Advanced Metrology for Ultrasound in Medicine 2010

(AMUM 2010)

Journal of Physics: Conference Series Volume 279

Teddington, United Kingdom
12 – 14 May 2010

ISBN: 978-1-61782-908-6
ISSN: 1742-6588

Advanced Metrology for Ultrasound in Medicine 2010

(AMUM 2010)

Journal of Physics: Conference Series – Volume 279

Teddington, UK
12 – 14 May 2010

ISBN: 978-1-61782-908-6
ISSN: 1742-6588

IOP Institute of Physics

Printed from e-media with permission by:

Curran Associates, Inc.
57 Morehouse Lane
Red Hook, NY 12571

Some format issues inherent in the e-media version may also appear in this print version.

Printed by Curran Associates, Inc. (2011)

For permission requests, please contact the Institute of Physics
at the address below.

Institute of Physics
Dirac House, Temple Back
Bristol BS1 6BE UK

Phone: 44 1 17 929 7481
Fax: 44 1 17 920 0979

techtracking@iop.org

Additional copies of this publication are available from:

Curran Associates, Inc.
57 Morehouse Lane
Red Hook, NY 12571 USA
Phone: 845-758-0400
Fax: 845-758-2634
Email: curran@proceedings.com
Web: www.proceedings.com

TABLE OF CONTENTS

Advanced Metrology for Ultrasound in Medicine (AMUM 2010) IOP Publishing
Journal of Physics: Conference Series **279** (2011) 011002 doi:10.1088/1742-6596/279/1/011002

Dear Colleague

On behalf of the UK National Physical Laboratory, I warmly welcome you to the second international conference on Advanced Metrology for Medical Ultrasound. It has been a difficult year with a worldwide recession and travel disruption over Europe due to volcanic dust from Iceland. So I am very grateful that so many of you have been able overcome these difficulties and make it Teddington

This second running of AMUM has once again attracted an extremely high standard of participant and I am sure that this will be an outstanding meeting. Our delegate list includes experts from the National Measurement Institutes of eight countries, and from many of as the leading companies as well as academics and medical physicists . So I anticipate plenty of stimulating discussion, and I hope that AMUM will promote new collaborations between industry, academia and medicine which will further advance the science of measurement for medical ultrasound.

The success of a meeting like AMUM depends, of course, on the quality of the presentations but even more important is the active participation of the other assembled experts. So I encourage everyone to take part in the discussions and debates that go on both during the scientific sessions and in the breaks.

I would like to thank you all – advisory board, presenters, audience, exhibitors and the AMUM secretariat - for your participation in this conference and I hope you have an interesting, informative and enjoyable three days at NPL.

Adam Shaw
Chair, AMUM Scientific Committee

PREFACE

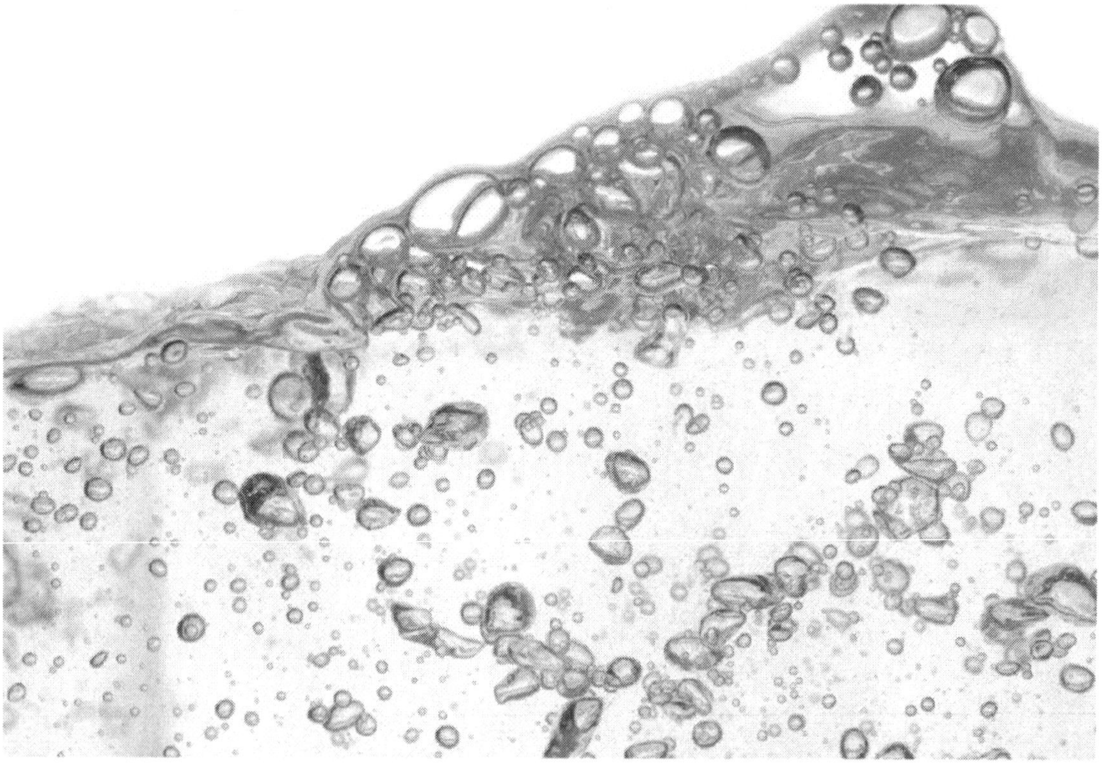

AMUM 2010 is the successor to the first AMUM conference held in 2004, which was the conference featured in the very first issue of the *Journal of Physics: Conference Series* (http://iopscience.iop.org/1742-6596/1/1). Like its predecessor, AMUM 2010 proved to be very successful and provided a fantastic opportunity for the world's ultrasound experts from medicine, industry and academia to explore the measurement challenges presented by new and emerging clinical ultrasound equipment. 2010 was a very difficult year economically for much of the world, and then air travel over Europe was thrown into chaos by prolonged ash clouds from the Icelandic volcano Eyjafjallajökull in April. So it was something of a relief to see such good attendance; this is a clear indication of the need for this conference with its focus on metrology in medical ultrasound.

There were a total of 70 attendees: of these 34 were from the UK, 16 from the rest of Europe, and 20 from outside Europe. National Metrology Institutes from no fewer than nine different countries were represented, some of them very new to the field, some with a recent growth of interest in ultrasound, and others with a well-established reputation. I was particularly pleased to see younger researchers from those NMIs new to ultrasound attend and have the chance to mix with such a wide grouping of their peers and present their work. There was also a strong attendance by industry including the major imaging companies such as GE Healthcare, Siemens, Philips Healthcare, Hitachi Medical, Aloka, Medison and the newer Supersonic Imagine; and more specialist companies such as Imasonic, Onda Corporation, Unisyn Medical Technologies, and Polytec and Precision Acoustics, our partner in organizing the event.

The conference was spread over 3 days with plenty of time for discussions over coffee; the afternoon of the 3rd day was dedicated to laboratory visits within NPL. We were fortunate to have some excellent Keynote speakers. Professor Khuri-Yakub from Stanford opened proceedings and gave us the lowdown on the potential of CMUT transducers in the generation and detection of ultrasound. Rob Hekkenberg from TNO in the Netherlands underlined the importance of international standardisation and gave us the benefit of his experience in helping set up new facilities around the world. Professor Ron Roy from Boston University gave us a personal view on how establishing well-founded measurement methods has been essential in driving forward the research in his department. Finally, Professor Markys Cain from NPL and Queen Mary, University of London, told us about some of the subtler issues around characterising the properties of piezoelectric, ferroelectric, multiferroic and magnetoelectric materials. Proffered papers covered topics from hydrophones and power measurement through to novel measurement methods, materials testing and even some biological studies.

Ultrasound in medicine goes from strength to strength, with new applications being developed almost monthly it seems. Metrology and Standards need to keep pace with these developments, which is what makes this conference important. I would like to thank everyone who attended or otherwise contributed to the success of AMUM 2010. I would particularly like to acknowledge the contributions of the guest speakers, the scientific advisory board members, NPL colleagues and the AMUM secretariat, who made the whole thing run so smoothly.

Adam Shaw
Chair, AMUM Scientific Committee

Journal of Physics: Conference Series **279** (2011) 011002 doi:10.1088/1742-6596/279/1/011002

ADVANCED METROLOGY FOR ULTRASOUND IN MEDICINE

National Physical Laboratory
Teddington
UK

12-14 May 2010

TIMETABLE and ABSTRACTS

CMUT Technology in Ultrasound

B. (Pierre) T. Khuri-Yakub
E. L. Ginzton Laboratory, room 11
Stanford University
Stanford, CA 94305-4088

Capacitive Micromachined Ultrasonic Transducers (CMUTs) have been developed in the past decade as alternative transducers for generating and detecting ultrasound. Capacitor transducers have been known for over 100 years; however, the advent of silicon micromachining has enabled the realization of the full potential of these transducers. Silicon micromachining allows the manufacture of capacitors with very thin gaps that can withstand electric fields of the order of 10^9 V/m. These fields enable performance that makes CMUTs competitive and superior to piezoelectric transducers.

In immersion applications, CMUTs are possible with fractional bandwidth of over 100 %, an electromechanical coupling coefficient close to unity; are made in the form of single element or one-dimensional (1D) or two-dimensional (2D) arrays of tens of thousand of elements, as well as annular arrays. They have been operated in the frequency range of 100 kHz to 50 MHz and have been included in imaging and therapy systems. Custom electronics have been developed and integrated with arrays of transducers to form compact catheter based medical ultrasound imaging systems. In airborne applications, CMUTs have been used as mass sensors with atto-gram resolution, and in pulse echo systems with large dynamic range including operation at high temperature (>600°C).

This presentation will review the operation of CMUTs, the technology used to make them, their characterization, transmit/receive RF electronics, and some airborne and medical imaging applications.

Journal of Physics: Conference Series **279** (2011) 011002 doi:10.1088/1742-6596/279/1/011002

Phase Characteristics for PVDF Membrane Hydrophones in the Frequency Range 1-100 MHz using Nonlinear Acoustics Approach

Gaurav Gandhi, Philip E. Bloomfield and Peter A Lewin

The School of Biomedical Engineering, Science and Health Systems, Drexel University, Philadelphia, PA, lewin@ece.drexel.edu

The purpose of this work was to develop phase calibration technique using an advanced semi-empirical computer model, which predicts the near-and the far field distributions making use of hyperbolic propagation operator, in contrast to the parabolic approximations used elsewhere. Knowledge of the relative phase shift is needed to deconvolve the pressure-time signal, and hence produce its faithful reconstruction, including the rise times and peak amplitudes. The phase responses of two PVDF membrane hydrophones – a 50 microns thick, 500 μm diameter, bilaminar Marconi-hydrophone and a custom made Precision Acoustics, 9 μm thick, 400μm diameter - were extracted from the complex frequency response of the nonlinear field simulated by the model. The simulation results were experimentally verified (uncertainties were determined at 95% confidence levels) and indicated that the nonlinear technique is well suited for membrane hydrophones having a uniform frequency response with variation in the range of ±10 %. The primary limitations of the proposed approach will be discussed. Phase calibration of needle hydrophone probe design was also attempted, however, in this case due to presence of intrinsic resonances, a fiber optic hydrophone with a virtually flat frequency response and hence zero phase-shift in the frequency range considered was needed.

Journal of Physics: Conference Series **279** (2011) 011002 doi:10.1088/1742-6596/279/1/011002

Uncertainty evaluation applied to calibration by deconvolution

Peter Harris and Pierre Gélat

Acoustics and Ionising Radiation
National Physical Laboratory
Hampton Road, Teddington TW11 0LW, UK
(peter.harris@npl.co.uk)

Abstract

The calibration of a sensor, such as an electroacoustic transducer, can be achieved through the mathematical operation of deconvolution. Two signals, a signal used to excite the sensor and the corresponding signal output by the sensor, are recorded, and the amplitude and phase responses of the sensor are obtained by deconvolving the signals. In practice, the signals are measured at discrete times, and the measured values defining the signals will have associated uncertainties. The operation of deconvolution is then realised digitally, and provides estimates of the amplitude and phase responses, which are to be reported with associated uncertainties.

In this talk, consideration is given to the problem of uncertainty evaluation for this calibration problem. The determination of the amplitude response is decomposed into the operations of (a) evaluating the coefficients of the discrete Fourier transforms of the signals, (b) forming the amplitudes of the coefficients, and (c) forming the quotient of the amplitudes, and the uncertainty evaluation for each operation is discussed. In particular, the extent to which the uncertainty evaluation for each operation can be treated using 'conventional uncertainty propagation', or more sophisticated approaches are needed, is considered. Some comments are made about the determination of the phase response of the sensor.

Journal of Physics: Conference Series **279** (2011) 011002 doi:10.1088/1742-6596/279/1/011002

Hydrophone area-averaging correction factors in nonlinearly generated ultrasonic beams

M P Cooling[1], V F Humphrey[1] and V Wilkens[2]

[1] *Institute of Sound and Vibration Research, University of Southampton, Southampton, SO17 1BJ, UK.*
[2] *Physikalisch-Technische Bundesanstalt, Bundesallee 100, 38116 Braunschweig, Germany*

The nonlinear propagation of an ultrasonic wave can be used to produce a wavefield rich in higher frequency components that is ideally suited to the calibration, or inter-calibration, of hydrophones. Such techniques usually use a tone-burst signal, limiting the measurements to harmonics of the fundamental calibration frequency. Alternatively, using a short pulse enables calibration at a continuous spectrum of frequencies. Such a technique is used at PTB in conjunction with an optical measurement technique to calibrate devices [1]. Experimental findings indicate that the area-averaging correction factor for a hydrophone in such a field demonstrates a complex behaviour, most notably varying periodically between frequencies that are harmonics of the centre frequency of the original pulse and frequencies that lie midway between these harmonics. The beam characteristics of such nonlinearly generated fields have been investigated using a finite difference solution to the nonlinear Khokhlov-Zabolotskaya-Kuznetsov (KZK) equation for a focused field. The simulation results are used to calculate the hydrophone area-averaging correction factors for 0.2 mm and 0.5 mm devices. The results clearly demonstrate a number of significant features observed in the experimental investigations, including the variation with frequency, drive level and hydrophone element size. An explanation for these effects is also proposed.

References

[1] V Wilkens and C Koch, 'Amplitude and phase calibration of hydrophones up to 70 MHz using broadband pulse excitation and an optical reference hydrophone', J. Acoust. Soc. Am. **115** (6), June 2004, pp 2892 – 2903.

Journal of Physics: Conference Series **279** (2011) 011002 doi:10.1088/1742-6596/279/1/011002

Quantifying acoustic emissions from cavitation events by passive detection

Jamie Collin, Constantin Coussios
Institute of BioMedical Engineering, University of Oxford
jamie.collin@eng.ox.ac.uk

Relative measures of the energy radiated as broadband noise emissions by inertially cavitating bubbles, such as the mean square voltage received using a passive cavitation detector (PCD), have been shown to be directly relatable to cavitation-mediated bioeffects. Though numerous techniques exist to detect inertial cavitation qualitatively, there is presently a lack of a unified approach that makes it possible to quantify and compare levels of cavitation activity across different experimental setups and conditions.

Using a substitution technique with a reference hydrophone, and a point scatterer embedded in a gel with low acoustic attenuation, the sensitivity on receive of a spherically focused PCD transducer can be obtained. This makes it possible to relate the PCD signal received during inertial cavitation activity to the acoustic power radiated from the focus of the detecting transducer. The technique is validated against PCD measurements of single-bubble inertial cavitation activity in a tissue-mimicking material and compared to theoretical predictions using the Keller-Miksis model. Calibrated measurements of acoustic power radiated as broadband noise emissions could thus provide a quantitative measure of cavitation dose that can be directly related to resulting bioeffects, such as increased sonoporation or enhanced heating.

Advanced Metrology for Ultrasound in Medicine (AMUM 2010) IOP Publishing
Journal of Physics: Conference Series **279** (2011) 011002 doi:10.1088/1742-6596/279/1/011002

Progress in the development of the Fabry Perot polymer film fibre optic hydrophone concept

P Morris[1], A Hurrell[1], A Jerling[2], P Beard[2]

1. *Precision Acoustics Ltd. Hampton Farm Business Park, Dorchester, DT2 8QH,* *paul@acoustics.co.uk*
2. *Dept. Medical Physics and Bioengineering, University College London, WC1E 6BT* *pbeard@medphys.ucl.ac.uk*

The Fabry-Perot fibre-optic hydrophone system has now matured from a proof of concept device into a commercially available product. During the maturation, various improvements have been made to the performance and functionality of the device. The frequency response of early sensors was dominated by resonances arising from diffraction of acoustic waves at the tip of the sensor. By using a rounded fibre tip, these effects have been significantly reduced, leading to a sensor with much improved frequency response.

In addition to the improved acoustic frequency response, development of the temperature measurement capability has led to a sensor with a temperature resolution better than 0.1 K with a sampling rate of approximately 200 Hz. The dual measurement capability enables the simultaneous acquisition of acoustic pressure and temperature rise data which can be used to correlate the heating produced by HIFU with the associated acoustic field parameters.

Development of the fibre-optic hydrophone system is still ongoing, supported by a PhD project at University College London. In particular new sensor fabrication techniques are being investigated, with a view to extending the upper limit of acoustic pressure measurement towards 100 MPa and the upper limit of temperature measurement towards 100°C.

Journal of Physics: Conference Series **279** (2011) 011002 doi:10.1088/1742-6596/279/1/011002

Pressure Pulse Measurements Using Optical Hydrophone Principles

Ueberle F., Jamshidi-Rad A.

University of Applied Sciences, Life Sciences / Medizintechnik, D-21105 Hamburg, Lohbrügger Kirchstrasse 65, friedrich.ueberle@haw-hamburg.de

Pressure pulses are used in extracorporeal lithotripsy, pain therapy other applications. Typical lithotripter pulses reach positive pressure amplitudes of more than 100 MPa and negative pressures of more than -20 MPa. These pulses cause several mechanical effects and cavitation, which are breaking stones, but are also a big challenge for the rigidity and integrity of the measurement sensors, which should at least be able to survive a complete measurement cycle of several thousand pulses.

On the other hand, due to the steep time signature of the focal pressure pulse as well as the small spot size of most pressure pulse sources, the properties of the sensors must both support extremely fast rise times in the nanosecond range, a broad and flat bandwidth of more than 10 MHz and sub – millimeter sensitive spot sizes.

During recent years, two main types of optical sensors were developed, which are based on the principle of measuring reflection changes of a laser beam at a glass-water surface: The fiber optic sensor using bare optical fibers and the "light spot" sensor using a thick glass block. The properties of these sensor types are discussed based on measurements of strong and lesser focusing pressure pulse sources.

Journal of Physics: Conference Series **279** (2011) 011002 doi:10.1088/1742-6596/279/1/011002

IEC 61949 - measuring acoustic output under quasilinear conditions.

FA Duck

Medical Physics and Bioengineering Dept, Royal United Hospital, Bath BA1 3QF, UK.
f.duck@bath.ac.uk

High amplitude acoustic pulses, as may be produced in diagnostic applications of ultrasound, can generate significant harmonic frequency components when propagating in water. As measurements of acoustic pressure and intensity in water are required by IEC Standards (IEC 60601-2-37 and IEC 62359) to estimate in situ or derated quantities, it is important to account for nonlinear propagation if derating is to provide accurate estimates.

IEC Technical Specification IEC 61949 describes a protocol to characterise an acoustic field using standard measurements in water but at quasilinear levels rather than maximum output. This paper will discuss the measurement protocols and implications of the TS.

Journal of Physics: Conference Series **279** (2011) 011002 doi:10.1088/1742-6596/279/1/011002

Evaluation of Stability and Repeatability of a Pulsed Check Source

S. Howard, D. Stoft, X. Chen, H. Morris and C. Zanelli

Onda Corporation, 592 E. Weddell Drive Suite 7, Sunnyvale CA USA 94089
(email: sh@ondacorp.com)

We examine the stability and repeatability of a portable and compact 3.5 MHz F/5 Pulsed Check Source (PCS). The PCS is used to verify the performance of hydrophone scanning systems and radiation force balances by providing acoustic waveforms similar to pulsed diagnostic medical equipment. The device also generates pulses in the quasi-linear range (to avoid high frequency components) and thus allows hydrophone calibration checks. It is therefore important to quantify how measurements with the PCS are affected by typical variations in experimental setups. A series of careful measurements have been made to quantify the device's dependence on temperature (less than 0.4% per ^{o}C in total power output over the range of 19 to 25°C). Additionally, a long-term study is underway to evaluate aging effects. Together with a discussion of design considerations, we discuss the effects of typical alignment errors in hydrophone scanning, as well as the errors due to dissolved oxygen content.

Journal of Physics: Conference Series **279** (2011) 011002 doi:10.1088/1742-6596/279/1/011002

Ultrasound- Standardisation – Metrology

Rob T. Hekkenberg

For the human society it is important that one can rely on information given. In science, technology and health care this concerns figures and statements presented in product specifications, in documentation discussing and determining safety arguments and in literature.
In the case of ultrasound, at the moment, only two types of parameters can reliably and traceably be quantified. Basic physical quantities like ultrasound pressure and power are identified by application specific parameters. But only in a limited range of application. To be able to make proper judgements, identification of uncertainties of quantities is a must.
It is well known that ultrasound energy can produce adverse biological effects in tissue. Activities have to be undertaken to ensure a safe application of ultrasound in medicine. For medical ultrasound equipment quality control is also an important issue. This can be performed using different types of test objects.
In the process of reliability and traceability, International Standardisation and the existence of standard laboratories and qualified test houses is very important. It is interesting to realise that in the last years institutes in the area of ultrasound stopped their activity, while others started.
As other physical phenomena due to an ultrasound field are of increasing importance, reliable methods of determining such phenomena need to be set up.
The presentation will briefly point to subjects listed and focus on the establishment of new standard laboratories.

Rob Hekkenberg
Senior Consultant Ultrasound Safety and Standards
Guest at TNO-Quality of Life
Wassenaarseweg 56, 2333 AL, Leiden, The Netherlands
P.O.Box 2215, 2301 CE, Leiden, The Netherlands
Tel: +31 (71) 3315695
Fax: +31 (71) 518 1902
Email: rob.hekkenberg@tno.nl

Advanced Metrology for Ultrasound in Medicine (AMUM 2010) IOP Publishing
Journal of Physics: Conference Series **279** (2011) 011002 doi:10.1088/1742-6596/279/1/011002

A fully automated system for ultrasonic power measurement and simulation accordingly to IEC 61161:2006

Rodrigo P.B. Costa-Felix[1], André V. Alvarenga[1], Rob Hekkenberg[2]

[1]*Laboratory of Ultrasound, Directory of Scientific and Industrial Metrology (DIMCI), National Institute of Metrology, Standardization, and Industrial Quality (Inmetro), Av. Nossa Sra das Graças, 50 (Prédio 1), Duque de Caxias, RJ, Brazil, ZIP 25250-020*

[2]*TNO Prevention and Health, Zernikedreef 9, 2333 CK Leiden, The Netherlands*

[1]*rpfelix@inmetro.gov.br*

Ultrasonic power measurement worldwide accepted standard is IEC 61161, presently in its 2nd edition (2006), under review. To fulfil its requirements, considering that a radiation force balance is to be used as ultrasonic power detector, a large amount of raw data (mass measurement) shall be collected as function of time to perform all necessaries calculations and corrections. Uncertainty determination demands huge calculation effort in raw and processed data as well. Although it is possible to be undertaken in an old-fashion way, using spread sheets and manual data collection, automation software are often used in metrology to provide a virtually error free environment concerning data acquisition and repetitive calculations and corrections. Considering that, a fully automate ultrasonic power measurement system was developed and comprehensively tested. A 0.1 mg of precision balance model CP224S (Sartorius, Germany) was used as measuring device and a calibrated continuous wave ultrasound check source (Precision Acoustics, UK) was the device under test. Over the balance plate, a 150 ml container filled with degassed water was laid and an absorbing sheet target was immersed in. Besides the feature of automation software, it was implemented a routine of power measurement simulation. It was idealized as a teaching tool of how ultrasonic power emission behaviour is with a radiation force balance with absorbing target. The automation software was considered as an effective tool for speeding up ultrasonic power measurement, while allowing accurate calculation and attractive graphical partial and final results.

Journal of Physics: Conference Series **279** (2011) 011002 doi:10.1088/1742-6596/279/1/011002

Three dimensional vibration testing in biomedical applications utilizing Laser Doppler Vibrometry

Roger Traynor[1], Joe Armstrong[2]

[1]Polytec Limited – Product Manager – Sales Department
Lambda House, Batford Mill, Harpenden, United Kingdom
r.traynor@polytec-ltd.co.uk

[2]Polytec Limited – Sales Manager – Sales Department
Lambda House, Batford Mill, Harpenden, United Kingdom
Joe.Armstrong@polytec-ltd.co.uk

Abstract

Scanning Laser Doppler Vibrometry enables fast and efficient measurements of vibration behavior at all optical accessible surfaces. After implementing the measurement object geometry in a teaching process the number of measurement points can be increased almost arbitrarily which would be completely impossible with conventional transducers. By means of importing a FEM grid and measuring the real object vibration at the defined positions essential data for the validation of the FE model can be provided.

Until now, measurements have been limited to the frequency spectrum up to approx. 30 MHz and to velocities of up to 10m/s. With its new innovative design UHF-120 Laser Doppler Vibrometer, Polytec has extended the frequency range to 1.2GHz and the velocity range to more than 150m/s. The new Polytec measurement system is thus ideal for examining high frequency devices, with high precision. Due to the small focal spot size of the laser beam, even tiny structures with short acoustic wavelengths, such as Surface Acoustic Waves (SAW), can be characterized with high lateral resolution. The ability to resolve high velocities now means that it is also possible to characterize high performance ultrasonic transducers.

A New Definition for Acoustic Dose

FA Duck

Medical Physics and Bioengineering Dept, Royal United Hospital, Bath BA1 3QF, UK.
f.duck@bath.ac.uk

This paper will discuss a recent proposal for definitions of acoustic dose and acoustic dose-rate [1]. Acoustic dose is defined as the energy deposited by absorption of an acoustic wave per unit mass of the medium supporting the wave. Expressed in this way, acoustic dose-rate, Q_m, in W kg^{-1}, is central to the prediction of both rate of temperature rise and radiation force. These quantities have spatial and temporal dependency, depending on the local field parameters (acoustic pressure, particle velocity, intensity) and local material properties (absorption coefficient, α_a, and mass density, ρ). Spatial and/or temporal averaging can be applied where appropriate. For plane-wave monochromatic conditions in a homogeneous medium, $Q_m=2\alpha_a I/\rho$, (I is the time-averaged intensity), and may incorporate temporal and frequency dependencies of energy deposition. Acoustic dose and acoustic does-rate are exact analogues for Specific Absorption and Specific Absorption Rate (SAR), quantities central to non-ionising electromagnetic radiation dosimetry. Acoustic dosimetry in the presence of tissue/gas interfaces remains a considerable challenge. Possible approaches will be discussed.

1. **Duck FA**. Acoustic dose and acoustic dose rate. Ultrasound in Med & Biol 2009;35:1679-1685.

Advanced Metrology for Ultrasound in Medicine (AMUM 2010) IOP Publishing
Journal of Physics: Conference Series **279** (2011) 011002 doi:10.1088/1742-6596/279/1/011002

Building and assessing anatomically relevant phantoms for neonatal transcranial ultrasound

G.Memoli[1], M.Gatto[2], N. Sadhoo[1], P. Gélat[1], R.A. Harris[2], A. Shaw[1]

[1] *Acoustics and Ionising Radiation Division, National Physical Laboratory, Teddington (UK)*
[2] *Wolfson School of Mechanical & Manufacturing Engineering, Loughborough University, Loughborough (UK)*
Contacts: gianluca.memoli@npl.co.uk, m.gatto@lboro.ac.uk

This study is part of a project aimed at the design of a validated, clinically relevant phantom to survey the temperature increase caused by ultrasound equipment currently used in neonatal head-scanning. This phantom, constituted of brain and skull mimicking materials, will allow for a more reliable evaluation of ultrasound heating than is currently possible.

Three Dimensional Printing (3DP) was employed for the fabrication of the skull phantom and a specific strategy was successfully pursued to embed a Fine-Wire Thermocouple (FWT) within the 3DP phantom during the manufacturing process. In this way, the FWT could be located at a known position inside the skull, close to the brain mimic. An *in-process* Non-Destructive Analysis (NDA), specifically developed by the investigators in Loughborough for 3DP manufacturing, was used to assess the correct position of the deposited FWT inside the fabricated phantom, later verified by ultrasound scanning.

Equally important was the selection of thermocouples for the more fluid brain-mimic: a discussion on the experimentally-observed differences between Thin Film Thermocouples (TFTs) and FWTs will also be reported here, with a discussion on the possible artefacts due to viscous heating. Modelling and measurements will be used to justify the final selection.

Carotid atherosclerotic plaque characterisation by measurement of ultrasound sound speed *in vitro* at high frequency, 20 MHz

Authors: M.P. Brewin, M.J. Birch

Address: Clinical Physics Department, Royal London Hospital, Barts and the London NHS Trust, LONDON, E1 2BL (email: m.p.brewin@qmul.ac.uk)

The shift in pulsed ultrasound pressure wave speed as a potential indicator of atherosclerotic plaque composition is under investigation. As part of an ethically approved trial, plaque is harvested from consenting adults undergoing carotid endarterectomy (CEA). It is then embedded within an IEC agar-based gel [1] of known acoustic properties [2]. The plaque is then scanned in a raster fashion using a tightly focussed 20 MHz ultrasound beam. The wave speed and pulse attenuation are calculated using both a transmission and a reflective substitution pulse-echo method [3]. The plots are repeated at both RTP and 37°C. The attenuation maps are matched to a digital photograph using Excel and the corresponding speed maps are then compared to the histology. By solution of simultaneous equations that link constituent percentages and composite speed, the individual speed values for five constituents are calculated; TMM, elastin, fibrin/collagen, calcification and lipid/necrosis.

Initial results have shown that the acoustic maps match well to the digital photographs. The results for speed in the TMM is consistently close to the expected value of average soft tissue (~1540 ms^{-1}) while fibrous tissue shows a higher value of ~1570 ms^{-1} and fat gives a lower value of 1470 ms^{-1}.

References

1. Development of an example flow test object and comparison of five of these test objects, constructed in various laboratories, Teirlinck et al, Ultrasonics (1998) 36 pp 653-660

2. The acoustic properties, centered on 20 MHz, of an agar-based tissue-mimicking material and its temperature, frequency and age dependence, Brewin et al., Ultrasound Med Biol (2008), 38(8) pp 1292-1306

3. Acoustical characteristics of biological media, J.Bamber, Chapter 141, Encyclopaedia of Acoustics 1997 (Edited by M.J.Crocker)

Journal of Physics: Conference Series **279** (2011) 011002 doi:10.1088/1742-6596/279/1/011002

The Resolution Integral as a Metric of Performance for Diagnostic Grey-Scale Imaging

S D Pye, W Ellis

Medical Physics, NHS Lothian, Royal Infirmary, Edinburgh UK
stephen.pye@luht.scot.nhs.uk

There is a strong clinical need for objective measurement of grey-scale ultrasound image performance. The *Resolution Integral* R is essentially a measurement of the ratio of the penetration of an ultrasound beam in soft tissue to the ultrasound beam width [1]. Highly sensitive transducers combine deep penetration with high resolution, and so the ratio penetration / beam-width is large. *Depth of Field Lr* and *Characteristic Resolution Dr* are defined such that R= Lr/Dr. Lr defines a region of optimum resolution, and Dr is representative of the lateral and azimuthal resolution within Lr. We have used an Edinburgh Pipe Phantom to measure values of R, Lr and Dr for 180 transducers with a range of ages (1-20 yrs), manufacturers (15), technologies (linear, curvy, mechanical sector) and applications (general radiology, small parts, endocavity). The characteristics of individual probes are concisely quantified and displayed using a plot of Lr vs Dr. This plot clearly discriminates transducers with different clinical applications and differing levels of performance, and we describe its use to inform the specification of equipment, pre-purchase evaluation, acceptance, assessment after repair, and twilight testing.

[1] Pye et al. J Physics Conf Series 2004 1:187-192

Journal of Physics: Conference Series **279** (2011) 011002 doi:10.1088/1742-6596/279/1/011002

The Edinburgh Pipe Phantom: characterising ultrasound scanners beyond 50 MHz

[1]C M Moran, [2]W Ellis, [2]A Janeczko, [3]D Bell, [2]S D Pye

[1] Medical Physics, University of Edinburgh, UK
[2] Medical Physics, NHS Lothian, Royal Infirmary, Edinburgh, UK
[3] Precision Acoustics Ltd, Dorchester, UK
stephen.pye@luht.scot.nhs.uk

The Edinburgh Pipe Phantom was originally developed to assess the technical performance of diagnostic scanners up to 15MHz. It comprises a series of anechoic cylinders with diameters 0.4 – 8mm embedded in agar-based tissue mimic. This design enables measurement of the characteristics (Resolution Integral R, Depth of Field Lr, Characteristic Resolution Dr) of grey-scale images with transducer centre frequencies from around 2.5 to 15MHz. We report the further development of the Edinburgh Pipe Phantom as a tool for characterising ultrasound scanners with centre frequencies up to at least 55MHz. This was done by moulding a series of anechoic pipe structures in a block of agar-based tissue mimic with pipe diameters 0.045 – 1.5mm. This paper presents measurements of R, Lr and Dr for a series of transducers assessed for pre-clinical scanning, with centre frequencies in the range 10 – 55 MHz. Resolution Integral values ranged from 35 (GE Vivid 5, i13Lv probe) to 72 (Philips iU22, L15-7io probe and VisualSonics Vevo 770, RMV704 40MHz probe) with corresponding Lr, Dr values of 25mm, 720μm; 43mm, 600μm; 6.7mm, 93μm respectively. In conclusion, the high frequency Pipe Phantom was able to successfully determine the imaging characteristics of ultrasound probes up to 55MHz.

Journal of Physics: Conference Series **279** (2011) 011002 doi:10.1088/1742-6596/279/1/011002

Apfel's laws and sound science: how physical measurements impact physical and medical acoustics research at Boston University

Ronald. A Roy, Robin O. Cleveland, R. Glynn Holt, Caleb H. Farny, James R. McLaughlan, Puxiang Lai, Nicholas J. Manzi (Dept. of Mechanical Engineering, Boston Univ., 110 Cummington St., Boston, MA 02215) and
Parag V. Chitnis (Riverside Research Institute, 156 William Street, 9th Floor, New York, NY 10038)

Robert E. Apfel once postulated the following three laws for acoustic cavitation science: know thy fluid, know thy sound field, and know when something happens. The Boston University Physical Acoustics Research Group is engaged in activity on problems as diverse as the use of light and sound to promote focused ultrasound surgery to cavitation mitigation in the mercury-filled target of the world's most powerful pulsed-neutron source. Through it all, the ability to monitor key physical parameters and make precise and accurate measurements of physical properties such as pressure and temperature can impact data interpretation and inform experimental design. We provide a capsule summary of these research activities cast in the context of Apfel's three laws and the ways that our measurement schemes enable and/or stimulate scientific inquiry. [Work supported by the National Institutes of Health, the US Army, the Office of Naval Research, the US SMDC, the Riverside Research Institute, the Oak Ridge National Laboratory, and NSF ERC Award No. EEC-9986821.]

Journal of Physics: Conference Series **279** (2011) 011002 doi:10.1088/1742-6596/279/1/011002

Evaluation of a Tissue Mimicking Material during HIFU Exposure

S. Maruvada, Y. Liu, B. A. Herman, and G. R. Harris

Food and Drug Administration, Center for Devices and Radiological Health, 10903 New Hampshire Ave., Bldg., Silver Spring, MD 20993.
subha.maruvada@fda.hhs.gov

In pre-clinical testing it is essential to characterize clinical HIFU devices using tissue-mimicking materials (TMMs) with well known characteristics, including predictable temperature rise and cavitation properties. The purpose of this study was to monitor cavitation behavior and determine its effect on temperature rise in a HIFU TMM containing an embedded thermocouple. A 75 µm fine wire thermocouple was embedded in a hydrogel-based TMM previously developed for HIFU. HIFU at 1.1 and 3.3 MHz was focused at the thermocouple junction. HIFU focal pressures from 1-5 MPa were applied and the temperature rise and decay were recorded. Three hydrophones were used to monitor cavitation activity during sonication. A hydrophone confocal with the HIFU transducer and a cylindrical hydrophone lateral to the HIFU beam were used as passive cavitation detectors for spectral analysis of cavitation signals, and a needle hydrophone placed beyond the HIFU focus was used to record changes in the pressure amplitude due to blockage by bubbles at or near the focus. B-mode imaging scans were employed to visualize bubble presence during sonication. In a separate measurement, schlieren imaging was used to monitor the change in field distribution behind the TMM. All methods produced a high level of correlation in monitoring cavitation in the TMM.

Journal of Physics: Conference Series **279** (2011) 011002 doi:10.1088/1742-6596/279/1/011002

Instantaneous input electrical power measurements of HIFU transducer

Baki Karaböce[1], Yakup Gülmez[1], Sirinath Rajagapol[2], Adam Shaw[2],

[1] : *TÜBİTAK UME P.K. 54 41470 Gebze Kocaeli TURKEY*
[2] : *National Physical Laboratory, Hampton Road Teddington Middlesex TW11 0LW, UK*
e-mail: baki.karaboce@ume.tubitak.gov.tr

Abstract

HIFU transducers have widely usage in therapeutic ultrasound in medicine. The output ultrasonic power of HIFU transducer can be measured in number of methods described in IEC 61161 standard. The ultrasound power radiated from a transducer is directly dependent on applied input electrical voltage and current and consequently power. But, up to now, no method were developed and adopted for the input electrical power measurements. Hence, a workpackage was carried out for the establishment of such method in the frequency range of 1 to 3 MHz as a part of EURAMET EMRP Era-net plus "External Beam Cancer Therapy" project.

The parameters of the voltage and current were characterized by employing four terminal impedance measurement methods for HIFU transducers at UME and NPL Ultrasonics Laboratories. The error parameters between the two and four terminal impedance measurement methods were investigated. Several current shunts were developed to determine the residual parameters and some improvements were made to reduce them. Current measurements were also realized with Philips current probe and preamplifier. The measurement results of UME and NPL will be compared to each other in order to understand the power reflection, impedance matching and harmonic problems.

In this paper, establishment and validation of a method for the measurement of instantaneous electrical power delivered to a reactive ultrasound transducer in the required frequency range with a target accuracy of ±5%, is presented.

Journal of Physics: Conference Series **279** (2011) 011002 doi:10.1088/1742-6596/279/1/011002

Calorimetric Method for Measuring High Ultrasonic Power Using Water as Heating Material

T. Kikuchi, T. Uchida

Acoustics and Vibration Metrology Division,
National Metrology Institute of Japan(NMIJ)/AIST,
1-1-1 Umezono, Tsukuba, Ibaraki, 305-8563 Japan. tsuneo.kikuchi@aist.go.jp

This study shows the calorimetric method for measuring high ultrasonic power by using water as heating material. In recent years, at the National Metrology Institute of Japan (NMIJ), an ultrasonic power primary standard from 1 mW to 15 W has been established by the radiation force balance (RFB) method. Conventionally, RFB method is widely used for the ultrasonic power measurement, but this method is not suitable for very high power measurement due to thermal damages of the absorbing targets. High power ultrasonic standards, however, are being required by medical HITU measurements or sonochemistry industries. In order to meet these requests, we have started to develop an ultrasonic power standard between 15 W and higher than 200 W. As calorimetric method is one of the alternatives to RFB, we have adopted this method by using water as the heating material. Water has excellent features as the "standard material", because the physical properties are well known and is easily available. We have already achieved the ultrasonic power measurement by the calorimetric method up to 100 W, and the measured power agreed well with the NMIJ primary standard up to 20 W.

Journal of Physics: Conference Series **279** (2011) 011002 doi:10.1088/1742-6596/279/1/011002

Buoyancy Method – A Potential New Primary Ultrasound Power Standard

Srinath Rajagopal and Adam Shaw
Acoustics and Ionizing Radiation Division, National Physical Laboratory, Teddington, UK
srinath.rajagopal@npl.co.uk

Abstract – The acoustic output power produced by medical ultrasound equipment is a key quantity representing the safety to the exposure of ultrasound during diagnostic or therapeutic applications. The acceptable method of measuring the total acoustic output power published by the International Electrotechnical Commission (IEC 61161: 2007) is measuring the radiation force using a gravimetric balance. The method described here is a non-radiation force method, which is previously been validated using a form of calorimetry in which the average temperature change in the medium is inferred from its volume change. The same method is now extended and validated up to 10 MHz and powers in the range of 200 mW and upwards is measured. An in-depth understanding of the systematic effects is carried out resulting in improved uncertainties in the measured power. The calculated systematic uncertainties for measured power @ 95 % confidence level is approximately ± 2.4 %. The improved technique and validated measurements up to 10 MHz offers an alternative to the radiation force method for power measurements.

Journal of Physics: Conference Series **279** (2011) 011002 doi:10.1088/1742-6596/279/1/011002

HIFU Ultrasound Power Measurements at INRiM

G. Durando, C.Guglielmone, C. Musacchio

INRIM strada delle cacce 91, 10135 Torino – ITALY (g.durando@inrim.it)

In this work the new system for the ultrasound power measurement of High Intensity Focused Ultrasound transducers realized at INRIM ultrasounds laboratory is presented. The system is based on a submersible load cell that takes the place of the balance. This solution presents essentially two advantages. The first one, of mechanical nature, is relevant to the fact that the target is directly connected to the force transducer, eliminating unwanted target motion at high power. The second, of electric nature, concerns the possibility to reduce the insonation time (the ON period of the electric driving signal to the HIFU transducer) under of 2 s, and is allowed for by the faster response of the force transducer (700 Hz bandwidth). The main components of uncertainty and the overall budget of the measurement system are presented together with the results of measures of conductance, *G*, carried on a HIFU transducer, at the work frequency 3,5 MHz, for values of power ranging from 10 W to 100 W. The results of the ultrasonic conductance, *G*, obtained with the new system are compared with values obtained using the traditional measuring system for low powers (*P* ? 20W).

Journal of Physics: Conference Series **279** (2011) 011002 doi:10.1088/1742-6596/279/1/011002

Ultrasound power measurements of HIFU transducer with a more stable radiation force balance

Baki Karaböce, Enver Sadıkoğlu, Eyüp Bilgiç

TÜBİTAK UME P.K. 54 41470 Gebze Kocaeli TURKEY,
e-mail: baki.karaboce@ume.tubitak.gov.tr

Abstract

A new radiation force balance (RFB) system was established at National Metrology Institute (UME) Ultrasonics Laboratory for High intensity focused ultrasound (HIFU) power measurements. The new system is highly stable at high power levels up to 500 watts. The measurement system consists of a Plexiglas cylindrical balance arm, target mounting scale disks, conical reflecting and absorbing targets, adjustment nuts, and a hanging wire. Both of the two sides of balance were mounted similar size and weight targets. The equilibrium of the balance arm can be adjusted with nuts on screws located at both sides of the balance arm. Transducer was mounted to bottom of water tank. Absorbers in the bottom and the near walls of the tank were used for reflecting target case. Ultrasound power was applied to one scale of the balance where the reflecting/absorbing target was mounted and corresponding force was measured on the other scale of balance where was connected to a balance with a thin wire while the thin rest standing on a support.

Ultrasound power of two HIFU transducers at 0.93, 1.1 and 3.3 MHz and were measured with conventional and new system, the values were compared and uncertainty components were assessed in this paper.

Journal of Physics: Conference Series **279** (2011) 011002 doi:10.1088/1742-6596/279/1/011002

Interlaboratory comparison of HITU power measurement capabilities

K-V Jenderka[1], G Durando[2], B Karaboce[3], S Rajagopal[4], A Shaw[4]

[1]*Physikalisch-Technische Bundesanstalt, Bundesallee 100, 38116 Braunschweig, Germany (E-mail: klaus-vitold.jenderka@ptb.de),*
[2]*Istituto Nazionale di Ricerca Metrologica, Strada delle Cacce 91, 10135 Torino, Italy,*
[3]*Tübitak Ulusal Metroloji Enstitüsü, P.K. 54 41470 Gebze–Kocaeli, Turkey,*
[4]*National Physical Laboratory, Hampton Road, Teddington, Middlesex, TW11 0LW, United Kingdom*

High Intensity Therapeutic Ultrasound (HITU) is gaining in importance among the spectrum of therapeutic options to combat cancer. HITU has already been approved and is in clinical use for the treatment of organs like the prostate, the liver and the uterus. Nevertheless, the metrology of the applied high power ultrasound fields, and in consequence, reliable treatment planning and monitoring, is still a challenge.

As part of European Metrology Research Programme project, the four National Metrology Institutes from the UK, Germany, Italy and Turkey conducted an interlaboratory comparison of their power measurement capabilities at power levels of 5, 25, 75 and 150 W each at frequencies of 1.1, 1.5 and 3.3 MHz.

The task was to measure the total, time-averaged ultrasonic output power, emitted by the circulated transducers under specified electrical excitation conditions into an anechoic water load, and the actual RMS transducer input voltage. The output value to be reported was the electro-acoustic radiation conductance including the associated standard and expanded uncertainties. Several different measurement techniques were applied to gain further insight into HITU power measurement.

The deviations from the calculated comparison reference value found for the different techniques are discussed and conclusions for the further improvement of measuring procedures are drawn.

Journal of Physics: Conference Series **279** (2011) 011002 doi:10.1088/1742-6596/279/1/011002

Progress in developing a solid-state sensor that responds to acoustic power or time-averaged intensity

B Zeqiri[1], M Hodnett[1], S Rajagopal[1], G Zauhar[2] and P N Gélat[1].

[1]National Physical Laboratory, Acoustics and Ionising Radiation Division, Hampton Road, Teddington, TW11 0LW, United Kingdom (bajram.zeqiri@npl.co.uk).

[2] University of Rijeka, Medical Faculty, Department of Physics, B. Branchetta 20, 51 000 Rijeka, Croatia.

This paper describes a solid-state thermal-technique developed at NPL to measure ultrasonic power at diagnostic and physiotherapy power levels. The technique exploits the pyroelectric properties of a membrane of polyvinylidene fluoride (*pvdf*), which is backed by a polyurethane material that is intrinsically extremely absorbing to ultrasound. Energy within the ultrasound beam is absorbed in this backing layer, leading to an increase in temperature of the *pvdf* and a consequent pyroelectric output voltage waveform whose shape partly depends on the applied power. The paper briefly presents the method, and describes research undertaken since AMUM 2004 that was targeted at understanding and optimizing the technique, through an evaluation of its performance within in a range of applied acoustic fields. Studies point to the importance of the acoustic and thermal properties of the absorbing backing layer in controlling sensor response and reveal several key features of the method: its sensitive nature, suggesting instruments capable of measuring powers as low as a few mW, and, intriguingly, the possibility of directly measuring acoustic intensity, through application of a sensor whose area is small in relation to the size of the applied acoustic beam. Currently, acoustic intensity is derived using hydrophones assuming a plane-wave acoustic field.

Journal of Physics: Conference Series **279** (2011) 011002 doi:10.1088/1742-6596/279/1/011002

A combined measurement and modeling method to characterize nonlinear HIFU fields

V.A. Khokhlova,[1,2] P.V. Yuldashev,[1] M.S. Canney,[3] M.R. Bailey,[1] O.A. Sapozhnikov,[1,2] and L.A. Crum[1]

[1]Center for Industrial and Medical Ultrasound, Applied Physics Laboratory, University of Washington, 1013 NE 40th St., Seattle, WA 98105 USA (E-mail: vera@apl.washington.edu)
[2]Department of Acoustics, Faculty of Physics, Moscow State University, Moscow 119991, Russia
[3]INSERM U556, 151 Cours Albert Thomas, Lyon Cedex 03, France

Acoustic characterization of nonlinear HIFU fields is important for the development of regulatory standards for clinical devices. In this work, a combined measurement and modeling characterization method is reported. Two modeling approaches are tested. The first approach includes 3D simulation of the nonlinear wave equation with the source boundary condition obtained experimentally using acoustic holography. This rigorous model can accurately simulate the entire nonlinear HIFU field, but it is computationally intensive. The second approach is based on simplified simulations using the KZK equation and an equivalent focused piston source boundary condition obtained by matching linear simulations to low amplitude measurements in the focal region. To compare the two models and experiment, characterization is performed for a 2 MHz transducer of 4.2 cm aperture and 4.5 cm focal length. Focal waveforms, simulated and measured using a fiber optic hydrophone, are compared. It is shown that the simplified model provides accurate results at the focus up to very high outputs when shocks are present. Preliminary characterization data are also obtained for a 2D random therapeutic phased array and show the feasibility of using the simplified piston model for therapeutic devices with different geometries. [Work supported by NIH EB007643, NSBRI SMST01601, and RFBR 09-02-01530.]

Advanced Metrology for Ultrasound in Medicine (AMUM 2010) IOP Publishing
Journal of Physics: Conference Series **279** (2011) 011002 doi:10.1088/1742-6596/279/1/011002

Measurement of Ultrasound Beam Angular Direction From Phase

J Civale, I Rivens, G ter Haar.

Joint Department of Physics, Royal Marsden Hospital, Institute of Cancer Research, Downs Road, Sutton, Surrey, UK

john.civale@icr.ac.uk

Acoustic field calibrations often use Fast Fourier Transforms (FFT) to quantify the spectral amplitude components of detected acoustic signals. FFTs can however also provide phase information which may be analysed to quantify the angular direction of wave propagation and hence determine beam mis-alignment.

The acoustic fields of a number of HIFU transducers (f numbers 2.8, 1.7, 1.3, 0.8) have been measured experimentally using a beam plotting system capable of waveform acquisition. Measurements were made using a fibre-optic hydrophone system (Precision Acoustics, UK) which provides excellent spatial resolution (<100 μm) and acoustic sensitivity (>100 mV/MPa). Theoretical pressures and phase distributions for these transducers were predicted using a linear acoustic field model. The phase data was used to compute plots which quantify the direction of propagation of the ultrasound beam, demonstrating the ability to detect beam mis-alignment. Spectral pressure and phase information can also be used to visualise instantaneous acoustic pressure distribution.

Further uses of beam angular direction information in focused fields includes potential for improved accuracy in determining pressure amplitude distributions with knowledge of the hydrophone's directional response, and the possibility of deriving parameters describing the effective curvature of acoustic wavefronts at a given spatial position.

Advanced Metrology for Ultrasound in Medicine (AMUM 2010) IOP Publishing
Journal of Physics: Conference Series **279** (2011) 011002 doi:10.1088/1742-6596/279/1/011002

Hydrophone Spatial Directivity and the induced difference in HIFU Acoustic Field Test

T Chen[1], J M Hu[1], W Huang[1], W Zhang[1] and D Zhang[2]

[1] *Jiangsu Province Institute for Medical Equipment Testing , China*
[2] *Nanjing University, China*

Email:ctgeorge@163.com

Hydrophone has been widely used in testing ultrasonic field distribution. But the hydrophone has a certain directivity, which may bring error in the actual acoustic field measurements. So it is necessary to meet some requirements about the directivity of hydrophone. In the measurement, the spatial directivities of needle hydrophone, membrane hydrophone and robust hydrophone is measured and a comparison about measured data of HIFU acoustic field with them is given. Experimental results indicate that the asymmetric special directivity of hydrophone is due to manufacturing processes and other reasons. As a result, it is not insufficient to test the directivity of hydrophone along one axis. It is necessary to adjust the direction of the hydrophone when characterizing the HIFU field.

Journal of Physics: Conference Series **279** (2011) 011002 doi:10.1088/1742-6596/279/1/011002

Beam profile measurement on HITU transducers using a thermal intensity sensor technique

V Wilkens, S Sonntag, K-V Jenderka

Physikalisch-Technische Bundesanstalt, Bundesallee 100, 38116 Braunschweig, Germany
E-mail: volker.wilkens@ptb.de

Thermal intensity sensors based on the transformation of the incident ultrasonic energy into heat inside a small cylindrical absorber have been developed at PTB in the past, in particular to determine the acoustic output of medical diagnostic ultrasound equipment. Currently, this sensor technique is being expanded to match the measurement challenges of high intensity therapeutic ultrasound (HITU) fields. At the high acoustic power levels as utilized in the clinical application of HITU transducers, beam characterization using hydrophones is critical due to the possible damage of the sensitive and expensive measurement devices. Therefore, the low-cost and robust thermal sensors developed offer a promising alternative for the determination of high intensity output beam profiles. A sensor prototype with a spatial resolution of 0.5 mm was applied to the beam characterization of an HITU transducer operated at several driving amplitude levels and burst length settings. Axial beam plots and lateral profiles at focus were acquired. The absolute overall output power was, in addition, determined using a radiation force balance. By comparison of the beam profiles obtained, the influence of the driving level on the location of the focus and on the beam dimensions is investigated.

Journal of Physics: Conference Series **279** (2011) 011002 doi:10.1088/1742-6596/279/1/011002

Calibration of HIFU intensity fields measured using an infra-red camera

S.M. Bobkova,[1] V.A. Khokhlova[1,2], A. Shaw,[3] L.R. Gavrilov,[4] and J.W. Hand[5]

[1]*Department of Acoustics, Physics Faculty, Moscow State University, Moscow 119991, Russia*
[2]*Center for Industrial and Medical Ultrasound, APL, University of Washington, Seattle 98105, USA*
[3]*Acoustics Division, National Physical Laboratory, Teddington TW11 0LW, UK*
[4]*N.N. Andreev Acoustics Institute, 117036 Moscow, Russia*
[5]*Radiological Sciences Unit, Imperial College Healthcare NHS Trust, London W12 0HS*

A trend in HIFU technologies is to use 2D phased arrays that offer electronic steering of a single focus and formation of patterns of multiple foci. Conventional methods to characterize array fields using scanned hydrophone would be prohibitively slow given the potentially large number of focusing conditions. An alternative technique for rapid qualitative assessment of intensity distributions was recently developed [Hand *et al.*, PMB 2009]. The method is based on infrared camera measurements of the temperature rises induced by low amplitude short ultrasonic bursts in a thin absorber. Here, the method is extended to estimate the absolute values of intensity. The temperature rise in an absorber positioned in the focal plane was measured and calculated from the modelled acoustic field of a 2D 1-MHz randomized phased array and the heat transfer equation. The modelling results were related to the measurements; a corresponding correction factor between the intensity and temperature was obtained and introduced in the conversion of temperature images to intensity. The method was applied to study various regimes for driving the array. Free fields in water and focusing through ribs were recorded and simulated. Good correlation between the measured and modeled results in both spatial distributions and the absolute values of intensity was demonstrated. [Work supported by Cancer Research UK-C18509/A78, NIHR BRC, RFBR 09-02-00066 and 09-02-01530, and NIH R01EB007643.]

Journal of Physics: Conference Series **279** (2011) 011002 doi:10.1088/1742-6596/279/1/011002

Properties of piezoelectric materials at NPL

Markys Cain
Materials Division, National Physical Laboratory, Teddington, UK
markys.cain@npl.co.uk

ABSTRACT

The Multifunctional Materials group at NPL develops new ways of measuring the multiphysics properties of piezoelectric, ferroelectric and more recently multiferroic and magnetoelectric materials. The metrology developed supports medium to long term industrial uptake of these materials for new sensing, actuation and telecommunications applications. In this talk, I will explore the principles of operation of these materials used as sensors or actuators and highlight several of the newer metrologies we have developed at NPL including non-contact and contact probe methods in the assessment of piezo activity.

Journal of Physics: Conference Series **279** (2011) 011002 doi:10.1088/1742-6596/279/1/011002

Estimation of tissue lesion created by a transmitter with lens using a virtual source approach

Dong Zhang, Tianbo Fan, Xiufen Gong

Institute of Acoustics, Key Laboratory of Modern Acoustics(Nanjing University), Ministry of Education, Nanjing University, Nanjing 210093, China; email: dzhang@nju.edu.cn

A virtual source model is developed to describe the nonlinear acoustic field generated by a transmitter with aluminous lens. This model assumes the geometry focal length of the virtual concave transducer to be equal to the acoustic focal length of the real transducer. Thus the acoustic field in the focal area can be accurately presented by that generated by the virtual source. To examine its validity, this theoretical model in combination with the bio-heat equation is applied to the size estimation of tissue lesion created by high intensity focused ultrasound. A transmitter with aluminous lens is used to create tissue lesions for bovine livers in vitro under three different focal acoustic intensities of 7, 13.1, 25.4 kW/cm^2. The predicted tissue lesion volumes coincide well with those previously reported measured results for 7 kw/cm^2, but are smaller those in the measurement for 13.1 and 25.4 kW/cm^2 owing to effect of vapor bubbles.

Journal of Physics: Conference Series **279** (2011) 011002 doi:10.1088/1742-6596/279/1/011002

Scattering of the field of a multi-element phased array used for ultrasonic surgery

P. Gélat[1], N. Saffari[2] and G. ter Haar[3]

(1) *Acoustics and Ionising Radiation Division, National Physical Laboratory, Hampton Road, Teddington TW11 0LW, UK*
(2) *Department of Mechanical Engineering, University College London, Torrington Place, London WC1E 7JE, UK*
(3) *Therapeutic Ultrasound Group, Physics Department, Institute of Cancer Research, Sutton, SM2 5NG, UK*
Pierre.Gelat@npl.co.uk

The efficacy of high intensity focused ultrasound (HIFU) for the non-invasive treatment of cancer has been clearly demonstrated for a range of different cancers including those of the liver, prostate and breast. As a non-invasive focused therapy, HIFU offers considerable advantages over other techniques such as chemotherapy and surgical resection, in terms of invasiveness and risk of harmful side effects. Despite its advantages, however, there are a number of significant challenges currently hindering its widespread clinical application. One of these challenges is the need to transmit sufficient energy through the ribcage to induce tissue necrosis at the required site whilst minimising the formation of side lobes, so sparing normal tissue.

Multi-element random arrays are currently showing great promise in overcoming limitations of single-element transducers. Nevertheless, successfully treating a patient for primary or secondary liver tumours requires a thorough understanding of the way in which the ultrasonic pressure field from a HIFU array is scattered by the ribcage. The acoustic transfer characteristics between the surface of each element of the array and the treatment location in the presence of the ribs may then be inferred. An approach based on the Kirchhoff-Helmholtz integral was investigated using distributed computing and its usefulness in predicting the scattered field at high *ka* values was assessed.

A passive acoustic system for extracorporeal lithotripsy to estimate early on treatment effectiveness

F Fedele[1], K Thomas[2], TG Leighton[3], S Ryves[2], AJ Coleman[1]

[1] *Medical Physics Department, Guy's and St Thomas NHS Foundation Trust, London, UK (fiammetta.fedele@gstt.nhs.uk)*
[2] *Urology Department, Guy's and St Thomas NHS Foundation Trust, London, UK*
[3] *Institute of Sound and Vibration Research, University of Southampton, UK*

Although extracorporeal shockwave lithotripsy has now been in the clinic for at least three decades, there has been little advance in efforts (i) to estimate the efficacy of the treatment whilst it is in progress, or (ii) to determine the end-point of a treatment session in terms of the degree of stone fragmentation achieved [1]. In vitro experimentation [2, 3] and clinical trials have shown that a passive diagnostic system, developed in a collaboration between Guy's and St Thomas NHS Trust and the University of Southampton, has the potential to provide evidence of effectiveness and end-point [2-6].

The system exploits secondary emissions generated during shock-tissue interaction, whose features depend on the stone targeting and stone fragmentation [2]. A clinical prototype passive acoustic sensor (PAS) has been in use since March 2009 by urologists at Guy's and St Thomas NHS Trust [7]. This paper presents the results for the first subset of 25 treatments monitored.

An X-Ray follow-up assessment could be performed only in 23 cases. In 15 of the 18 treatments that generated usable signals, the PAS device correctly predicted the outcome, as judged by clinical follow-up. The system failed to classify two successful treatments, whilst one additional successful treatment was classified as borderline. In the remaining 5 of the 23 treatments, it was possible to determine from the results that lack of acoustic coupling and other technical issues resulted in a data-set that could not be analysed. The device is being redesigned to reduce the likelihood of such problems recurring. Interestingly, the measure of effectiveness provided by the PAS device after 500 shocks was predictive of the final treatment outcome ($p< 0.001$). This suggests that the PAS system could be used in pre-assessment; indicating if the stone is susceptible to ESWL or if the patient should be sent for surgery.

[1] Skolarikos A., Alivizatos G. and de la Rosette J.: Extracorporeal shock wave lithotripsy 25 years later: complications and their prevention. *European. Urology*, 2006, 50 (5), 981-990.

[2] Fedele F. Acoustic sensing of renal stone fragmentation in extracorporeal shockwave lithotripsy. PhD Thesis, University of Southampton, 2008.

[3]Leighton T. G., Fedele F., Coleman A., McCarthy C., Jamaluddin A. R., Turangan C. K., Ball G J., Ryves S., Hurrell A., De Stefano A., White P. R.. The development of a passive acoustic device for monitoring the effectiveness of shockwave lithotripsy in real time. *Hydroacoustics*, 2008, 11, 159-180.

[4]Leighton T. G., Fedele F., Coleman A., McCarthy C., Ryves S., Hurrell A., De Stefano A., White P. R. Clinical Studies Of Real-Time Monitoring Of Lithotripter Performance Using Passive Acoustic Sensors. *Proceedings of the 2nd International Urolithiasis Research Symposium*, 2008, Indianapolis, 256-277.

[5]Leighton T. G., Fedele F., Coleman A. J., McCarthy C., Ryves S., Hurrell A. M., De Stefano A. and White P. R.: A passive acoustic device for real-time monitoring of the efficacy of shockwave lithotripsy treatment. *Ultrasound Medicine and Biology*, 2008, 34(10), 1651-1665.

[6] Turangan C.K., Jamaluddin A.R., Ball G.J. and Leighton, T.G.: Free-Lagrange simulations of the expansion and jetting collapse of air bubbles in water, *Journal of Fluid Mechanics*, 2008, 598, 1-25

[7] Smith N.C., Fedele F., Ryves S., Leighton T.G., Coleman A.J., Thomas K.: The 'Smart Stethoscope': Predicting the outcome of lithotripsy, *BJU International*, 2009,103(4), 32.

Journal of Physics: Conference Series **279** (2011) 011002 doi:10.1088/1742-6596/279/1/011002

Biological response *in vitro* of skeletal muscle cells treated with different intensity continuous and pulsed ultrasound fields

Viviane M. Abrunhosa[1,2], Laise M. Campos[2], Claudia S. Mermelstein[2], Manoel L. Costa[2], Rodrigo P.B. Costa-Felix[1]

[1] *Laboratory of Ultrasound, Directory of Scientific and Industrial Metrology (DIMCI), National Institute of Metrology, Standardization, and Industrial Quality (Inmetro), Av. Nossa Sra das Graças, 50 (Prédio 1), Duque de Caxias, RJ, Brazil, ZIP 25250-020*

[2] *Laboratory of Muscle Differentiation and Cytoskeleton, Biomedical Sciences Institute, Federal University of Rio de Janeiro (UFRJ), Cidade Universitária, Rio de Janeiro, RJ, Brazil, ZIP 21949-590*

rpfelix@inmetro.gov.br

Therapeutic ultrasound has been used in physiotherapy to accelerate tissue healing. Although the ultrasonic wave is widely used in clinical practice, not much is known about the biological effects of ultrasound on cells and tissues. This study aims to evaluate the biological response of ultrasound in primary cultures of chick myogenic cells. To ensure the metrological reliability of whole measurement process, the ultrasound equipment was calibrated in accordance with IEC 61689:2007. The skeletal muscle cells were divided in four samples. One sample was used as a control group and the others were submitted to different time and intensity and operation mode of ultrasound: 1) 0.5 W/cm_2 continuous for 5 minutes, 2) 0.5 W/cm_2 pulsed for 5 minutes, 3) 1.0 W/cm_2 pulsed for 10 minutes. The samples were analyzed with phase contrast optical microscopy before and after the treatment. The results showed alignment of myogenic cells in the sample treated with 0.5 W/cm_2 continuous during 5 minutes when compared with the control group and the other samples. This study is a first step towards a metrological and scientific based protocol to cells and tissues treatment under different ultrasound field exposures.

Journal of Physics: Conference Series **279** (2011) 011002 doi:10.1088/1742-6596/279/1/011002

Speed of sound and density of ethanol-water mixtures across the temperature range 10 to 50 Celsius degrees

J Tong *, M J W Povey [1], X Zou [1], B Ward, C P Oates

Regional Medical Physics Department, Freeman Hospital, Newcastle upon Tyne, NE7 7DN, UK

[1] *Procter Department of Food Science, University of Leeds, LS2 9JT, UK*

* *jie.tong@nuth.nhs.uk*

Ethanol-water mixtures are often used in ultrasound test objects as a sound propagation medium for assessing the measurement tools provided by ultrasound imaging systems, as part of quality assurance practice [1]. They are easy to prepare with a range of speed of sound to fit different applications. However, due to the fact that there are only limited data available in the literature [2-4], it is desirable to provide a comprehensive set of speed of sound and density data at commonly used temperatures for reference purposes. These data should also be useful in the study of molecular interactions in binary solutions [5, 6].

In this paper, measurements of the speed of sound and density in ethanol-water mixtures are reported from concentrations of 3% to 40% (w/w) across the temperature range 10 to 50 °C. The speed of sound was measured using a `ResoScan Research System` (TF Instruments GmbH, Germany), based on the ultrasound resonator method; the density was measured using an `Anton Paar DMA 4500 M` densitometer (Anton Paar GmbH, Austria), based on the oscillating U-tube method. The accuracies of the two measurement types were ± 0.25 m/s and ± 0.05 kg/m^3, respectively, which were better than those previously published. The measurement data were interpolated and presented analytically in a muti-polynomial form as functions of the temperature and the concentration. Based on these results, parameters such as the temperature coefficient of the speed of sound, the thermal expansion coefficient, and the isentropic compressibility of the mixture were obtained. Potential applications of these parameters in terms of molecular acoustics are also discussed.

[1] International Electrotechnical Commission (IEC), "IEC 61391-1 Ed. 1.0: Ultrasonics – pulse-echo scanners – part 1: techniques for calibrating spatial measurement systems and measurement of system point-spread function response", IEC, Geneva, Switzerland, 2006

[2] K Martin and D Spinks, "Measurement of the speed of sound in ethanol/water mixtures", Ultrasound in Med. & Biol., Vol. 27(2), 289-291, 2001

[3] A Gilacomini, "Ultrasound velocity in ethanol-water mixtures", J. Acoust. Soc. Am., Vol. 19(4), 701-702, 1947

[4] D R Lide, et al., "CRC Handbook of chemistry and physics", 82th Ed., CRC Press Inc., Florida, 2000.

[5] G A'Arrigo, et al., "Acoustic properties of water-ethanol mixtures at low temperature', J de Physique, Col. C7 (9), 185-193, 1984

[6] A P Sarvazyan, "New frontiers of molecular acoustics in liquids", J. Acoust. Soc. Am., Vol. 96(5), 3225-3230, 1994.

Journal of Physics: Conference Series **279** (2011) 011002 doi:10.1088/1742-6596/279/1/011002

The effects of focused transducer geometry and sample size on the measurement of ultrasonic transmission properties

T J Atkins[1], F A Duck[1] and, V F Humphrey[2]

[1]*Department of Medical Physics and Bioengineering, Royal United Hospital, Bath, BA1 3NG UK.*
[2]*Institute of Sound and Vibration Research, University of Southampton, Southampton, SO17 1BJ, UK.*
Timothy.Atkins@ruh-bath.nhs.uk

The response of two coaxially aligned weakly focused ultrasonic transducers, typical of those used for measuring the attenuation of small samples using the immersion method, has been investigated. A theoretical expression has been used to examine how the transducer separation affects the receiver output. The effects of the sample size on transmission measurements have also been analyzed by integrating the sound pressure distribution functions of the radiator and receiver over different limits to determine the size of the region that contributes to the system response. The results enable the errors introduced in to measurements of attenuation to be estimated as a function of sample size. Smaller samples are required if the measurements are made with unpaired transducers rather than a pair of identical transducers. The calculations are compared with an experimental study of the axial response of three unpaired transducers in water. The separation of each transducer pair giving the maximum response was determined, and compared with the field characteristics of the individual transducers. The optimum transducer separation, for accurate estimation of sample properties, was found to fall between the sum of the focal distances and the sum of the geometric focal lengths as this reduced diffraction errors.

Journal of Physics: Conference Series **279** (2011) 011002 doi:10.1088/1742-6596/279/1/011002

A systematic study of uncertainties associated with the Finite Amplitude method for determining the acoustic nonlinearity parameter

Bajram Zeqiri[1], Lise Retat[2] and Pierre Gélat[1].

[1]National Physical Laboratory, Acoustics and Ionising Radiation, Hampton Road, Teddington, TW11 0LW, United Kingdom (bajram.zeqiri@npl.co.uk).

[2]Therapeutic Ultrasound Group, Physics Department, Institute of Cancer Research, Sutton, SM2 5NG, United Kingdom.

The nonlinearity parameter (*B*/*A*) plays a key role in controlling finite amplitude propagation and consequent acoustic waveform distortion in any medium. Conventionally, the most convenient measurement method for *B*/*A* has been the Finite Amplitude Insertion Substitution (FAIS) technique where distortion in a test medium is compared to that through an equivalent thickness of water. This paper describes a systematic study of the FAIS method made using a 30 mm diameter hydrophone as a detector. The use of such a large area receiver ensures that effects due to diffraction are minimised, allowing measurements to be made at reduced transducer-receiver distances. An attenuator, closely matched to water in terms of acoustic impedance, positioned at the entry window of the test cell, ensures the amount of second harmonic signal level entering the cell is minimised.

The performance of the measurement system has been investigated through characterisation of two reference fluids: Corn Oil and Ethylene Glycol. Derived values of *B*/*A* will be compared with literature and presented alongside an assessment of uncertainties. The activity described is part of a programme of work at NPL to develop standardised methods of measurement of key ultrasonic properties of materials, including speed of sound, attenuation and backscatter coefficient.

POSTERS

Advanced Metrology for Ultrasound in Medicine (AMUM 2010) IOP Publishing
Journal of Physics: Conference Series **279** (2011) 011002 doi:10.1088/1742-6596/279/1/011002

Measurement system for experimental determination of acoustic properties of gels at INRIM

C.Musacchio, G.Durando, A.Bernardi, A.Troia

INRIM - Istituto Nazionale di Ricerca Metrologica-, Strada delle Cacce 91, 10135 –Torino, Italy; c.musacchio@inrim.it

Due to the large diffusion of ultrasonic technique in medicine, use of tissue mimicking materials to understand physical interaction of ultrasound has grown in importance. From this view measuring acoustic properties of such materials is crucial.

This work shows the settlement at INRIM laboratory of a simple system for acoustic properties measurements with particular interest in attenuation of ultrasound.

The system is based on substitution method and Fourier analysis and allow the evaluation of speed of sound, impedance and attenuation in the frequency range between 3 MHz and 12 MHz.

Measurements have been done on gel samples. In particular the system ah been used to check the characteristics and the behavior of different preparations of polyacrylamide gels realized in the laboratory.

Journal of Physics: Conference Series **279** (2011) 011002 doi:10.1088/1742-6596/279/1/011002

Ultrasound quality control in Italy: present and future

S. Balbis[1], C. Musacchio[2], C. Guiot[1], R. Spagnolo[2]

[1]*Dipartimento di Neuroscienze, Università di Torino*
[2]*INRIM – Istituto Nazionale di Ricerca Metrologica -, Strada delle Cacce 91, 10135 –Torino, Italy;* c.musacchio@inrim.it

Ultrasound (US) diagnostic equipments are widely diffused in Italy but, in spite of the progressive obsolescence of "aged" scanner and the increased complexity of the more recent instrumentation, a national quality assurance program for testing the performances of the US equipments is still missing.

According to ISPESL-Ministry of Health (1999) and SIRM (Società Italiana di Radiologia Medica, 2004), US quality controls are recommended, but such "good practise" is only restricted to a few public sanitary structures, ASLs (e.g. Aosta, Varese).

A joint Research Centre among the three Piedmontese Universities and INRIM, partially funded by Regione Piemonte, has been established in 2009 as Reference Center for Medical Ultrasounds. In addition to research, development and training tasks, the Center aims to the diffusion of the quality assurance in clinical US equipments. According to data from Italian National Ministry of Health (2006), around 7 % of US diagnostic equipment (946 over 13526) are located in Piedmont: 75.6% in public hospitals, 9.3 % in conventionated hospitals, 4.3% in public and 10.8% in private territorial structures.

Our goal is the formation of a regional database, which progressively includes data related to acceptance test, constancy and Quality Control tests and maintenance, in order to drive equipment turnover and carefully monitoring equipment efficiency.

The Importance of Expression of Uncertainty of Acoustical Parameters of Ultrasonic Phantoms

A. B. B. Souza[1], L. E. Maggi[1,2,3], R. M. Ichinose[1], W. C. A. Pereira[1], M. A. von Kruger[1]
[1]Programa de Engenharia Biomédica / COPPE – UFRJ, Rio de Janeiro, Brasil.
[2] Departamento de Fisioterapia / Universidade Estadual de Goiás, Goiânia, Brasil.
[3]Ultrasound Laboratory, Diavi/Dimci/Inmetro, Duque de Caxias, RJ, Brasil.

luis.maggi@gmail.com

Uncertainty is defined as a means to estimate value dispersion. It is a concept being increasingly applied in scientific experiments aiming at determining the main sources of results variability. The longitudinal velocity and attenuation coefficient are acoustic parameters commonly used to characterize materials and biological tissues. In this work it is studied the uncertainty in experiments designed to evaluate these parameters on two different materials (silicone rubber and PVCP). The uncertainties were studied following the Guide to the Expression of Uncertainty in Measurement and calculated by a program in Labview8.6. One setup was developed to measure the acoustic parameters by a transmission/reception technique. Five signals of each medium (water and materials) were collected. The attenuation coefficient was calculated using the relation between the amplitude spectrum peak of the water signal and its respective point on the spectrum of the material signal. The longitudinal velocity was calculated using the time delay between signal peaks (from water and from the material). The individual uncertainties of each part of the setup were estimated and these values allowed the identification of the sources that most contributed to increase the value of associated uncertainty. It, thus, permitted to improve experiment quality and reliability.

Advanced Metrology for Ultrasound in Medicine (AMUM 2010) IOP Publishing
Journal of Physics: Conference Series **279** (2011) 011002 doi:10.1088/1742-6596/279/1/011002

Ultrasound Metrology in Mexico: a round robin test for medical diagnostics

R. Amezola Luna, A.L. López Sánchez, A.A. Elías Juárez

Centro Nacional de Metrología (CENAM, México)
Carretera a los Cués km.4.5, El Marqués, Querétaro, México. 76246.
e-mail: ramezola@cenam.mx

Abstract

This paper presents preliminary statistical results from an on-going imaging medical ultrasound study, of particular relevance for gynecology and obstetrics areas. Its scope is twofold, firstly to compile the medical ultrasound infrastructure available in cities of Queretaro-Mexico, and second to promote the use of traceable measurement standards as a key aspect to assure quality of ultrasound examinations performed by medical specialists. The experimental methodology is based on a round robin test using a characterized ultrasound phantom for medical imaging. The physician, using its own ultrasound machine, couplant and facilities, measures the size and depth of a set of pre-defined reflecting and absorbing targets of the reference phantom, which simulate human illnesses. Measurements performed give the medical specialist an objective feedback regarding some performance characteristics of their ultrasound examination systems, such as measurement accuracy, dead zone, axial-lateral resolution, depth of penetration and anechoic targets detection. By the end of March 2010, close to one hundred entities with medical ultrasound facilities, from both public and private institutions, have performed measurements. A network of medical ultrasound calibration laboratories in Mexico, with traceability to national measurement standards, may indeed contribute to reduce measurement deviations and thus attain better diagnostics.

Journal of Physics: Conference Series **279** (2011) 011002 doi:10.1088/1742-6596/279/1/011002

Sensibility of hydrous ethanol adulteration detection using ultrasonic parameters validated in a metrological base

Monique K-K Figueiredo[1,2], Luis E. Maggi[1,3], André V. Alvarenga[1], Marcio F. Portilho[4], Gilberto A. Romeiro[2], Rodrigo P.B. Costa-Felix[1]

[1]*Laboratory of Ultrasound, Directory of Scientific and Industrial Metrology (DIMCI), National Institute of Metrology, Standardization, and Industrial Quality (Inmetro), Av. Nossa Sra das Graças, 50 (Prédio 1), Duque de Caxias, RJ, Brazil, ZIP 25250-020*

[2]*Depto of Organic Chemistry, Institute of Chemistry, Fluminense Federal University (UFF), Campus of Valonguinho,Outeiro de São João Batista, s/n, Niteroi, RJ, Brazil, ZIP 24020-150*

[3]*Biomedical Engineering Program, COPPE/Federal University of Rio de Janeiro (UFRJ), Horácio Macedo Ave., n.º 2030, Center of Technology, Block H – room/327 Cidade Universitária, Rio de Janeiro, RJ, Brazil, ZIP 21941-914*

[4]*Renewable Energy Division, Petrobras Research and Development Center (CENPES/ PETROBRAS), Horácio Macedo Ave., n.º 950, Cidade Universitária, Rio de Janeiro, RJ, Brazil, ZIP 21941-598*

[1]*rpfelix@inmetro.gov.br*

The aim of this study is to identify possible changes in fuels, in this case hydrous ethanol, through ultrasonic parameters such as attenuation and propagation speed. The system setup and method were implemented at Inmetro's Laboratory of Ultrasound. Experiments and method uncertainties were assessed accordingly to ISO/IEC Guide 98 1:2009 (Uncertainty of measurement – Part 1: Introduction to the expression of uncertainty in measurement). Mixtures of ethanol and water varying from 90% to 98% of alcohol in mass were used as testing samples. Attenuation and propagation speed were accurately measured and uncertainties evaluated. A calibration chart were the final result, and illegal (accordingly to Brazilian biofuel regulations) concentration of water could be determined with a probability higher than 90% close to permit limit of 92% of alcohol in mass. To achieve that figure, a functional combination of tested parameters should be implemented. Those results could be used as a tool to identify adulteration of biofuel, even in analysis performed on site.

Ultrasonic immersion probes characterization for use in Non Destructive Testing accordingly to EN 12668-2:2001

Cristiane E.R. Silva, André V. Alvarenga, Rodrigo P.B. Costa-Felix

Laboratory of Ultrasound, Directory of Scientific and Industrial Metrology (DIMCI), National Institute of Metrology, Standardization, and Industrial Quality (Inmetro), Av. Nossa Sra das Graças, 50 (Prédio 1), Duque de Caxias, RJ, Brazil, ZIP 25250-020

cesilva@inmetro.gov.br

Ultrasound is often used as a Non Destructive Testing (NDT) technique to analyze components and structures to detect internal and surface flaws. To guarantee reliable measurements, it is necessary to calibrate instruments and properly assess related uncertainties. An important device of an ultrasonic instrument system is its probe, which characterization should be performed accordingly to EN 12668-2. Concerning immersion probes beam profile, parameters to be assessed are beam divergence, focal distance, width, and zone length. Those parameters are determined scanning a reflector or a hydrophone throughout the transducer beam. Within the present work, a methodology developed at Inmetro's Laboratory of Ultrasound to evaluate relevant beam parameters is presented, based on hydrophone scan. A water bath and a positioning system to move the hydrophone were used to perform the scan. Studied probes were excited by a signal generator, and the waterborne signals were detected by the hydrophone and acquired using an oscilloscope. An user-friendly virtual instrument was developed in LabVIEW to automate the system. The initial tests were performed using a 1 MHz-ultrasonic unfocused probe (Ø = 1.27 cm), and results are consistent with manufacturer's specifications. Besides, expanded uncertainties are lower than 7% for all parameters under consideration.

Advanced Metrology for Ultrasound in Medicine (AMUM 2010) IOP Publishing
Journal of Physics: Conference Series **279** (2011) 011002 doi:10.1088/1742-6596/279/1/011002

Ultrasonic calibration and certification of reference standard blocks types V1 and V2 for use in Non Destructive Testing

Cristiane E.R. Silva[1], Luis E. Maggi [1,2], Rodrigo P.B. Costa-Felix[1]

[1] *Laboratory of Ultrasound, Directory of Scientific and Industrial Metrology (DIMCI), National Institute of Metrology, Standardization, and Industrial Quality (Inmetro), Av. Nossa Sra das Graças, 50 (Prédio 1), Duque de Caxias, RJ, Brazil, ZIP 25250-020*

[2] *Biomedical Engineering Program, COPPE/Federal University of Rio de Janeiro (UFRJ), Horácio Macedo Ave., n.º 2030, Center of Technology, Block H – room/327 Cidade Universitária, Rio de Janeiro, RJ, Brazil, ZIP 21941-914*

[1]*cesilva@inmetro.gov.br*

Ultrasonic Non Destructive Testing (US-NDT) has many industrial applications, such as flaw detection, welding integrity evaluation and thickness measurement of solid material. To evaluate the instrumentation conditions, prior to a measurement begins a "calibration" should be undertaken with the aid of a reference standard block (RSB). Types V1 (ISO 2400) and V2 (ISO 7963) are the most widely used in US-NDT. Due their characteristics, V1 and V2 RSB can be classified as Reference Material, so that they can be certified, resulting in a Certified Reference Material (CRM). This paper describes technical aspects of a V1 and V2 RSB certification based on ISO 2400, ISO 7963, EN 12223 and EN 27963. Accomplishing to the certification protocol established at Inmetro's Laboratory of Ultrasound, RSB relevant dimensions were determined through calibration. Ultrasound speed and attenuation were assessed using a scope and signal generator at 5 MHz. Uncertainties were determined including measurement precision under repeatability conditions. Attenuation was calculated after an exponential best fitting of the envelope peaks of each echo. A complete uncertainty budget was determined for both parameters. Within the specimens used, velocities were in accordance with specifications 5920 ± 30 m/s), but measured attenuation of 0.1 dB/cm was not.

Journal of Physics: Conference Series **279** (2011) 011002 doi:10.1088/1742-6596/279/1/011002

The Ultrasound effects on non tumoral cell line at 1 MHz therapeutic frequency.

L. Di Giambattista[(1,2)], P. Grimaldi[(1)], I. Udroiu[(3)], D. Pozzi[(4)], G. Cinque[(5)], M.D. Frogley[(5)], S. Buogo[(6)] and A. Congiu Castellano[(1)].

[(1)] *Physics Department, Sapienza, University of Rome (IT).*
[(2)] *CISB, Sapienza, University of Rome (IT).*
[(3)] *DIPIA, ISPESL, via Urbana 167, Rome (IT).*
[(4)] *Experimental Medicine and Pathology Department, Sapienza, University of Rome (IT).*
[(5)] *Diamond Light Source Ltd, Didcot, Oxfordshire (UK).*
[(6)] *Institute of Acoustics O. M. Corbino, via del Fosso del Cavaliere 100, Rome (IT).*

Presenting author:*l.digiambattista@caspur.it*

The aim of this research is to investigate some bioeffects due to *Therapeutic Ultrasound* (1 MHz and $50<I_{PA}<60$ W/cm^2) which could allow to enhance drugs or genes delivery in non tumoral cells [1]. *Ultrasound* (US) has been demonstrated to alter the cell membrane permeability due to a biophysical mechanism (*Sonoporation*) and exploited as a promising non-invasive gene transfer method. We have used the NIH-3T3 cell line as a model system and exposed it to US medical equipment for 15, 30, 45, 60 minutes at distances of 10 and 15 cm from the source transducer, corresponding to the far field region where $z > z_{last} = \frac{a^2}{4\lambda} = 4.0 \pm 0.4$ cm. We work with the maximum power in pulsed system with 75% duty cycle. Characterization of the unfocused, planar and with a circular geometry 1 MHz source transducer, was performed by the Institute of Acoustics O. M. Corbino. The acoustics pressure was measured by a calibrated 0.5 mm needle hydrophone. The US effects on cells were assessed by *FTIR Imaging* with focal plane array FPA detector and *cytokinesis-block micronucleus (CBMN)* assay. The combination of IR analysis results with the cytogenitic assay outcomes has shown that these experimental conditions do not cause DNA mutations in the NIH-3T3 cell line.

[1] L. Conti, P. Grimaldi, I. Udroiu, A. Bedini, C. Giliberti, L. Giuliani, R. Palomba, A. Congiu Castellano. Effects induced in cells by ultrasound revealed by ATR-FTIR spectroscopy. *Vibrat. Spectrosc. (2010)* 52:79-84.

Nonlinear acoustics determination of phase characteristics of PVDF membrane hydrophones

Philip E Bloomfield[1], Gaurav Gandhi[2] and Peter A Lewin
Drexel University School of Biomedical Engineering, Science, and Health Systems, Philadelphia, PA 19104-2875, USA

Abstract. When an ultrasonic pressure wave propagates through a nonlinear medium, the relative phasing of the generated harmonics causes a distinct asymmetry between the positive and negative pressure levels and between the rise and fall time of examined waveforms. A faithful quantitative reproduction of the source transducer's pressure field requires amplitude and phase measurements by calibrated hydrophone probes. Nonlinear hydrophone calibration provides amplitude and phase information at discrete multiples of an acoustic source's fundamental frequency. Two PVDF bilaminar membrane hydrophones were first calibrated in terms of their amplitude sensitivity to the pressure levels generated by two different HIFU (High Intensity Focused Ultrasound) circular source transducers operating at 5 MHz and 10 MHz, enabling phase studies up to 105 and 100 MHz, respectively. Introducing two newly-developed phase-dispersion representations, the phase responses of the two membrane hydrophones were determined with respect to the phase of the complex frequency response extracted from the nonlinear field simulated by a semi-empirical computer model which predicts the near and the far field pressure distributions. These phase differences compared favorably with the results obtained from the commercially available *PiezoCAD* simulation model. The protocol for specifying the complex pressure field of source transducers through measurements using the calibrated hydrophones is described. The results obtained indicate that the membranes exhibit close to linear decay of phase against the frequency.

1. Introduction and Background

In an ultrasound field where the acoustic spectrum contains a significant spread of frequencies the output voltage waveform, which is the acoustic spectrum convolved with the complex frequency response of the measurement system, will differ in shape from the true pressure waveform since the amplitude and phase response of PVDF membrane hydrophones are frequency dependent. The determination of peak rarefactional and peak compressional pressure amplitudes requires phase response data in the frequency range considered. Hence, a complete description of a source transducer's pressure field should consist of the phase transfer function (PTF) as well as the amplitude transfer function (ATF). To recover the true pressure-time waveform of a source transducer, the measured signal's spectrum must be deconvolved from the hydrophone's spectral response by means of its calibration data available as complex sensitivity versus frequency response. Cooling and Humphrey [1] proposed utilizing nonlinear wave propagation and a time-shifted relative phases approach to phase characterize membrane hydrophones. This work verifies their approach and provides evidence that their approach can be replaced by a direct comparison of the phases relative to the phase values at the fundamental. In the following sections, the semi-empirical methodology used to extract the characteristic phase spectra of two PVDF membrane hydrophones from their voltage-time measurements of the nonlinear pressure-time waveform generated by HIFU sources is presented. The amplitude and phase spectra of a 5 MHz and a 10 MHz HIFU source were measured by a *Marconi* and a *Sonora* bilaminar hydrophone, respectively, whose phase characteristics were obtained and displayed in the 100 MHz bandwidth by utilizing two different schemes for extracting hydrophone

[1] Any correspondence should be addressed to: bloomfpe@drexel.edu
[2] Current Address: Artann Laboratories Inc., Ewing, NJ

Journal of Physics: Conference Series **279** (2011) 012001 doi:10.1088/1742-6596/279/1/012001

phase spectra characteristics. This work supports using the nonlinear propagation method can be used to calibrate PVDF membrane hydrophones allowing faithful reproduction of the measured pressure-time waveform, providing the hydrophone's complex sensitivity is known at discrete frequencies, which are multiples of the fundamental of an acoustic source.

2. Methodology

Two different schemes for extracting hydrophone phase spectra characteristics from nonlinear acoustic field measurements were used. In section 2.2 the steps to obtain the results of section 3 are given.

2.1. Hydrophone phase characteristics and phase spectrum of source transducer.

The measured voltage signal is the response of the source's pressure output convolved with the hydrophone's response characteristic. The JW non-linear program, summarized in section 2.3, predicts the complex pressure spectrum produced by the transducer source. Thus, the hydrophone sensitivity spectrum is obtained from the ratio of the measured voltage spectrum and the non-linear spectrum.

$$A_H e^{i\phi_H}{}_{[mV/MPa]} = A_M e^{i\phi_M}{}_{[mV]} / A_J e^{i\phi_J}{}_{[MPa]} \tag{1}$$

$$\frac{A_M}{A_J} e^{i(\phi_M - \phi_J)} = \frac{A_M^R + iA_M^I}{A_J^R + iA_J^I} = \frac{A_M}{A_J}\left[\left(\frac{A_M^R}{A_M}\frac{A_J^R}{A_J} + \frac{A_M^I}{A_M}\frac{A_J^I}{A_J}\right) + i\left(\frac{A_M^I}{A_M}\frac{A_J^R}{A_J} - \frac{A_M^R}{A_M}\frac{A_J^I}{A_J}\right)\right] \tag{2}$$

Then, with **(N+H)** symbolizing the hydrophone measurement of a new source transducer, its pressure output spectrum can be determined from the hydrophone's response sensitivity spectrum via:

$$A_N e^{i\phi_N}{}_{[MPa]} = A_{N+H} e^{i\phi_{N+H}}{}_{[mV]} / A_H e^{i\phi_H}{}_{[mV/MPa]} \tag{3}$$

The quantities in the brackets in (2) are identified as the cosine and sine difference formulae. Frequency spectra are provided by the F.T. programs in terms of the Real and Imaginary parts of the amplitude or the Amplitude and phase. Thus hydrophone characteristic phase spectra can be specified in two equivalent ways: via spectra of the intensive measures Re/Ampl and Im/Ampl or by the phase spectrum. Either characterization may be used to deconvolve the pressure field of a source transducer.

2.2. Step by step semi-empirical methodology for acquiring characteristic hydrophone phase spectrum from pressure-time waveform

Acquire Pressure-Time Waveform & Run JW nonlinear model program ⇒

Obtain Fourier transform spectra of: measured voltage signal (M), the JW nonlinear field prediction (J), and hydrophone modeling (PC), for verification.

⇨ $R + i*I$ ⇨ Amplitude = $A = (R^2 + I^2)^{1/2}$ at source harmonics, f_n.

Two alternative pathways* [1a)-1c) or 2a)-2c)] *to obtain the hydrophone phase spectrum ⇒

1a) Phase (ϕ) = $\tan^{-1}(I/R)$ or $\sin^{-1}(I/A)$ or $\cos^{-1}(I/A)$; assign ϕ to quadrant determined by $R + i*I$

1b) Calculate hydrophone (H) phase spectrum as: $\phi_H = \phi_{M\text{-}J} = \phi_{Meas} - \phi_J$.

1c) Compare the calculated differences between the FFT measured (M) phases and the nonlinear JW model phases: $\phi_{M\text{-}J}(f_n) - \phi_{M\text{-}J}(f_1)$, to the PiezoCAD predicted phases: $\phi_{PC}(f_n) - \phi_{PC}(f_1)$

2a) Calculate: $\cos(\phi) = R/A$ and $\text{sine}(\phi) = I/A$ for M, and JW

2b) Determine hydrophone phase spectrum directly via cosine and sine phase difference formulae applied to the measured and the JW nonlinear program's predicted phase.

2c) Compare the above 2b) responses to the PiezoCAD hydrophone cosine and sine predictions.

It is appropriate to note that the spatial averaging errors due to the underestimate of the pressure amplitude do not affect the phase measures which are calculated from the ratio of extensive quantities.

2.3 Non-linear Propagation Model

The "JW" nonlinear acoustic simulation model, used as a reference to compare with the experimental values, requires as a boundary condition the apodization curve, which is a mathematical reproduction of the surface-pressure amplitude profile of the acoustic source that is measured by a 1D scan at the surface of the source along its diameter. This semi-empirically determined representation of the pressure distribution at the surface of the source transducer is distinct from the Gaussian distribution, as in the parabolic solution to the KZK model used by some other simulation models [1,2,3]. After the needed parameters have been input, the JW program theoretically propagates the sound field away from the transducer's surface to an arbitrary field point. The output is available in the form of the complex Fourier spectra of the field generated by the source. The desired phase and amplitude information is extracted from this complex spectrum. The details of the JW model, which uses the hyperbolic propagation operator to calculate the acoustic field generated in a nonlinear and lossy medium, were explained in [4,5,6]. In this work, the JW model calculated the amplitude and phase spectra at the focal points of the 5 MHz, 20 mm diameter, focal number 1.9, *Sonic Concepts* acoustic source and the 10 MHz, 10 mm diameter focal number 4.21 *Panametrics* acoustic source. The FFT programs provided real and imaginary amplitudes as well.

3. Results

As outlined in section 2.2 the hydrophone phase spectrum is acquired from selected differences between the discrete Fourier transforms of the acquired pressure-time waveform and the nonlinear program's calculated pressure field. Selected experimental results are shown in figures 1 to 4.

The phase spectrum, $\phi_H = \phi_{Meas\text{-}JW}$ (◇), shown in figure 1 specifies the Sonora hydrophone's phase characteristic. Figure 2 shows the characteristic phase spectra for both the Marconi and Sonora bilaminar hydrophones in comparison to the PiezoCAD modeled phase spectra in comparison to the PiezoCAD modeled phase spectra in terms of the phases themselves.

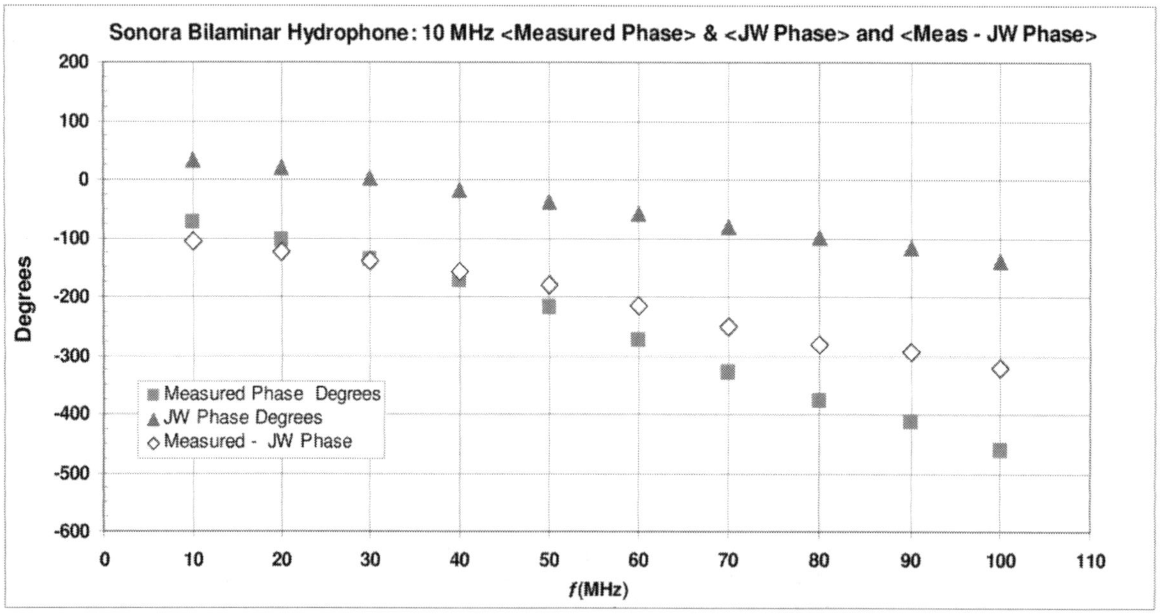

Figure 1. The 10 MHz JW and measured phases and their calculated difference, $\phi_{Meas\text{-}JW}$

The characteristic hydrophone phase spectra $\phi_H(f_n) = \phi_M(f_n) - \phi_{JW}(f_n)$ and $\phi_{PC}(f_n)$ were normalized to zero phase at the principal frequency as indicated in figure 2 where the comparison of the measured bilaminar hydrophone characteristic phase spectra to the PiezoCAD modeled phase spectra shows reasonable validation of the phase dispersion representation presented here.

The PiezoCAD model amplitude and phase spectra were determined at the harmonic frequencies. The PVDF thicknesses of the (2*9μm) Sonora and the (2*25μm) Marconi bilaminar hydrophones were modeled as (2*10μm) and (2*25.5μm) membranes to account for the adhesive thicknesses.

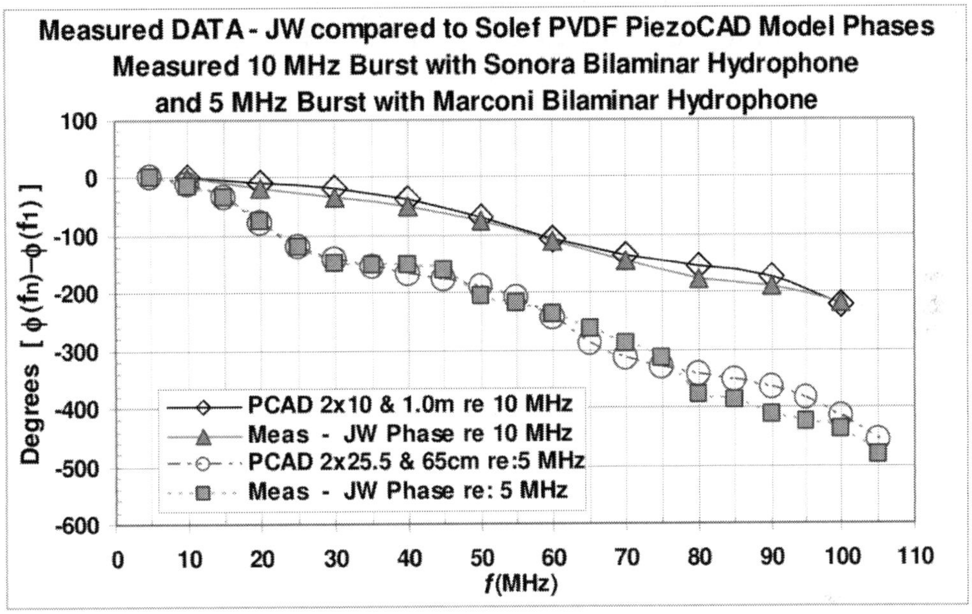

Figure 2. The bilaminar hydrophone characteristic phase spectra $\phi_H = \phi_M - \phi_{JW}$ compared to ϕ_{PC}.

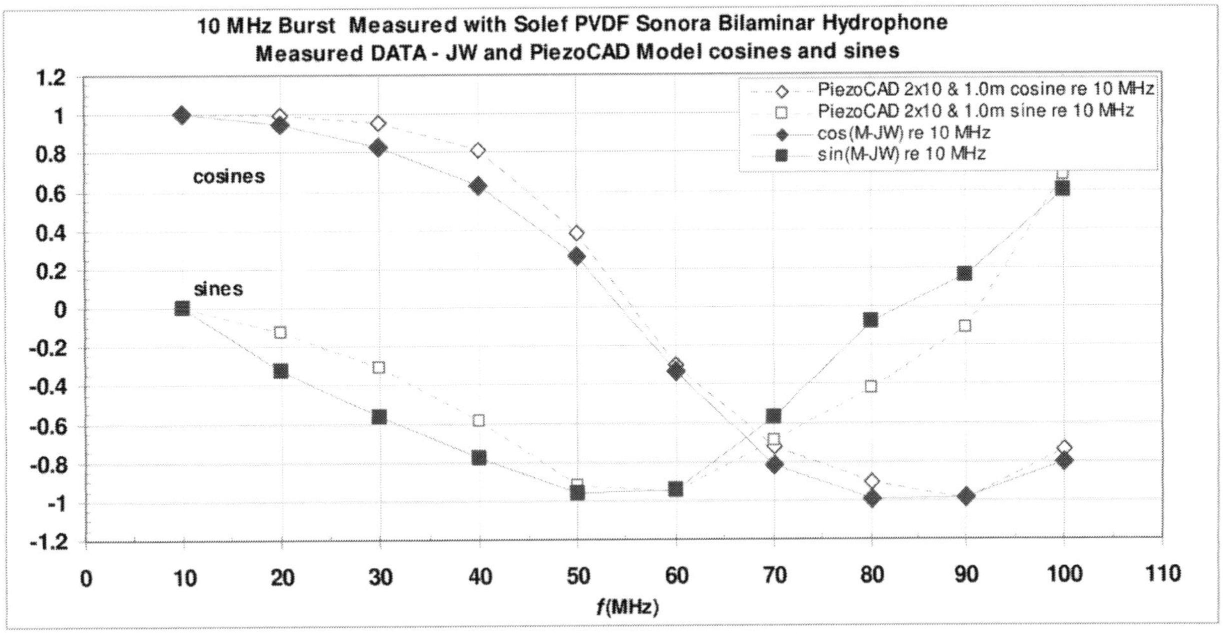

Figure 3. The Sonora hydrophone $\cos(\phi_H) = \cos(\phi_{M\text{-}J})$ and $\sin(\phi_H) = \sin(\phi_{M\text{-}J})$ phase spectra compared to the PiezoCAD calculated cosine and sine, $\cos(\phi_{PC})$ and $\sin(\phi_{PC})$.

Journal of Physics: Conference Series **279** (2011) 012001 doi:10.1088/1742-6596/279/1/012001

The phase representations in figures 3 and 4 (including the phase shift to zero phase at 10 and 5 MHz, respectively) were calculated using the cosine and sine difference formulae as discussed in sections 2.1 and 2.2. The figure comparing the Marconi spectra, $\sin(\phi_H)$ to $\sin(\phi_{PC})$, has the format of figure 4 shifted 20 MHz lower; i.e., the 45 and 85 MHz min and max for the cosine phase representations occur at 25 and 65 MHz, respectively, for the sine phase representations.

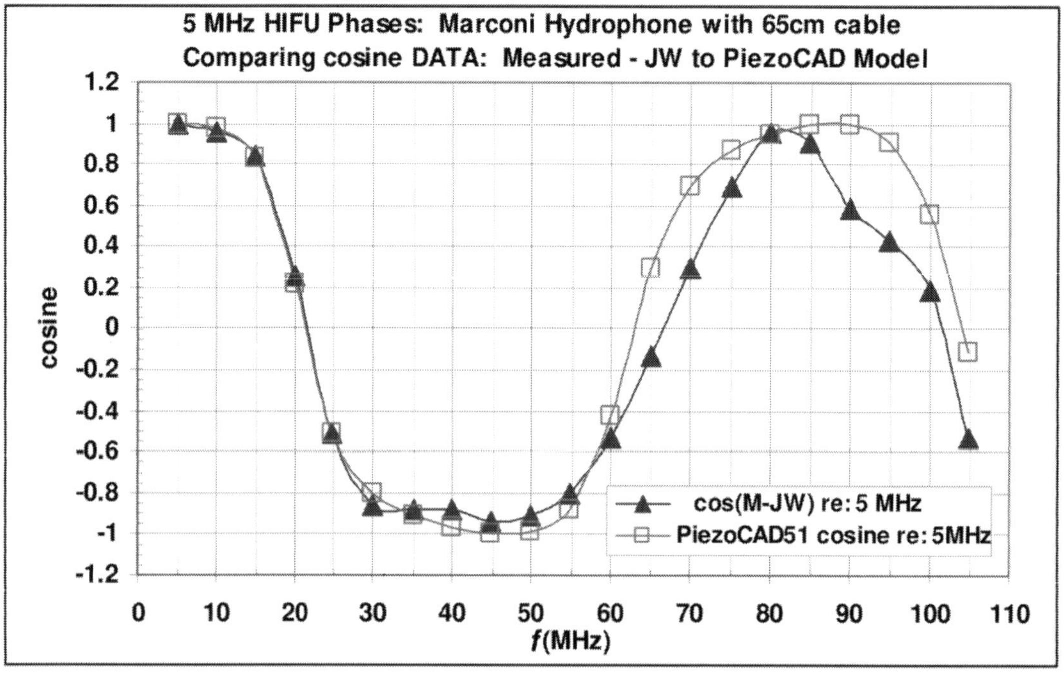

Figure 4. Marconi hydrophone $\cos(\phi_H)$ phase spectrum compared to the PC calculated $\cos(\phi_{PC})$.

4. Summary and Conclusions

A complete description of a transducer's pressure field should consist of the phase transfer function (PTF) as well as the amplitude transfer function (ATF). The faithful quantitative reproduction of a source transducer's pressure field requires amplitude and phase measurements by calibrated hydrophone probes. An ideal hydrophone should exhibit behavior close to that of an ideal point receiver. However, the available probes are not fully adequate for this purpose because the probe's frequency response is usually insufficient to cover the whole range of needed frequencies and the probe's finite aperture introduces spatial averaging errors. A probe area larger than the pressure field area leads to an underestimate of the pressure amplitude. To recover the true pressure-time waveform of a source transducer, the measured signal's spectrum must be deconvolved from the hydrophone's spectral response by means of its calibration data available as complex sensitivity versus frequency response. The nonlinear propagation method of hydrophone calibration, which provides amplitude and phase calibration information at discrete frequencies which are multiples of the fundamental of an acoustic source, was examined. The results presented were limited to examining and verifying the applicability of the non-linear propagation method to determine the phase characteristics of membrane hydrophones. Detailed algorithms for extracting phase values from hydrophone pressure measurements were given. The data presented were based on the analysis and measurement of the pressure field at the focal point of two different HIFU circular source transducers energized by 5 and 10 MHz 10-cycle burst signals, enabling phase studies up to 105 and 100 MHz, respectively by two different bilaminar PVDF membrane hydrophones. It was pointed out that the measured response included the phase characteristics of the hydrophone plus the phase of the acoustic (HIFU) source. In order to phase calibrate the membrane hydrophone, one must subtract the source's non-linear acoustic contribution to the measured field. This could be done by using a fiber-optic hydrophone implemented as a point receiver in 100 MHz bandwidth and exhibiting virtually uniform amplitude response and a zero phase shift [7]. Then the phase of the tested hydrophone signal could be directly compared with the signal measured by the fiber optic hydrophone. However, in this study the phase responses of the two membrane hydrophones were determined with respect to the phase of the complex frequency response extracted from a computer-simulated nonlinear field by subtracting the calculated harmonic frequency phase values from the corresponding measured Fourier phase values. The JW finite amplitude computer model, which assumes the hydrophone in the field behaves as a point receiver, incorporated the characteristics of the HIFU transducer as measured by a pressure scan at the surface of the acoustic source by a needle hydrophone.The PVDF hydrophone phase characteristics obtained from the difference between the calculated (finite amplitude computer model) harmonic frequency phase values and the corresponding measured Fourier phase values were compared to the phase output of the commercial program, *PiezoCAD*, which predicted the hydrophone's frequency characteristics. Alternatively, the real and imaginary parts of the Fourier spectra directly yielded the cosines and sines corresponding to the phase differences. The earlier discussed time-shifted relative phases approach of Cooling and Humphrey [32] was replaced here by a direct comparison of the phases relative to the phase values at the fundamental. The *Measured–JW* phase values (as well as the cosine and sine values) compared very well with the corresponding *PiezoCAD* model values. Based on this outcome it can be concluded that reliable PVDF membrane hydrophone phase calibration characteristics can be obtained by the non-linear propagation method using the algorithm presented. The algorithm allows the determination of the complex pressure field of source transducers through measurements using the calibrated hydrophone. The overall uncertainty of the amplitude sensitivity calibration was dependent on frequency and determined to be about ±12% up to 40 MHz, ±20% from 40 to 60 MHz and ±25% from 60 to 100 MHz.

Fundamental limitations of the approach were also discussed and these included discrete versus continuous method. The results obtained indicate that the PVDF membrane hydrophones exhibit close to linear decay of phase against the frequency. In a subsequent work the influence of the attached cable length on the hydrophone's phase and amplitude will be presented.

Acknowledgment: This research was partially supported by the National Institute of Health through R01EB007117 grant.

References

[1] Cooling M P and Humphrey V F 2008 A nonlinear propagation model-based phase calibration technique for membrane hydrophones *IEEE Trans. Ultrason. Ferroelect. Freq. Contr.* **55** 84-93

[2] Gélat P N, Wright L, Bickley C and Sadhoo N 2007 Magnitude and phase characterisation of a reference broadband membrane hydrophone up to 100 MHz using nonlinear field modelling *NPL Report AC 3* – ISSN 1754-2936

[3] Humphrey V F 2000 Nonlinear propagation in ultrasonic fields: measurements, modelling and harmonic imaging *Ultrasonics* **38** 267-72

[4] Wojcik, J 1998 Conservation of energy and absorption in acoustic fields for linear and nonlinear propagation *J. Acoust. Soc. Amer.* **104**(5) 2654-63

[5] Radulescu E G, Wojcik J, Lewin P A, Nowicki A 2003 Nonlinear propagation model for ultrasound hydrophones calibration in the frequency range up to 100 MHz *Ultrasonics* **41** 239-45

[6] Wojcik, J, Nowicki A, Lewin P A, Bloomfield P E, Kujawska T and Filipczynski L, 2006 Wave envelopes method for description of nonlinear acoustic wave propagation *Ultrasonics*, **44**(3) 310-29

[7] Umchid S, Gopinath R, Srinivasan K, Lewin P A, Daryoush A S, Bansal L and El-Sherif, M 2009 Development of calibration techniques for ultrasonic hydrophone probes in the frequency range from 1 to 100 MHz *Ultrasonics*, **49**(3) 306-11

Journal of Physics: Conference Series **279** (2011) 012002 doi:10.1088/1742-6596/279/1/012002

Hydrophone area-averaging correction factors in nonlinearly generated ultrasonic beams

M P Cooling[1], V F Humphrey[1] and V Wilkens[2]

[1] Institute of Sound and Vibration Research, University of Southampton, Southampton SO17 1BJ, UK
[2] Physikalisch-Technische Bundesanstalt, Bundesallee 100, 38116 Braunschweig, Germany

vh@isvr.soton.ac.uk

Abstract. The nonlinear propagation of an ultrasonic wave can be used to produce a wavefield rich in higher frequency components that is ideally suited to the calibration, or inter-calibration, of hydrophones. These techniques usually use a tone-burst signal, limiting the measurements to harmonics of the fundamental calibration frequency. Alternatively, using a short pulse enables calibration at a continuous spectrum of frequencies. Such a technique is used at PTB in conjunction with an optical measurement technique to calibrate devices. Experimental findings indicate that the area-averaging correction factor for a hydrophone in such a field demonstrates a complex behaviour, most notably varying periodically between frequencies that are harmonics of the centre frequency of the original pulse and frequencies that lie midway between these harmonics. The beam characteristics of such nonlinearly generated fields have been investigated using a finite difference solution to the nonlinear Khokhlov-Zabolotskaya-Kuznetsov (KZK) equation for a focused field. The simulation results are used to calculate the hydrophone area-averaging correction factors for 0.2 mm and 0.5 mm devices. The results clearly demonstrate a number of significant features observed in the experimental investigations, including the variation with frequency, drive level and hydrophone element size. An explanation for these effects is also proposed.

1. Introduction

The propagation of a high amplitude ultrasound wave is a fundamentally nonlinear process that can result in waveform distortion with the resultant generation of harmonics of the frequencies initially present in the wave. For suitable combinations of pressure and propagation distance, this eventually leads to the formation of shocks within the waveform, where the pressure changes very rapidly with position or time. Such frequency-rich waveforms are very useful in ultrasonic metrology.

For example a primary standard for ultrasound in water in the frequency range 1 to 40 MHz can be implemented using a laser interferometer to measure the displacements of a membrane hydrophone in a nonlinearly distorted wave field [1]. In this way the hydrophone output can be related to the pressure amplitude via the particle displacement measured with the interferometer. The use of a nonlinearly distorted wavefield removes the need to use a number of individual transducers.

Calibrations can also be transferred from a standard hydrophone to secondary devices by inter-calibration in a nonlinearly distorted wavefield [2]. If a tone-burst drive is used, and a quasi-CW section of the resultant nonlinearly distorted waveform is analysed, then results are obtained at harmonics of the source frequency. Calibration is then achieved at these discrete frequencies, with the frequency sampling determined by the fundamental frequency used (typically 1 MHz).

A range of developments are driving the need to perform hydrophone calibrations at higher ultrasonic frequencies (up to 100 MHz). This implies the use of a higher frequency transducer (about 5 MHz) as the source for the nonlinearly distorted wavefield. However, if a 5 MHz tone burst is used then the calibrations will only be obtained with a low frequency resolution due to the harmonic spacing of 5 MHz. An alternative approach, that has been used at Physikalisch-Technische Bundesanstalt (PTB) in Germany, is to use a very short acoustic pulse generated as the result of a pulsed discharge into a transducer (see figure 1) [3]. The initial pulse has a broadband spectrum centred on a peak frequency determined by the resonant frequency of the source. Consequently, the waveform generated by nonlinear propagation has a continuous spectrum extending up to high frequencies. The resulting spectrum does however show maxima at frequencies that are close to harmonics of the centre frequency of the source waveform, with dips in between at 'intermediate' frequencies.

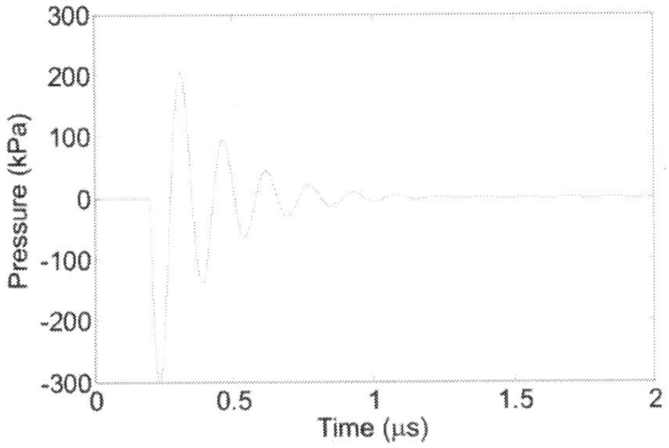

Figure 1. Example source waveform used in numerical model; this is based on that used for the experimental measurements performed at PTB in reference [3].

A hydrophone will in general give an output that is related to the average pressure across its active area. If the acoustic field varies in amplitude (or phase) across this area then the hydrophone output will be different from that which would have been obtained with a point-like device. The ratio of the spatial average to axial pressure is termed the area-averaging factor. Since spatial averaging has the potential to affect calibration, it is important to know this factor. It will, in general, depend on both the device and the field.

While calibrating hydrophones in the nonlinear pulsed wavefields described, it was noted at PTB that the area-averaging factor varied significantly with frequency (see figure 2). This was calculated by using an optical system to measure the pressure amplitude across the area of the hydrophone element and numerically averaging the resultant values. In particular it was noted that the area-averaging factor varied periodically between the 'harmonic' and 'intermediate' frequencies.

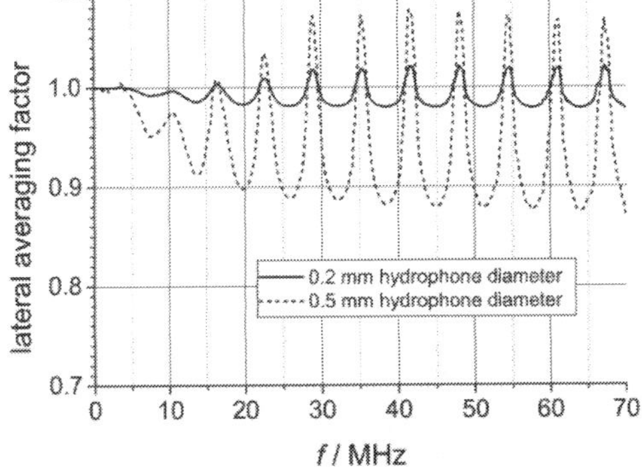

Figure 2. Area-averaging factor for planar hydrophones of 0.2 and 0.5 mm diameter calculated from experimental measurements with an optical probe at PTB [3].

This paper investigates these phenomena by use of numerical modelling. In particular it aims to understand the factors that determine the sensitivity of the area-averaging factor to the pulse characteristics. The modelling was used to identify the causes of the phenomena observed rather than exactly reproduce the experimental situation.

2. Model

A finite difference code, known as the Bergen Code, was used to solve the nonlinear Khokhlov-Zabolotskaya-Kuznetsov (KZK) equation and hence model the propagation of the finite amplitude beam generated by a focusing transducer [4]. The KZK equation is a parabolic approximation to the full nonlinear wave equation that consistently accounts for absorption, nonlinearity and diffraction in sound beams. The parabolic or paraxial approximation assumes that the energy propagates at small angles to the beam axis; this approximation is valid for acoustic sources that are many wavelengths across and not strongly focusing, and for field points that are not too far from the beam axis or too near the source plane [5]. The code used works entirely in the frequency domain, representing the waveform as a Fourier series. For the runs reported an axial step (Δz) of 0.005 mm and radial step (Δr) of 0.031 mm were used.

In order to model the propagation of a short pulse the fundamental frequency in the model was set to be 500 kHz, whereas the centre frequency of the pulse (f_c) was 6.5 MHz. The initial pressure pulse was modelled by

$$p(t) = \begin{cases} 0, & 0 < t \le t_s, \\ p_0 \exp(-t/\tau)\sin\left[2\pi f_c (t - t_s)\right], & t > t_s, \end{cases} \qquad (1)$$

where for this work $t_s = \tau = 0.2$ µs. The value of p_0 was chosen to give peak rarefactional pressures at the source of 300 kPa or 500 kPa for the results presented here.

The field of an ideal circular focusing transducer with radius 6.2 mm and geometrical focal length 56 mm was modelled; results are presented for a transverse plane at an axial distance $z = 47.5$ mm from the source where the peak rarefactional pressure was a maximum for the 500 kPa run.

In order to calculate the area-averaging effect the average pressure over the surface of a planar circular hydrophone, assumed perfectly aligned on-axis, was first calculated as a function of frequency, f. This is given by

$$p_{\text{ave}}(z, f) = \frac{1}{\pi b^2} \int_0^b p(z, r, f)\, 2\pi r \mathrm{d}r, \qquad (2)$$

where b is the hydrophone element radius and $p(r, z, f)$ is the complex pressure distribution as a function of radial position, r. The area-averaging factor (α) is then given by:

$$\alpha = \left| \frac{p_{\text{ave}}(z, f)}{p(z, 0, f)} \right|. \qquad (3)$$

The calculations were performed for hydrophones with a diameter of 0.2 and 0.5 mm. The integral of equation (2) was numerically evaluated by cubic spline interpolation of the amplitude and phase radial profiles, followed by trapezoidal numerical integration.

3. Results

Figure 3(a) shows the calculated area-averaging factor for a relatively low initial pressure at the source of 300 kPa, as a function of frequency for the two hydrophone diameters. This shows that the averaging effect is much less significant for the smaller hydrophone, and only results in a small reduction relative to the on-axis pressure, reaching a minimum value of 0.94 by 100 MHz. In contrast, the 0.5 mm device shows a much more significant effect that increases dramatically with frequency such that α is 0.47 by 100 MHz. This follows from the expected behaviour of nonlinearly propagating beams, with the higher harmonics having narrower beams. Figure 3(a) also shows periodic fluctuations in the area-averaging factor that are more pronounced for the larger hydrophone. Comparison with the pulse spectrum confirms that the minima in α correspond to frequencies that are harmonics of the centre frequency, that is where $f \approx Nf_c$ for integer N. The maxima correspond to the intermediate frequencies where $f \approx (N+1/2)\,f_c$.

In contrast figure 3(b) shows results for a significantly higher drive level of 500 kPa. For the 0.2 mm device the fluctuations are now much more pronounced, although the minima of the area-averaging factor are not significantly different in amplitude from the lower drive case. However, the peaks are much more pronounced and are visible across the whole of the frequency band. The higher amplitude of the oscillations is such that the area-averaging factor is greater than unity for some frequencies. This indicates that the average pressure over the hydrophone is larger than the pressure on axis, implying that the pressure increases with radial displacement from the axis.

These effects are clearly much more pronounced in Figure 3(b) for the larger hydrophone (0.5 mm diameter), which displays large periodic fluctuations in the area-averaging factor. The factor varies significantly in magnitude, ranging from 0.62 to over 1.24. It should be noted that the maximum values are probably not accurately reproduced due to the limited resolution of the model in the frequency domain ($\Delta f = 500$ kHz). It is interesting to note that the minimum α values are actually larger for the high drive case than for the low drive case. This can be attributed to saturation effects resulting in a flattening of the beam (as has previously been observed during farfield calibration [6]).

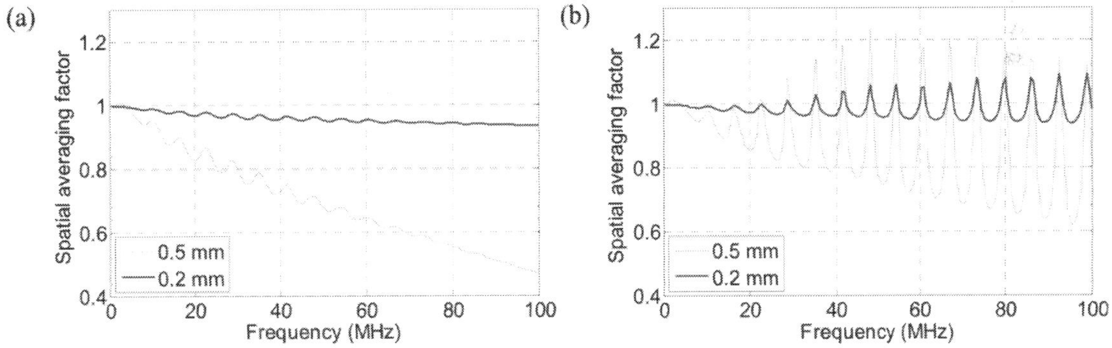

Figure 3. Area-averaging factor for hydrophones of 0.2 and 0.5 mm diameter calculated from the numerical model at a range of 47.5 mm. Results are calculated for initial peak rarefactional pressures at the source of (a) 300 kPa and (b) 500 kPa.

In order to confirm the origins of the area-averaging factor fluctuations, the radial variation of pressure amplitude is plotted in figure 4 for two particular frequency components. In figure 4(a) the results are shown for the 300 kPa case, while in figure 4(b) the corresponding results for the higher drive level are presented. For the low drive case the pressure amplitude is a maximum on-axis for both a frequency of 39 MHz, the 6th harmonic of the centre frequency, and 35.5 MHz, an intermediate frequency (half way between the 5th and 6th harmonics). Notably the two levels are not that dissimilar on axis and rapidly approach each other off axis. In contrast the corresponding curves for the higher drive case (figure 4(b)) are very different. In particular the curve for the intermediate frequency now has a minimum on axis, with the pressure amplitude increasing with radial displacement. This causes the area-averaging factor to be greater than unity. It is also noticeable that the pressure amplitude on axis is much lower than that for the harmonic frequency.

As presented so far, the area averaging factor is usually considered purely an amplitude scaling factor. There is, however, also a phase shift associated with spatial averaging, and this too can be calculated using the numerical model. Such phase changes could be potentially significant if this type of wavefield was used for phase calibration of hydrophones. Figure 5 illustrates an example of the phase change due to area-averaging for the 500 kPa case. This shows that for both the 0.2 mm and 0.5 mm hydrophones, the phase change varies periodically, with rapid changes associated with the intermediate frequencies. These effects can be minimised by making measurements at lower drive levels. In addition, the measurements can be repeated with different matching conditions between the generator and transducer so that the fundamental and harmonic frequencies are shifted and the complete frequency range is covered with high strength signals [7]. Averaging the resulting calibrations reduces the effect of the area-averaging factor oscillations with frequency.

(a) 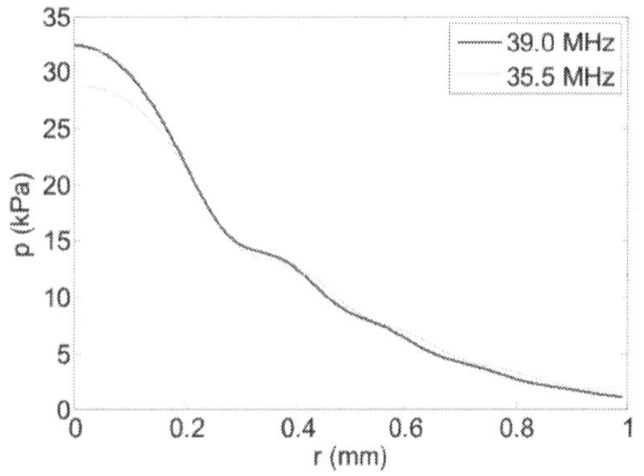

(b) 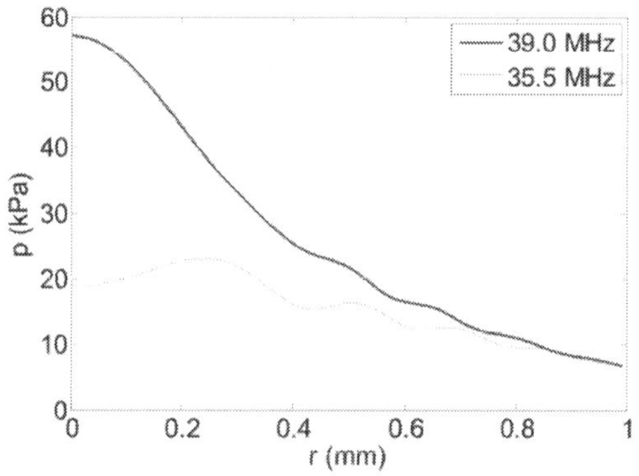

Figure 4. Radial variation of the pressure amplitude for the 6th harmonic of the centre frequency (39 MHz) and a nearby intermediate frequency (35.5 MHz) at a range of 47.5 mm. Results are presented for initial peak rarefactional pressures at the source of (a) 300 kPa and (b) 500 kPa

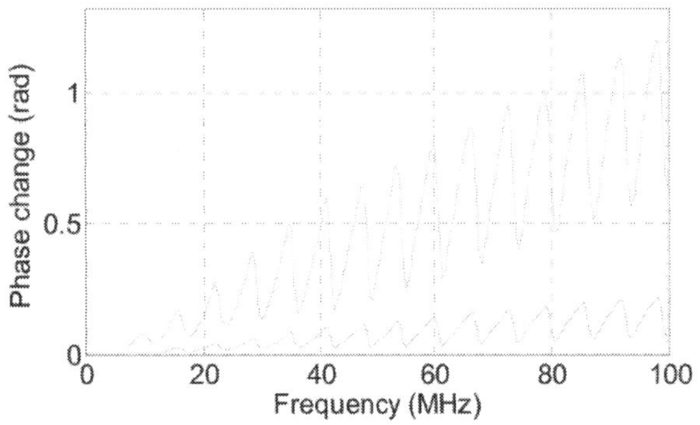

Figure 5. Phase change due to area-averaging for hydrophones of 0.2 and 0.5 mm diameter, calculated from the numerical model for a peak rarefactional pressure at the source of 500 kPa, at a range of 47.5 mm.

4. Discussion and conclusions

The results presented in section 3 have demonstrated all of the phenomena reported in reference [3]. They show that the area-averaging factor is very sensitive to frequency, drive level and hydrophone diameter. The similarity of the response shown in figure 3(b) to that shown in figure 2 is very encouraging and indicates that the modelling technique is capable of reproducing the characteristics of the behaviour.

In order to understand the behaviour presented in section 3 it is informative to look at the distorted waveforms that occur on-axis in the observation plane at a range of 47.5 mm as a result of nonlinear propagation (see figure 6). These waveforms were calculated from the Fourier components in the model and then passed through an FIR filter with a cut off at 200 MHz to avoid oscillations due to Gibbs phenomenon. Comparison of the two graphs shows that for the low drive case only the first cycle is of sufficient amplitude to have significantly distorted and produced a shock front at this range. The resulting rapid rise in pressure results in frequency components across the whole spectrum. In contrast the higher drive case shows significant distortion of both the first two cycles with a shock front present on the second cycle as well as the first. It can be considered that each shock gives rise to a wide spectrum of frequencies. As the second shock occurs at a time $1/f_c$ later than the first, frequencies that are multiples of f_c will add constructively in the overall pulse spectrum. In contrast, the contributions at intermediate frequencies from the two shocks will be out of phase and tend to cancel. Hence the reduction of the level on axis at the intermediate frequencies, seen in figure 4(b), can be attributed to cancellation effects associated with the second cycle going in to shock. As the observation position is moved away from the acoustic axis and the amplitude of the pulse reduces, the second cycle no longer distorts as significantly and the cancellation effect reduces; this results in less cancellation so that the signal amplitude at intermediate frequencies actually increases.

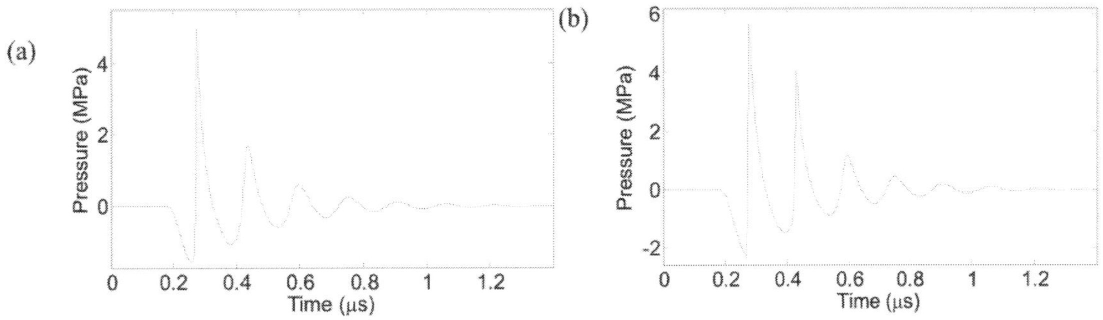

Figure 6. Example on-axis pressure waveforms at 47.5 mm from the source, for two source pressure amplitudes: (a) 300 kPa and (b) 500 kPa.

An implication of these observations is that the area-averaging effect will be very sensitive to the exact shape of the initial pulse waveform. For example, inverting the pulse results in a very different response with larger fluctuations (not presented here); the polarity shown here in figure 1 was actually chosen experimentally because of this fact.

A broad conclusion of this work is that if pulsed nonlinear waveforms are to be used in calibration work then the drive level should be adjusted to give rise to just one shock front in the distorted pulse at the observation position. This will ensure that the spectrum is broad but will minimise fluctuations in the spectral level, avoiding rapid fluctuations in the area-averaging factor such as those shown in figure 3(b).

References

[1] Bacon D R (1988) IEEE Trans. Ultrason. Ferroelectr. Freq. Control 35, 152 - 161.
[2] Smith R A and Bacon D R (1990) J. Acoust. Soc. Am. 87, 2231 - 2243.
[3] Wilkens V and Koch C (2004) J. Acoust. Soc. Am. 115, June 2004, pp 2892 - 2903.
[4] Aanonsen S I, Barkve T, Naze Tjotta J and Tjotta S (1984) J. Acoust. Soc. Am. 75, 749 - 768.
[5] Naze Tjotta J, Tjotta S and Vefring E H (1991) J. Acoust. Soc. Am. 89, 1017 - 1027.
[6] Cooling M P and Humphrey V F (2008) IEEE Trans. Ultrason. Ferroelectr. Freq. Control 55, 84-93.
[7] Wilkens V and Molkenstruck W (2007) IEEE Trans. Ultrason., Ferroelect, Freq. Contr. 54, 1784 - 1791.

Pressure Pulse Measurements Using Optical Hydrophone Principles

Friedrich Ueberle, Abtin Jamshidi-Rad

Hamburg University of Applied Sciences, Life Sciences / Biomedical Engineering
Lohbrügger Kirchstraße 65, D-21033 Hamburg, Germany

Friedrich.ueberle@haw-hamburg.de

Abstract. Pressure pulses are used in extracorporeal lithotripsy, pain therapy and other medical applications. Typical lithotripter pulses reach positive pressure amplitudes of ca. 20 to more than 100 MPa and negative pressures of -5 to more than -20 MPa, depending on the focusing properties and energy settings of the source. The IEC standard 61846, which defines the acoustic parameters of pressure pulse fields, describes the properties of "Focus-" and "Field-" type hydrophones, which were originally specified as PVDF sensors.
During recent years, two types of optical sensors were developed, which are based on the principle of measuring reflection changes of a laser beam at a glass-water surface: The fiber optic sensor using bare optical fibers and the "light spot" sensor using a thick glass block. Measurements with both hydrophone types were made with a low pressure transducer ($p+_{max}$=3 MPa), and two electromagnetic lithotripter sources with the same total acoustic energy (E_{5MPa}=90mJ), one with a wide focus (FWHM = 11 mm, $p+_{max}$= 30 MPa) and the other with a small focus (FWHM = 3,5 mm, $p+_{max}$= 83 MPa). The results show that both optical sensor types provide high pressure-time signal fidelity comparable to PVDF membrane sensors. Both optical hydrophones can serve as "Focus-" and "Field-" hydrophones as defined in the lithotripsy measurement standard IEC 61846.

1. Development and Application of Pressure Pulse Technology

In February 1980 the first patient was relieved from his kidney stones with a pressure pulse lithotripter. During the last 30 years, extracorporeal lithotripsy has evolved and remains the standard treatment method for kidney stone patients. In 2003 70% of all German kidney stones were treated by pressure pulse lithotripsy alone, additional 15% received pressure pulses combined with other endourological treatments [1].

New applications of pressure pulses were developed, like pain therapies for heel spur, tennis elbow and other orthopedic pain diseases [2, 3]. Most recent applications target the heart muscle [4] and seek for methods to target gene therapies by focused pressure pulses [5].

In clinical use Pressure Pulses are generated by electrohydraulic, electromagnetic or piezoelectric sources [6]. Due to the nonlinearity of the medium, the pressure pulses in the focus of a lithotripter usually have steep shock-like pressure fronts, which require measurements with high bandwidth.

2. Measurement of Pressure Pulses

2.1. History

The first measurements of pressure pulses were attempted with piezoelectric hydrophones, which were mainly designed for the measurement of shockwaves. The signals from these sensors allowed only estimations of single parameters like peak positive pressure or rise time; therefore reliable measurements of pressure, energy parameters, negative pulse portions and lateral and axial field distributions had errors >100% at first. In clinical papers of the early years, the usual statements about shockwave parameters were based on the kV setting of the electrical circuits of the lithotripters.

PVDF membrane hydrophones could measure the pressure pulses with high fidelity, but they are expensive and often damaged after only few pulses, mainly by cavitation. A significant step towards longer service life was the introduction of PVDF needle hydrophones [7], but some pulse parameters, in particular those of negative pressure portions and most energy parameters are still grossly underestimated due to refraction and resonance effects of the needle. In the 80s and 90s of the last century, most lithotripters on the market were characterized using these hydrophones [8]. In consequence a lot of biologic research data on effects and side effects used these data [9] and are therefore imprecise.

2.2. IEC 61846 measurement standard

In 1998, the IEC standard 61846 was published, defining the measurement parameters which describe both the temporal waveforms and the acoustic field of pressure pulse[1] sources [10]. From this date the approval of newly designed lithotripters for clinical use was based on the measurement methods described in this standard. Although the standard originally was written for lithotripters, it is also used to describe pain therapy (ESWT) devices and their effects [11].

The standard 61846 allows the use of "focus" and "field" hydrophones. Focus hydrophones are specified to have a performance comparable to PVDF membrane hydrophones with 25µm membranes and small sensitive spots. As these hydrophones are very fragile, they may just be used to characterize the pressure waveforms in the focus, so they only have to survive a few pulses. More robust "Field hydrophones" have reduced specifications (e.g. the PVDF needle hydrophones). They may be used in the pressure pulse field regions outside the focus, where the signals usually have longer rise times and less high frequency components.

In IEC 61846 two sets of pressure pulse parameters are defined; the first set is derived from the pressure-time curves measured at the focus (e.g. Figures 4, 5): Positive and negative peak pressures P+ and P-, rise time TR (10% to 90% of P+), positive pulse duration TD (50% of P+ from rising edge to falling edge) and the Pulse Intensity Integral $\mathrm{PII}(r,\theta) = \frac{1}{Z_W}\int_{(T)} p^2(r,\theta,t)dt$ which determines the energy flux density of the pressure pulse. The second set is field parameters: axial and lateral pressure distributions (FWHM = Full Width Half Maximum) and focus energy2 $E_R = \int_S \mathrm{PII}(r,\theta)dS$. Correlations of the parameters to stone comminuting and biological effects are discussed in [6].

---

[1] Before 1998 the term of lithotripter waves was „shockwave", as the electrohydraulic generated pulses in the focus usually had very short rise times in the range of a few nanoseconds. This term is still in widespread use. After the occurrence of other sources like electromagnetic and piezoelectric, it showed that pulses with slower rise times, which may occur at lower energy settings, also could disintegrate stones. In most pressure pulse sources, the rise time of the signals outside the focal area increases by some orders of magnitude [6].

[2] S is a circular area in the focus plane delimited by a diameter R, which, according to IEC 61846, shall be the diameter of the -6dB pressure amplitude isobar. As this value has not much significance for the evaluation of stone disintegration efficacy or biological (side) effects, other "effective" diameters like typical kidney stone sizes (12mm) or pressure isobars (e.g. 5 MPa, which is about the minimum pressure required for stone comminution), should be used [6], [7], [8], [9], [11], [15]

3. Optical Hydrophones – Physics and Technology

The physical principle of both the fiber optic hydrophone [12] and the light spot hydrophone [13] is based on the change of the refractive index of a fluid, when a passing pressure wave instantaneously changes the local density of the fluid, which is described for water by the Tait equation. This change of refractive index modulates the intensity of reflected light at a glass-fluid interface, which can be calculated with the Gladstone-Dale equation, relating the optical refractive index to the media density.

The fiberoptic hydrophone was invented by Eisenmenger in 1988. A thin optic fiber transports an infrared laser beam. The bare end of the fiber is submerged in the fluid, usually water. The fraction of light, which is reflected at the fiber – fluid interface and transmitted back to the top end of the fiber, is detected by a photo diode via a directional fiber coupler. When the fluid density is changed by the passing pressure pulse, the reflected light intensity is modulated proportional to the acoustic pressure p(t) [12]. The hydrophone used in this study (FASO = FaserOptische SOnde, the same principle as Eisenmengers FOPH = FiberOPtic Hydrophone) uses a 812nm diode laser at 120 mW through a 125 μm optical fiber.

The light spot hydrophone (LSHD) uses a 90x60x30 mm glass block, which is submerged in the fluid at the lower side. The 785 nm, 35 mW diode laser beam is passed through a coupling fiber to a lens, which directs the beam from the air above through the glass block under an angle (ca. 15°). The beam is focused to 50 μm spot at the glass-fluid interface by the optic lens. A photo diode is placed analogous above the glass block to receive the laser light, which is reflected (R_{GW} = 0,16%) at the glass-water interface. The intensity of the reflected laser light is modulated by the pressure pulse induced change of refractive index of the water beneath the glass block. As a portion of the pressure pulse also causes solid waves traveling through the glass block, only short pressure pulses can be measured unobstructed before the acoustic reflections from the upper (air) side of the glass interfere with impinging acoustic signals. For lithotripter and ESWT pressure pulses, 30 mm thickness of the glass block is chosen for undisturbed measurement of ca. 10μs pulse signals.

3.1. Calibration of the optical hydrophones

The in-situ calibration of both optical hydrophones does not need acoustic measurements. The calibration procedure of the LSHD only requires measuring the static reflected light intensity $R_{GW|p=0}$ under two different interface conditions: Glass-air and glass-water. From these values the sensitivity can be calculated [13]. The typical sensitivity of the LSHD is -9,6 mV/MPa. If the glass surface is damaged by the pressure pulses, the glass block can be displaced horizontally by some centimetres by a precision screw without changing the positions of the laser and detector. Thus the position of the active spot in the pressure field is conserved and measurements can continue in an instant without the need to change the position of the hydrophone for re-calibration.

The FASO is calibrated by comparing the static reflected light intensities by submerging the fiber in isopropanol and in water. The typical sensitivity of the FASO is -0,27mV/MPa. If the fiber is damaged or broken, it can be cleaved to achieve a new surface, and then recalibrated using the two fluids. When steep pulses, impinging oblique in the fiber are measured, the reflected pressure adds a short time pressure peak, which is compensated by an edge diffraction wave, emerging at the periphery of the fiber (Figure 1). When this diffraction wave has travelled over the flat fiber surface to the center, it completely compensates the reflected pressure pulse signal. In order to display the correct pressure – time curve of the signal and to avoid over-estimation of the pressure pulse amplitude P+, this additional spike needs to be removed, which requires an additional signal processing step.

3.2. Signal Processing

The signal processing serves two purposes: For the FASO, the removal of the additional spike can be achieved by deconvolution or by filtering the measured signal (Figure 1). As a reference signal for deconvolution, the hydrophone transfer function is needed.

For both FASO and LSHD, the photon noise of the laser source, which restricts the (non-averaged) use of the hydrophones to signals of P+ > 1..2 MPa, can be smoothed by additional low pass filtering.

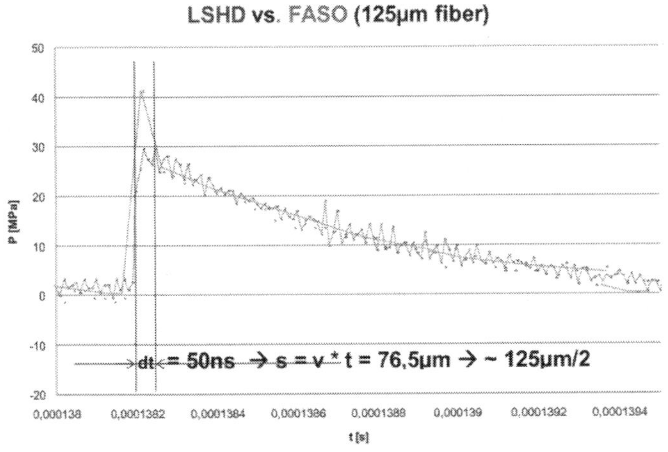

Figure 1: The leading edge of the fiberoptic hydrophone signal (red dots) shows a short peak, which is caused by the reflection of the impinging pressure pulse and is compensated after ca. 50 ns when the edge diffraction wave at the 125 µm fiber diameter has reached the center of the fiber tip. The LSHD signal (blue dots) doesn't show this peak because the glass surface is 60*90 mm wide.

After filtering, additional linearization may be applied in order to correct the slight (<10% up to 120 MPa) non-linearity, which is caused by the non-linearity of the pressure induced density change of the fluid (Figure 2). As this non-linearity only leads to significant errors at pressures above ca. 50 MPa, this correction step may be omitted for smaller signals.

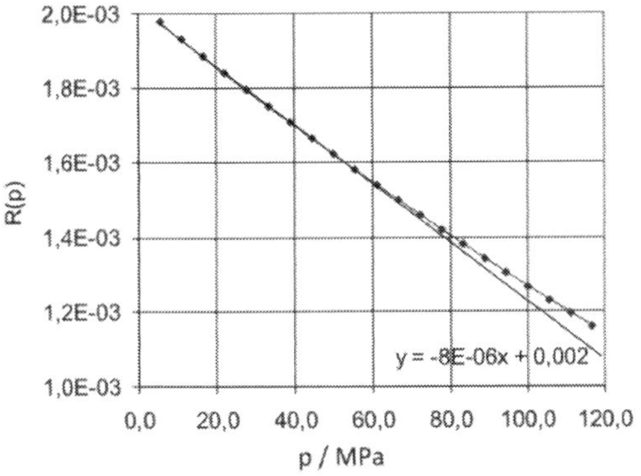

Figure 2: The optical refractive index R(p) is slightly non-linear (<10% in the usual therapeutic range up to P+ of 120 MPa), which should be corrected if pressures above ca. 50 MPa are measured (Data based on the LSHD glass parameters)

4. Measurements using Optical Hydrophones

4.1. Materials and Methods

The measurements were made in a prototype lithotripter source, which was originally built at the University of Stuttgart. The highly stable pressure pulse source is an electromagnetic type with a concave focusing membrane of 120 mm diameter and a focal distance of 200 mm (EMSE-01), giving a focus spot size of 11 mm (FWHM, at 13 kV energy setting). Additionally, it can be equipped with a Perspex lens, which reduces the focal distance to 100 mm and thus concentrates the acoustic pressure to a significantly smaller focus spot of 3,5 mm (FWHM). The system with additional lens is called EMSE-02. The lens allows an easy comparison of lesser and stronger focused pressure pulse fields at the same overall energy. Between our measurements, a damaged membrane of the EMSE had to be replaced; only measurements with the same membrane (marked "old" and "new") are compared in this paper.

The pressure pulse source is mounted at the bottom of a tank, filled with degassed water (< 2mg/l O_2) at 21+-2°C temperature. The hydrophones can be moved by computer-controlled 3-axis gimbals with a resolution of 100µm.

In order to determine the transfer functions of LSHD and FASO, the focus signals of a piezoelectric transducer (Panametrics: V307-SU, 5.0 MHz, diameter: 25 mm, focus distance: 95 mm, driven by a Panametrics Pulser-Receiver at 100V, 200V, 300V and 400V settings) were measured in the laboratories of the Fachbereich 1.62 Ultraschall, Physikalisch-Technische Bundesanstalt PTB,

Braunschweig with the PTB reference interferometric hydrophone [14]. From these measurements and measurements of the FASO at the same focus distance, the transfer function of FASO was calculated.

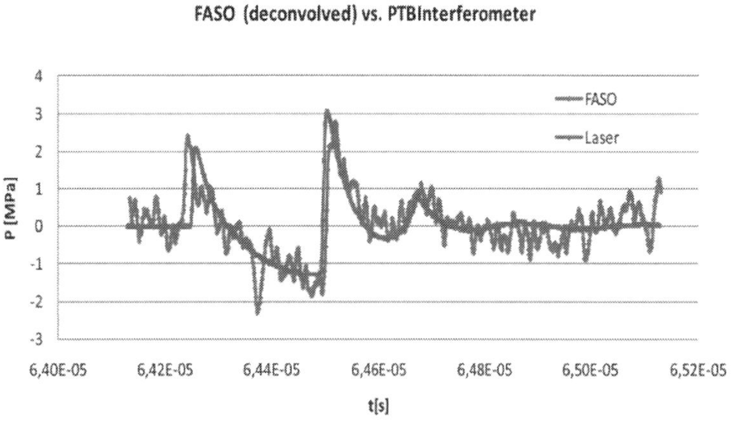

Figure 3: Deconvolved 5 MHz Panametrics (300 Volt) signal of FASO vs. PTB reference interferometer signal. At these low pressures, even averaged (N=64) FASO signals are noisy due to the laser photon noise. Parameters:

	Reference	FASO
P+ (1st/2nd)	2,1/3,06MPa	2,4/2,75MPa
P-	-1,27 MPa	(peak:-2,3)/ -1,85MPa
T_R (1st/2nd)	7 / 10 ns	7 / 7 ns
PII (64,7µs)	0,309µJ/mm²	0,307µJ/mm²

4.2. Results

The results of the FASO calibration with the Panametrics 5 MHz transducer at 3,06 MPa are demonstrated in Figure 3. Figures 4 compares the FASO vs. LSHD focal pressure pulse signals of EMSE-01 at 13 kV setting. The EMSE-01 vs. EMSE-02 13 kV focus and field parameters (Figure 5) were measured with LSHD. IEC 61846 Pressure pulse parameters are given beside the figures.

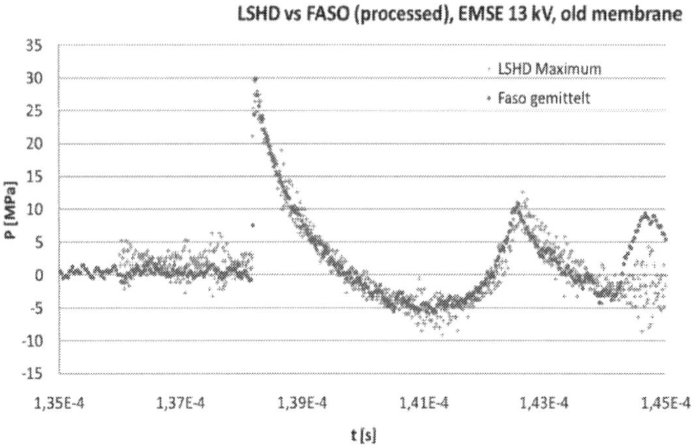

Figure 4: Comparison of LSHD and FASO measurements in the focal maximum of EMSE-01 (13 kV, old membrane). While the FASO signal was deconvolved; the LSHD signal was not filtered.

	FASO	LSHD
P+	29,8 MPa	29,6 MPa
P-	-5,3 MPa	-5,3±1MPa
T_R	20 ns	20 ns
T_D	760 ns	780±20 ns
PII+	0,40mJ/mm²	0,41mJ/mm²
PII_{ges}	0,50mJ/mm²	0,53 mJ/mm²

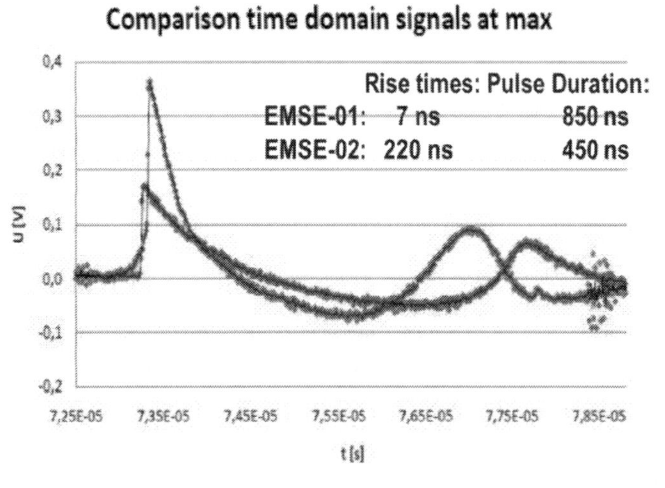

Figure 5: LSHD measurements of the focal pressure pulses of EMSE-01 and EMSE-02 (both at 13 kV; new membrane). Pressure pulse parameters according to IEC 61846:

	EMSE-01	EMSE-02
P+	36,4 MPa	76,1 MPa
P-	-10,3 MPa	-15,0 MPa
T_R	7 ns	220 ns
T_D	850 ns	450 ns
PII_{ges}	0,43mJ/mm²	1,04mJ/mm²

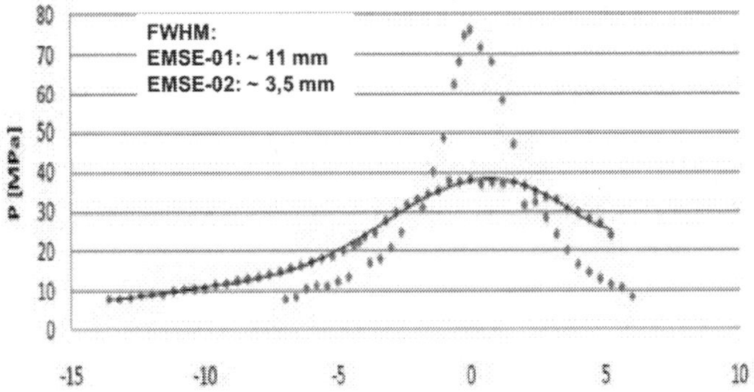

Figure 6: LSHD measurements of the lateral focal pressure distributions of EMSE-01 and EMSE-02 (both at 13 kV; new membrane).
Pressure pulse parameters according to IEC 61846 and DGSL [11,15] are:

	EMSE-01	EMSE-02
FWHM	11 mm	3,5 mm
E_{FWHM}	40 mJ	25 mJ
E_{12mm}	36 mJ	36 mJ
E_{5MPa}	~90 mJ	~90 mJ

5. Discussion

With the optical hydrophones LSHD and FASO, high fidelity measurements of lesser and stronger focused pressure pulse sources could be made. Processed pressure – time signals of both hydrophones at the same position in the pressure pulse fields are in very good agreement for pressures > 2 MPa.

Both hydrophones have the following advantageous properties: (+) good signal reproduction; (+) fast repair; (+) no acoustic calibration necessary; (+) high bandwidth; (+)long service life with low-cost spare parts (+); can be used as "Focus" and "Field" hydrophones according to IEC 61846.

The active part of the fiberoptic hydrophone (FOPH, FASO) additionally is (+) very small and flexible, but it (-) easily breaks due to cavitation and thus needs frequent cleaving and readjustment of the fiber, which may extend the measurement time at high energy settings and less-degassed water.

Disadvantages of the FOPH are its (-) large sensor head and the need for (-) upright positioning, as laser and photodiode need to be in air. The glass block LSHD is (+) robust and cavitation resistant, and in case of cracks in the glass it (+) can be repaired in an instant by shifting the position of the glass block without loss of the active spot position in the pressure pulse field, thus enabling fast and reliable measurements of pressure pulse fields.

We thank Dr. Wilkens (Physikalisch-Technische Bundesanstalt Braunschweig PTB, Fachbereich 1.62 Ultraschall) for the reference measurements and many competent and helpful discussions.

We thank Siemens Medical Solutions for supporting this project with a research grant.

References

[1] Interventionelle Therapie von Harnsteinen, T. Knoll, L. Trojan, A. Haecker, M. S. Michel, K. U. Köhrmann, P. Alken, NOTFALLMEDIZIN 2003; 29(12): 506—511

[2] Evaluation of Extracorporeal Shock Wave Therapy for Chronic Lateral Epicondylitis. Pettrone FA, et al. Paper 271, AAOS Annual Meeting, 2002

[3] Extrakorporale StoßWellen-therapie (ESWT) aus orthopädischer und traumatologischer Sicht, V. Auersperg, R. Dorotka, M. Sabeti-Aschraf, Ch. Dohnalek, St. Wanke, W. Schaden, J. Mineralstoffwechsel 2004 (4), 19-28

[4] Extracorporeal Cardiac Shock Wave Therapy Markedly Ameliorates Ischemia-Induced Myocardial Dysfunction in Pigs in Vivo, Nishida, Shimokawa, Oi et al., Circulation 2004, 110: 2977

[5] Gentherapie des Prostatakarzinoms durch extrakorporale akustische Energie: Erste In-vitro- und In-vivo-Ergebnisse, Michel, Erben, Köhrmann, Siegsmund, Alken, Aktuel Urol 2002; 33(3): 213-218

[6] Pressure Pulses in Medicine, F.Ueberle in: Srivastava, Leutloff, Takayama,Groenig (eds) Shock Focussing Effects in Medical Science and Sonoluminescence (2003), Springer, Heidelberg New York

[7] Einsatz einer breitbandigen Piezodrucksonde auf PVDF-Basis zur Untersuchung konvergierender Stoßwellen in Wasser; Müller, Platte (1985) Acustica 58

[8] Dornier-Lithotripter im Vergleich: Vermessung der Stoßwellenfelder und Fragmentationswirkungen; M.Müller (1990) Biomed. Technik 35(11): 250 – 262

[9] Determination of energy- dependent extent of vascular damage caused by high-energy shock waves in an umbilical cord model; Steinbach, Hofstaedter, Roessler, Wieland; Urological Research (1993), 21: 279–282

[10] IEC 61846 Pressure Pulse Lithotripters, Standard; International Electrotechnical Commission, Genf 1998

[11] www.digest-ev.de; Goto: "Gütesiegel" – "Vergleich" (Page lookup 31.5.2010)

[12] Fiberoptic Probe Hydrophone for Ultrasonic and Shock-Wave Measurements in Water; Staudenraus, J., Eisenmenger, W. (1993) Ultrasonics 31(4):267-273

[13] Light Spot Hydrophone, Innovation in Lithotripsy; Granz, Nanke, Fehre, Pfister, Engelbrecht, Medical Solutions, June 2004, 86–87

[14] Amplitude and phase calibration of hydrophones up to 70 MHz using broadband pulse excitation and an optical reference hydrophone; V. Wilkens, C. Koch, J. Acoust. Soc. Am 115 (2004), 2892–2903

[15] Working Group Technical Developments – Consensus Report; Wess, Ueberle, Dührßen, Hilcken, Reuner, Schultheiß, Staudenraus, Rattner, Haaks, Granz in Chaussy et al (eds) High Energy Shock Waves in Medicine. Thieme, Stuttgart New York, (1997) pp 59–71

A fully automated system for ultrasonic power measurement and simulation accordingly to IEC 61161:2006

Rodrigo P.B. Costa-Felix[1], **André V. Alvarenga**[1], **Rob Hekkenberg**[2]

[1]*Laboratory of Ultrasound, Directory of Scientific and Industrial Metrology (DIMCI), National Institute of Metrology, Standardization, and Industrial Quality (Inmetro), Av. Nossa Sra das Graças, 50 (Prédio 1), Duque de Caxias, RJ, Brazil, ZIP 25250-020*

[2] *TNO Prevention and Health, Zernikedreef 9, 2333 CK Leiden, The Netherlands*

[1]rpfelix@inmetro.gov.br

Abstract. The ultrasonic power measurement, worldwide accepted, standard is the IEC 61161, presently in its 2nd edition (2006), but under review. To fulfil its requirements, considering that a radiation force balance is to be used as ultrasonic power detector, a large amount of raw data (mass measurement) shall be collected as function of time to perform all necessary calculations and corrections. Uncertainty determination demands calculation effort of raw and processed data. Although it is possible to be undertaken in an old-fashion way, using spread sheets and manual data collection, automation software are often used in metrology to provide a virtually error free environment concerning data acquisition and repetitive calculations and corrections. Considering that, a fully automate ultrasonic power measurement system was developed and comprehensively tested. A 0,1 mg of precision balance model CP224S (Sartorius, Germany) was used as measuring device and a calibrated continuous wave ultrasound check source (Precision Acoustics, UK) was the device under test. A 150 ml container filled with degassed water and containing an absorbing target at the bottom was placed on the balance pan. Besides the feature of automation software, a routine of power measurement simulation was implemented. It was idealized as a teaching tool of how ultrasonic power emission behaviour is with a radiation force balance equipped with an absorbing target. Automation software was considered as an effective tool for speeding up ultrasonic power measurement, while allowing accurate calculation and attractive graphical partial and final results.

1. Introduction

Automated systems (software) are to be used to help data acquisition, conditioning and manipulation, as well as instruments (hardware) controlling and setting up. This paper presents software to be used in Ultrasonic Power and Radiation Conductance Measurement (USPRCM) automation. It was developed in LabVIEW 2009 (National Instruments, Austin, TX, USA), and tested in both simulation and real condition data acquisitions.

The software was developed in compliance with IEC 61161 [1]. It has been tested using absorbing and reflecting targets, so that it is prepared to calculate power measurement for both targets. The software can be freely distributed after a proper agreement of use.

2. Basic requirements and specifications

2.1. Power measurement

Ultrasound propagation in a medium will result in a radiation force which can be measured, for instance, using a regular microbalance. However, as the level of the force in many cases is very low, and also because heating due to ultrasound absorption of some propagation media (water, target for instance), the reading during ultrasound radiation is usually not stable. Further, ultrasonic radiation could diverge regarding different distances from the acoustic center of the radiation device, its characteristic, and reception target. All those aspects are detailed in [1] and other references referred therein.

2.2. An acceptable approach

Any medium where ultrasound propagates through will absorb some energy and so ultrasound will heat up that medium. The same happens with the target. As a consequence buoyancy may vary significantly during sonification. The microbalance which is used to measure the radiation force of an ultrasound wave observes continuously during the ultrasonic power measurement and shows the variation in its reading. The force could be measured just by observing and/or "manually" collecting the output data from the balance, but due to the instabilities this is not worthwhile. If measurements were undertaken automatically, especially if data is processed at a glance, it would be even better.

USPRCM software developed has the main feature of being able to process data from a raw mass measurement into a power measurement (assessment). Figure 1 shows results of a final measurement with that software. It shows "raw" (already processed) data in the graph in its upper right corner. It shows mass converted into power, and also regression points considered. The fat spots in the graph are those points where power sonification would start, disregarding flutter of "power on".

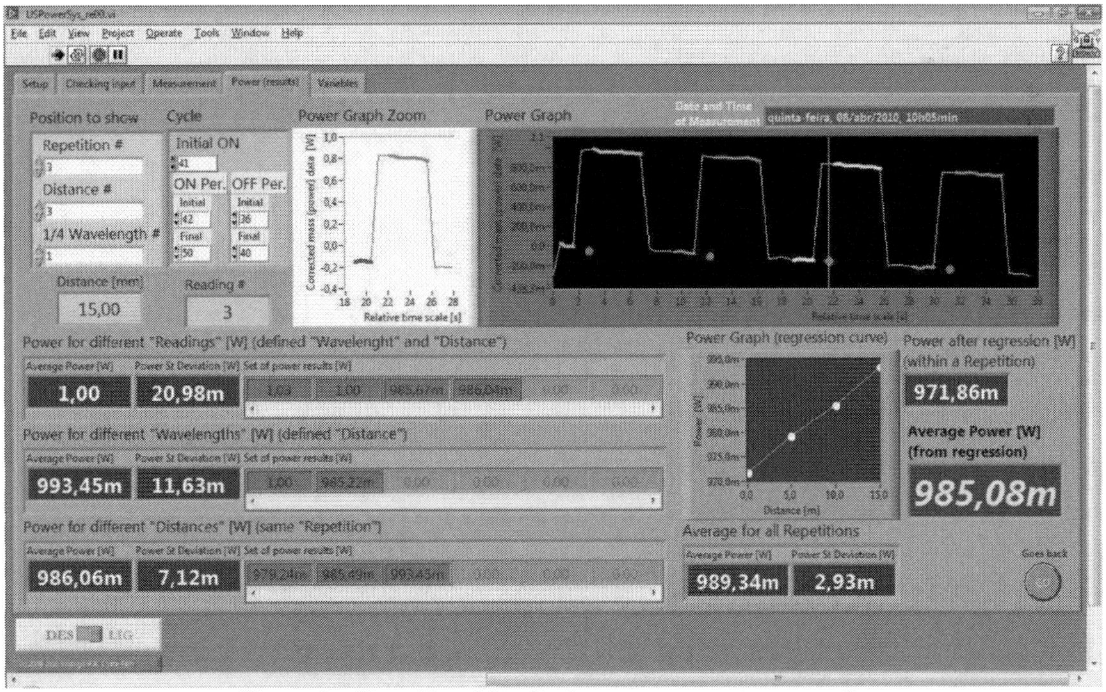

Figure 1 – Front Panel for final power measurement results

Numerical values are also depicted in that same front panel window. One can observe that both "average" and "regression" final results are disclosed. The first one, in the lower part of right hand side of that screen, is basically used for absorbing target or if close different distances were used. The "from regression" power is also disclosed to be used if reflecting target was used or if absorbing due propagation are of concern. On the lower left hand side of the screen of Figure 1, there are 3 "Average Power" boxes. The topmost one is for a fixed distance from the target to the source and for a defined "wavelength". Basically, the values depicted there are obtained from the graph (top right in the figure). The second one is the average of different "wavelength", if there is more than one. In the example illustrated, there are two different "wavelengths". The lowest one is the average within a repetition, i.e., for different distances.

The approach used herein was to have the possibility of performing measurements in at least 2 different separations from target to output device. For each distance, it is possible to perform up to 3 different measurements in a micro adjustment, called "1/4 wavelength", to avoid, for instance, incorrect measurement due to standing waves. Each measurement is to be repeated, so a number of "Repetitions" shall be undertaken. All those settings are to be adjusted in the window as disclosed in Figure 2.

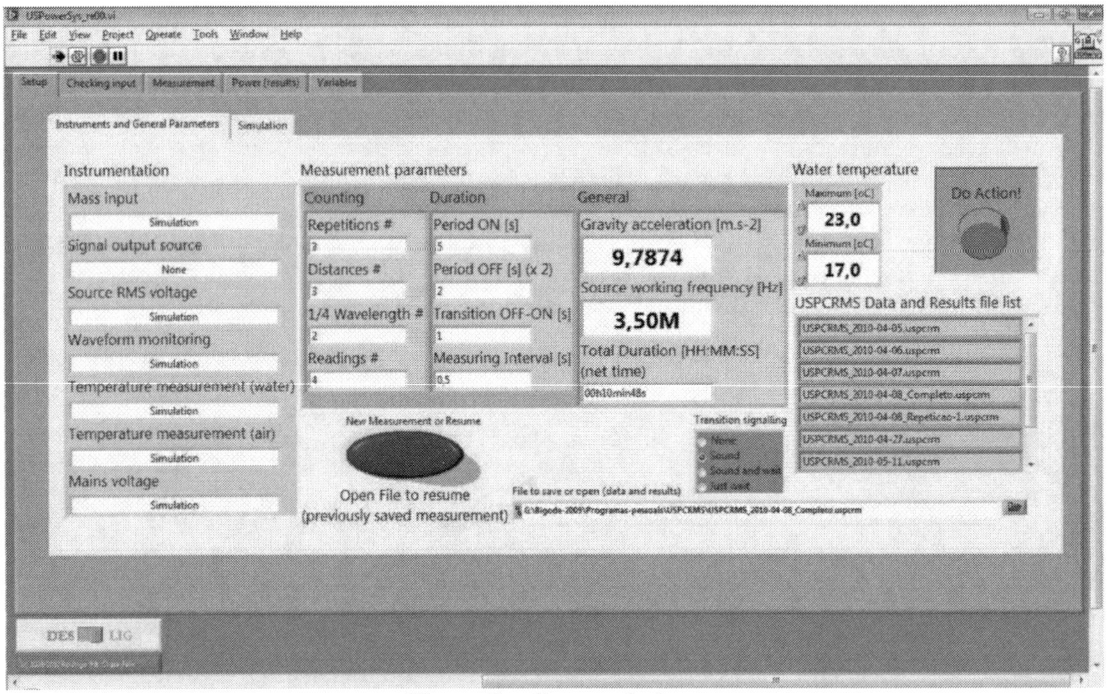

Figure 2 – Setup window

All accessories inputs shall be done in that window (Fig. 2). "Instrumentation" is to be settled to adjust what and which measurement instruments are to be used.

2.3. Measurement Front Panel

Figure 3 discloses the Front Panel that is shown during a measurement preset. It is possible to evaluate values of Mains voltage, Temperature (room and inside the water bath), Source output voltage (rms) as well as its waveform. All those measurement are available if proper instrumentation is plugged and settled in the window presented in Fig. 2.

In this same window (see Fig. 3), it is possible to adjust the actual distances where the measurements were done, and also to confirm (check-up) the total number of measurements done and those to be done.

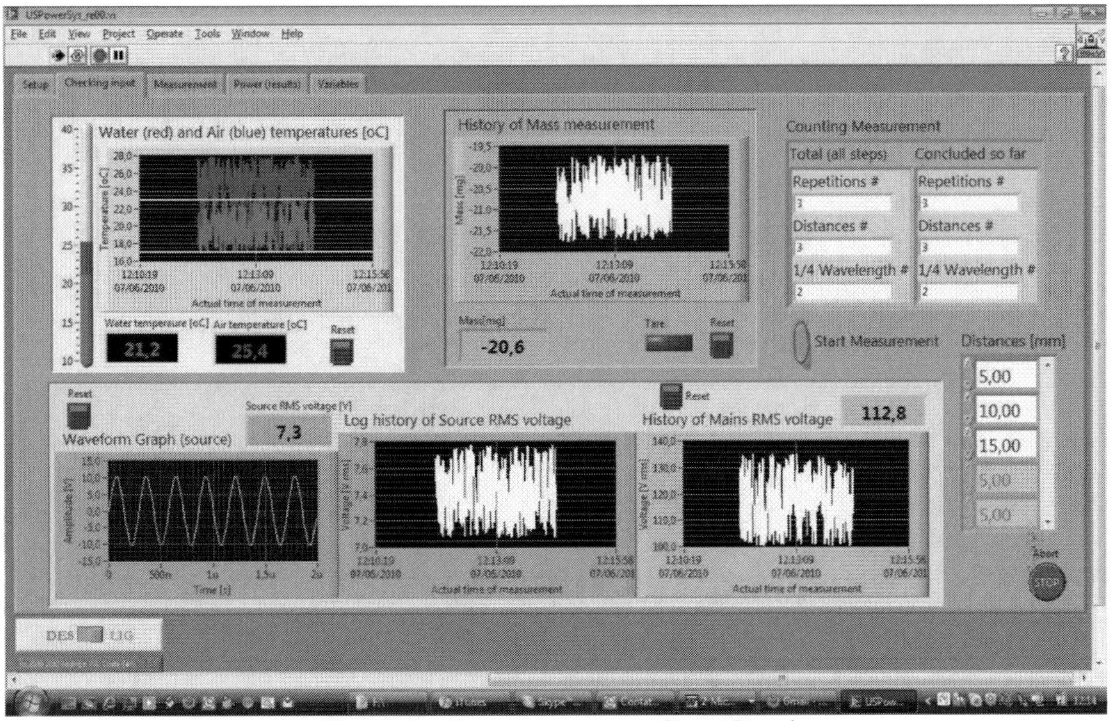

Figure 3 – Pre-measurement Front Panel

3. Conclusion

The system for USPRCM, as developed, is able to be used in a wide range of instruments and standards (balances). It is rapidly adaptable to different power measurement approaches and mountings, as far as they were accordingly to [1]. Because of its simplicity on being used, it can be explored as a teaching tool. That feature could be welcome for newborn laboratories dealing with power measurement. It can be freely distributed for partners of the owners.

References

[1] IEC 61161:2006, *Ultrasonics – Power measurement – Radiation force balances and performance requirements* , *International Electrotechnical Comission*, Geneva, Switzerland, 97pp.

Journal of Physics: Conference Series **279** (2011) 012005 doi:10.1088/1742-6596/279/1/012005

US quality control in Italy: present and future

S Balbis[1], C Musacchio[2], C Guiot[1], R Spagnolo[2]

[1]Dip. di Neuroscienze, Università di Torino, Corso Raffaello 30, 10125 Torino, Italy;
[2]INRIM, Istituto Nazionale di Ricerca Metrologica, Strada delle Cacce 91, 10135 Torino, Italy

E-mail: sonia.balbis@unito.it

Abstract. US diagnostic equipments are widely diffused in Italy but, in spite of recommendations (e.g. ISPESL-Ministry of Health (1999) and SIRM (Società Italiana di Radiologia Medica, 2004), US quality controls are restricted to only a few public sanitary structure and a national (or even regional) quality assurance program for testing the performances of the US equipments is still missing.

A joint Research Centre among the three Piedmontese Universities and INRIM , partially funded by Regione Piemonte, has been established in 2009 as Reference Centre for Medical Ultrasounds (CRUM). In addition to research, development and training tasks, the Centre aims at the local diffusion of the quality assurance in clinical US equipments . According to data from the Ministry of Health (2006), around 7 % of the Italian US diagnostic equipments (946 over 13526) are located in Piedmont: mostly (75.6%) in public hospitals, 9.3 % in conventionated hospitals, 4.3% in public and 10.8% in private territorial structures. The goal is the provision of a regional database, which progressively includes data related to acceptance test, status and QC tests and maintenance, in order to drive equipment turnover and carefully monitoring the overall equipment efficiency. Moreover, facilities are available at CRUM for monitoring both beam geometry and acoustic power and performing quantitative assessment of the delivered energy intensity.

1. Introduction

Due to the wide diffusion of the US diagnostic equipments and the advances in computational techniques applied to ultrasound technology, systematic monitoring of the US scanners performances could be very important to assure efficiency during clinical examinations.

The elaboration of a common quality control (QC) program is therefore mandatory to protect (or better serve) patients and, on the other hand, to provide an objective assessment to optimize the equipment turnover.

In particular, in our region, Piemonte, according to the data published by the Italian Ministry of Health (2006) [6], is present about the 7% of the whole national US instrumentation: 75.6% in public hospitals, 9.3 % in conventionated hospitals, 4.3% in public and 10.8% in private territorial structures.

QC in Italy have been recommended (see ISPESL-Ministry of Health, 1999, and SIRM, Società Italiana di Radiologia Medica, 2004) [11] but are not compulsory, and as a consequence only few public structures practice them.

A joint Research Centre among the three Piedmontese Universities and INRIM , partially funded by Regione Piemonte, has been established in 2009 as Reference Center for Medical Ultrasounds (CRUM) and, in addition to perform research and education in the field, it has the aim to diffuse quality assurance in medical US equipments and promoting a regional database.

1.1 The starting point: the international experiences

The importance of testing the performances of US imaging has been detailed by many authors over the last three decades, and specific procedures, from tests based on the simple use of plexiglass block to more sophisticated tissue equivalent test objects, have been so far proposed.

In the USA, since the 70s of the twentieth century, many papers on this topic have been published: see for instance P L Carson et al, 'Calibration and quality control in diagnostic ultrasound', *Phys. Med. Biol.* (1974), 19, 229-230 [3], who first defined which parameters are to be checked and which procedures have to be followed, according to the indications of the AIUM (American Institute of Ultrasound in Medicine).

Successively, a number of guidelines have been published by professional bodies, that are periodically updated as technology develops. In 1995, Price recommended tests to be performed in routine quality assurance of ultrasound imaging systems [10]. Three years after, in 1998, Goodsit et al published the report of AAPM (American Association of Physicists in Medicine) "Ultrasound Task Group N° 1 about the Real-time B-mode Ultrasound quality control test procedures", which is one of the most important document on this topic: in this paper a set of limit values to follow are suggested and the use of computerized image analysis is recommended, to provide more accurate and reproducible results [4].

In the same years the scientific community faced the problem also in France, and Vallbe and Donadey, starting from the most recent developments of the technology, described the main QC international experiences , the parameters to analyze and the procedures to follow, with a particular attention to the image resolution and contrast of the US equipments [16].

In most recent years, due to the rapid evolution of the equipments features and performances, several International Committees updated their guidelines. For example, AIUM published the documents: 'Routine Quality Assurance for Diagnostic Ultrasound Equipment' (2008) [2] and 'Standard Methods for Calibration of 2-Dimensional and 3-Dimensional Spatial Measurement Capabilities of Pulse Echo Ultrasound Imaging Systems' (2004) [1]. In the first paper, a basic QA program for clinical ultrasound facilities is outlined and a particular attention is devoted to the frequency of tests and the role of the personnel, with a number of examples of procedures fully described in the references paragraph. In the second document, the problem of the spatial accuracy in 2-dimensional (2D) and 3-dimensional (3D) imaging is investigated and calibration of measurements systems techniques are described. In particular, the following types of measurements are analyzed in details: 1) linear dimensions, 2) perimeters of areas, 3) areas, 4) curved dimensions, 5) volumes and surface areas, 6) position registration, 7) linear dimensions, areas, curved dimensions and volumes and surface areas in complex orientations and defined only by their difference in backscatter from the background, 8) representation of time, velocity and acceleration in M-scans and a whole chapter is dedicated to the test objects and their features.

In Italy, the most important document ,'Controlli di qualità in ecografia', has been published in 2004 by Società Italiana di Radiologia Medica, and collects the proposed procedures for US equipments QA, comparing them to the ones currently performed for NMR and x-rays equipments [11].

Such guidelines are actually based on the previous experience of Novario and his collaborators on QC in both Doppler [7] [8] and B-Mode [9], focusing on test objects, while, in most recent years, the same research group at the Ospedale di Circolo of Varese devoted his attention to the definition of ranges of acceptability for the parameters currently measured in QC routines[10].

Also in the Valle d'Aosta region, QA procedures are a solid experience, as described in [13]. In this paper the results in the 2007-2008 years are described and their positive aspects are shown: the authors conclude that the quality of US equipments has improved over the years and the overall costs have decreased, because QC made it possible to identify malfunctioning probes and to provide specific repairs instead of expensive interventions on the equipments. Finally, these positive results were clearly perceived by the users, as confirmed by a survey among the sonographers, which shows the congruence between the phantom results and the clinical perception.

2. The protocol

Starting from the international and Italian experiences previously described, CRUM is elaborating a QA protocol, to establish the main parameters and the corresponding limit values and tolerances. In particular, we have considered as a starting point the AMUM, AAPM and SIRM guidelines, in addition to some significant national experiences of clinical routine. The protocol is currently under examination of a group of expert for overall assessment, and will shortly be submitted for the national accreditation procedures by INRIM.

In this protocol B-Mode imaging and Doppler are taken into account, which are the most common diagnostic US modalities performed in medical practice, the protocol is composed by different sections: Italian and international regulations, measurement procedures, dedicated phantoms, test objects and software for imaging analysis, with an overview on the commercial products available.

Such QC program is particularly focused on the evaluation of progressive deterioration the US equipments over time, which is responsible of a slow but gradual deterioration of the image quality, which is hardly perceived by operators working daily with the same apparatus, and need therefore effective technical tools, based on objective parameters and reproducible measurements.

Data collection starts from the first check of the equipment (the so called acceptance tests), and continues with a series of annual 'status' or 'constancy' tests, where parameter values are checked and compared with the initial values recorded at the acceptance test (see Fig. 1)

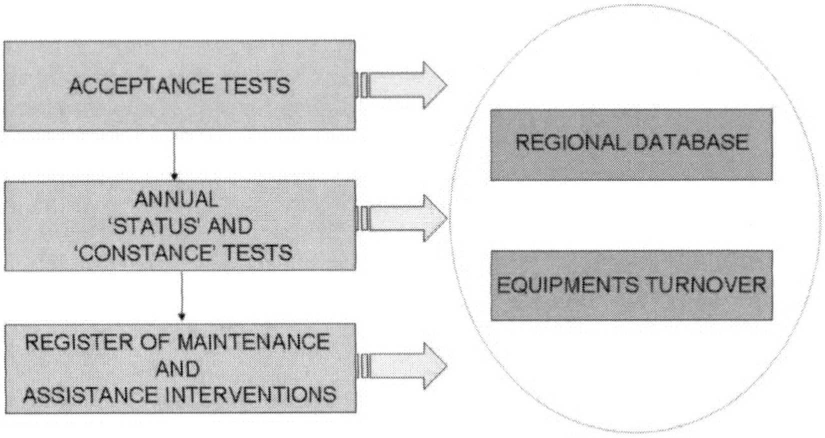

Figure 1. Sketch of the protocol proposed by CRUM for the US diagnostic equipment.

The US parameters both in gray scale and echo-colour Doppler which are evaluated are those refer red by the American Association of Physics in Medicine (Goodsitt et al. 1998) and recognized by the AIUM as reference standard.

In the following tables, the main QC parameters are shown:

Table 1. QC parameters in B-Mode modality

B-Mode modality
Image uniformity
Distance accuracy
Axial resolution
Lateral resolution
Anechoic –iperechoic object imaging
Dead zone
Depth of visualization
Focalization
Contrast resolution
Physical and mechanical inspection
Hardcopy and display monitor fidelity

Table 2. QC parameters in Doppler and Color-Doppler modality

Doppler modality	**Color-Doppler modality**
Doppler signal sensitivity	Color flow sensitivity
Flow sensitivity at depth	Color flow sensitivity at depth
Accuracy of flow velocity readout	Accuracy of the Color Doppler velocity
Accuracy of sample gate positioning	Color flow B-Mode image congruency
Wall filter	Colored noise
	Directional discrimination
	Motion discrimination

The above parameters have to be checked by the responsible of the equipment and/or by the medical physicists with a constant periodicity, but at least annually in the case of multidisciplinary US instruments heavily used daily in Radiological Departments, or even more frequently if required by the Radiologist.

In particular, the monitor fidelity, the image uniformity, the hard copy and display monitor fidelity, the distance accuracy and the depth of visualization are very easy to be tested, and their check should be performed more often: the time estimated for it is about 15 minutes and the time necessary to perform the whole QA protocol is about one hour / probe.

Moreover, a section of the protocol is devoted to the more recent diagnostic US modalities, e.g. 3D and 4D equipments and endocavitary probes, that need special phantoms and finally, to the available software to perform the automatic analysis of B-Mode images.

Another important problem still open is that of assessing the acoustic power emitted by the probes. As a matter of facts, although the use of ultrasound is commonly considered devoid of collateral effects, the widespread use of ultrasound examinations and the increase in output intensity level required by new devices (e.g. contrast agents modalities) settle the issue of defining the levels of US exposure which ensure both patients safety and best devices performance. Furthermore, an increasing interest in using ultrasound as a surgical and therapeutic tool, e.g. due to the use of High Intensity Focused Ultrasound (HIFU) for tissue ablation in the treatment of cancers and benign prostate hyperplasia, require a direct and quantitative assessment of the acoustic power delivered by the US beam in biological tissues.

For these reasons, at INRIM., many efforts have been done to develop effective techniques for ultrasonic sources and fields characterization, measurements set-ups and techniques for beam characterizations as required by national and international standards. Preliminary investigations of new technologies directed to the extension of the measurements capabilities to the field of HIFU devices are in progress.

In particular, the measurement of the total power emitted by an ultrasonic transducer based on the principle of the radiation force balance and following the guideline of the standard IEC 61161 can be performed using the apparatus described in Fig. 2.

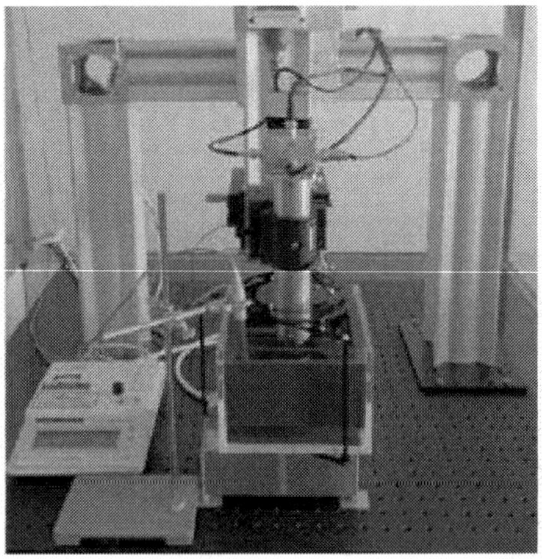

Figure 2. Radiation force balance

The ultrasonic power is determined by measuring the force exerted by the sound field generated by an ultrasonic source on a target, which is connected to a balance that estimates the apparent mass variation of the target itself due to ultrasonic field when the source is alternately switched on and off.

Two different radiation force balance set-ups are available at the INRIM ultrasound laboratory, which are used for different ultrasonic power ranges: the first set-up uses an absorbing target and is suitable for power less than 2 W, while the second set-up, used for higher power levels P>2W, adopts a reflecting target.

Such instrumentation is very bulky and at the moment measurements cannot be performed in hospitals, but a portable version is currently under study, making it feasible to include acoustic power control in QA protocol in the next future.

3. Conclusions

CRUM, Reference Center for Ultrasound in Medicine in Regione Piemonte, has the aim to promote and standardize US QC procedures in the regional territory, to create a regional database including all the diagnostic medical US equipments in Piemonte and monitoring their operational conditions.

Moreover, the training activity of CRUM aims at providing specific and updated skills in the field of medical US , in cooperation with the local Universities, not only to medical professionals and radiologists but also to graduates in Obstetrics, Radiology Technicians and Nurses following specific university training courses.

Finally, CRUM is interested in create interest and sensitizing all the professional figures involved with Medical US, with the aim of extending QC practice also to the US therapy equipments, developing appropriate protocols and procedures.

References

[1] AIUM, *Standard methods for calibration of 2-dimensional and 3-dimensional spatial measurements capabilities of pulse echo ultrasound imaging systems*, 2004.
[2] AIUM, *Routine Quality Assurance for Diagnostic Ultrasound Equipment*, 2008.
[3] Carson PL et al, *Calibration and quality control in diagnostic ultrasound*, Phys Med Biol 19: 229-230, 1974.
[4] Goodsitt A, Carson PL et al, *Real time B-mode ultrasound quality control test procedures: report of AAPM task group n°1.* Med Phys 25: 1385, 1998.
[5] IEC Technical, Report n°1390. *Real time pulse echo system: test procedure to determine performance specifications.* 1st Ed. 1996-1997.
[6] Ministero della Salute, Direzione Generale del Sistema Informativo - Ufficio di Direzione Statistica, *Attività Gestionali ed Economiche delle A.S.L. e Aziende Ospedaliere. Annuario Statistico del Servizio Sanitario Nazionale Anno 2006*, giugno 2008.
[7] Novario R, Goddi A, Conte L, Crespi A, *Use of Doppler phantom to evaluate color coded flow equipments*, Physica Medica, 5 (I): 224, 1989.
[8] Novario R, Goddi A, Crespi A, Conte L, *A new phantom for quality assurance of color-coded ultrasound flow equipment*, Physica Medica, 10 (3): 101-106, 1994.
[9] Novario R, Campani R, Conte L, Giribona P, Bazzocchi M, *Controllo di qualità delle apparecchiature ecografiche Omicron*, Genova, 1996.
[10] Novario R, Nicolini G, Tanzi F, Lorusso R, Conte L, Sharma SD, Sansotta C, Vermiglio G, Goddi A, Quality control in sonography: possibility of defining acceptability criteria, Physica Medica, 20 (3): 91-98, 2004.
[11] Price R, editor, *Routin quality assurance of ultrasound imaging systems.* York: The Institute of Physical Sciences in Medicine, 1995.
[12] SIRM, *Controlli di qualità in ecografia*, Supplemento de 'Il Radiologo' 1/2004.
[13] Tengattini A, Richetta E, Balbis S, Tofani S, Meloni T, *Aspetti fisici della qualità in ecografia clinica*, Rivista Italiana di Acustica, 32(4): 37-42, 2008.
[14] Thijssen JM et al, *Performance testing of medical echo-Doppler equipment*, European Journal of Ultrasound (2002), 15: 151-164.
[15] Thijssen JM et al, *Objective performance testing and quality assurance of medical ultrasound equipment*, Ultrasound in Medicine & Biology (2007), 33: 460-471.
[16] Vallbe M, Donadey A, *Technologies emergentes et controle de la qualité en échographie*, RBM 19, 1: 36-39 Elsevier, Paris, 1997.

A new definition for acoustic dose

FA Duck

Medical Physics and Bioengineering Department, Royal United Hospital, Bath BA1 3QF, UK.

Email: f.duck@bath.ac.uk

Abstract. This paper discusses a recent proposal for definitions of acoustic dose and acoustic dose-rate. Acoustic dose is defined as the energy deposited by absorption of an acoustic wave per unit mass of the medium supporting the wave. Its time-derivative, acoustic dose-rate, Q_m, in W kg^{-1}, is central to the prediction of both rate of temperature rise and radiation force. These quantities have spatial and temporal dependency, depending on the local field parameters (acoustic pressure, particle velocity, intensity) and local material properties (absorption coefficient, α_a, and mass density, ρ_0). Spatial and/or temporal averaging can be applied where appropriate. For plane-wave monochromatic conditions in a homogeneous medium, $Q_m=2\alpha_a I/\rho_0$, (I is the time-averaged intensity), a simple expression which may also incorporate frequency dependencies of energy deposition. Acoustic dose and acoustic does-rate are exact analogues for Specific Absorption and Specific Absorption Rate (SAR), quantities central to radiofrequency (RF) and microwave dosimetry. Acoustic dosimetry in the presence of tissue/gas interfaces remains a considerable challenge.

1. Introduction

The term 'acoustic dose' has never gained clear definition and usage in the context of the biological effects of ultrasound. This has often led to confusion in the description of the conditions pertaining to such effects and the interactions that give rise to them. Instead, a form of dosimetry terminology has evolved in connection with two of the mechanisms, heating and cavitation, rather than being associated with any more fundamental physical quantity. Concepts and definitions of thermal dose [1] and cavitation dose have evolved. Thermal dose is the time to achieve, on average, a particular biological outcome at a specified temperature rise (or time-temperature-rise integral). If the thermal dose is known for one temperature, it is possible to estimate, from the assumed Arrhenius relationship between time and temperature rise, the time to achieve the same outcome at a different temperature rise [2]. Thermal dose is really a threshold, however, for which scaling of dose has no meaning. It is not possible to predict the outcome of halving or doubling the thermal dose (that is halving or doubling the heating time) from knowledge of a particular thermal dose, nor does this thermal dose have any independent meaning relating it to another cell line or tissue, or biological endpoint. The term 'cavitation dose' has also been used by several authors, usually derived from the time-integral of the signal from a passive cavitation detector [3,4]. Such approaches do not give an absolute quantity that may be compared between laboratories, however, nor do they relate quantitatively and directly to the interaction process.

New definitions for acoustic dose, and its time-derivative acoustic dose-rate, based upon energy deposition per unit mass, have recently been proposed [5,6]. The present paper reiterates these definitions and develops the underlying concepts. The definitions have arisen in part from consideration of the way in which those concerned with dosimetry for electromagnetic radiations, both ionizing and non-ionizing, have dealt with the need to quantify the energy deposited in tissues, from which subsequent biological outcomes may be predicted. For ionizing radiations, absorbed dose is defined as the energy deposited per unit mass of a medium, because the biological effects of ionizing radiation depend strongly on this quantity. For RF and microwave radiation, Specific Absorption similarly quantifies the energy deposited per unit mass. The Specific Absorption Rate (SAR) is much more useful, however, since it is related directly to temperature rise and hence thermal biological effects. The definitions of acoustic dose and acoustic dose-rate are formulated to align with these dosimetric quantities, which are in widely-accepted use for other radiations. As will be seen below, the acoustic dose-rate relates more closely to biological effects mechanisms than does acoustic dose.

Acoustic dose and its time-derivative acoustic dose-rate are both spatially-dependent and time-dependent quantities. As with more conventional acoustic exposure quantities, it would be possible (and perhaps in some cases appropriate) to use a derived value as a descriptor of the whole field. For example the spatial average or spatial peak value, or the value at a specified point, could be calculated, being the dosimetric analogues of, for example, the commonly-used ultrasound safety parameters for acoustic exposure such as Thermal Index, or estimated *in-situ* spatial-peak temporal-average intensity. It is not the primary purpose, however, to develop here such derived values, but rather to set the general dosimetric principles in place, and to leave such possible developments to follow as and when they may be found to be useful.

2. Definitions of acoustic dose and acoustic dose-rate.

Acoustic dose, Φ, is defined as the energy deposited per unit mass of a medium supporting an acoustic wave. The SI unit for acoustic dose is joule per kilogramme (J kg^{-1}). Acoustic dose-rate, q_m, is defined as the instantaneous temporal rate of deposition of energy per unit mass of a medium supporting an acoustic wave. The SI unit for acoustic dose-rate is joule per second per kilogramme (J s^{-1} kg^{-1}) or, equivalently, watt per kilogramme (W kg^{-1}). Both acoustic dose and acoustic dose-rate may vary with position, and take values that are location-dependant. Acoustic dose-rate may be time-varying whilst acoustic dose relates to an acoustic exposure over a specified time period.

3. Expressions for acoustic dose and acoustic dose-rate

A closely analogous quantity to the acoustic dose-rate, the dissipation function, q_v, was used by Nyborg as the starting point for the calculation of the temperature rise in a relaxing medium [7]. The dissipation function is the instantaneous rate of energy deposition per unit volume. The local time-average of q_v over an appropriate time interval is

$$\langle q_v \rangle = -\nabla \cdot I = -\nabla \cdot \langle pu \rangle = -\nabla \cdot \langle i \rangle \tag{1}$$

where i and I are the instantaneous and time-averaged intensities respectively, p is the local acoustic pressure and u is the particle velocity. Thus, equivalently, the local time-average of acoustic dose-rate is

$$\langle q_m \rangle = -\frac{\nabla \cdot I}{\rho_0} = -\frac{\nabla \cdot \langle pu \rangle}{\rho_0} = -\frac{\nabla \cdot \langle i \rangle}{\rho_0} \tag{2}$$

where ρ_0 is the mean mass density.

Under the specific conditions of the uniform deposition of energy in a homogeneous absorbing medium from a plane travelling wave, the dissipation factor is

$$Q_v = 2\alpha_a I \tag{3}$$

where $2\alpha_a$ is the intensity absorption coefficient of the medium in nepers, and the notation Q_v is used for the time-average $\langle q_v \rangle$. This expression may be also applied, to good approximation in the far field of a piston source and in the focal region: more general expressions may be derived for other positions on the axis of symmetry of plane and focused transducers [8].

It follows that the rate of energy absorption per unit mass, or acoustic dose-rate, Q_m, (that is, $\langle q_m \rangle$) is

$$Q_m = \frac{2\alpha_a I}{\rho_0}. \tag{4}$$

The total energy deposited per unit mass of a medium, acoustic dose Φ, during a time period t, is

$$\Phi = \frac{2\alpha_a It}{\rho_0}. \tag{5}$$

Equations 4 and 5 assumed an acoustic wave of a single frequency, but may be easily rewritten to accommodate broadband exposure by integrating the product $2\alpha_a I$ over frequency, for which equation 4 may be re-written

$$Q_m = \frac{2\int_f (\alpha_a I)df}{\rho_0}. \tag{6}$$

4. Acoustically inhomogeneous media: scattering and reflection

Real materials supporting an ultrasound wave are generally not homogeneous. To be useful, concepts of acoustic dose and acoustic dose-rate must retain definition and meaning within an acoustically inhomogeneous medium. Acoustic phenomena such as scattering, reflection and refraction must be accommodated. None of these phenomena alter the material properties of the medium, α_a and ρ_0, however, only changing the intensity I (or more fundamentally the instantaneous intensity i). Exact acoustic dose estimates therefore require the local intensity to be estimated or measured. Soft tissue scattering is weak, so conventional approximations to estimating soft-tissue exposure *in-situ* may often be sufficient. Close to strong scatterers and reflecting surfaces, such as at soft-tissue/bone interfaces, and soft tissue/gas interfaces, the influence of the surface on the local intensity would require quantification. Amongst the conditions previously considered for the dissipation factor [8], the particular case of the superposition of two travelling waves was evaluated. This case is relevant for conditions close to tissue/bone or tissue/air interface.

5. Acoustic dose-rate and mechanical and thermal effects

The relationships between the product $2\alpha_a I$ and specific thermal and mechanical responses are well described in the literature. The initial rate of increase in temperature caused by the absorption of ultrasound is

$$\frac{dT}{dt} = \frac{2\alpha_a I}{\rho_0 C} = \frac{Q_m}{C} \tag{7}$$

where C is the specific heat capacity of the medium. Standard methods for estimating temperature rise from the absorption of ultrasound use this expression as the energy source term, depending on assumed thermal conduction and convection and 3D intensity distributions (ie 3D acoustic dose-rate distributions for uniform media) in order to predict the time-course and final steady-state increase in temperature.

In a travelling wave, the medium also experiences a volumetric force, F_v, in the direction of wave propagation, associated with radiation pressure. This force, expressed per unit mass, F_m, is

$$F_m = \frac{F_v}{\rho_0} = \frac{2\alpha_a I}{\rho_0 c_0} = \frac{Q_m}{c_0} \tag{8}$$

where c_0 is the sound speed. Comparison between equations 7 and 8 demonstrates the close relationship between heating and body force caused by acoustic absorption. In particular, the central role played by acoustic dose-rate in both these mechanisms demonstrates explicitly that both thermal and mechanical mechanisms are present under all exposure conditions, and that one cannot exist in the absence of the other. Pulsing regimes may be expected to elicit different relative outcomes, however. The recovery times may be expected to be much shorter for stress effects than for thermal effects: to a first approximation stress effects may be expected to depend on the pulse-average intensity, and thermal effects to the time-averaged intensity.

6. Dose estimation in the presence of gas bubbles

For the majority of situations the principles of estimation of acoustic dose-rate outlined above should lead to reasonable estimates of distributions of energy deposition, from the point-by-point product of absorption coefficient and intensity and knowledge of the mean mass density. Considerable challenges remain, however, in the measurement or prediction of acoustic dose-rate in the presence of bubbles. These may be introduced into the medium (assumed to be tissue), for example, either as the result of cavitation or boiling from high intensity ultrasound exposure, or from the introduction of gas-bubble contrast agents. In both these cases, the introduction of bubbles adds a new time-variation for all the three quantities in the dose-rate expressions, α_a, ρ_0 and I. Q_m is inherently a time-varying quantity, and in principle to should be possible to quantify this time variation through the time-variation of these three quantities as the bubbles form, grow, coalesce, move and disperse. Clearly, however, there would be a huge challenges arising from the wide range of bubble behaviour, characteristic of acoustic cavitation, and also from the variety of forms into which energy may be converted, including thermal, radiative, potential, kinetic and chemical.

One outcome of the presence of bubbles is to increase the effective absorption coefficient of the medium above its native value in the absence of bubbles. This in turn enhances heating. Under some conditions it may be possible to assign an average effective absorption coefficient to the bubbly medium. An alternative approach might enable the power deposited per unit mass to be estimated from calculation of the power deposited per bubble, and knowledge of the bubble concentration [5].

7. Discussion

This paper brings together several known aspects related to energy deposition from sound into its supporting medium in order to develop a coherent definition of the terms "acoustic dose" and "acoustic dose-rate". This is done with the intention of introducing uniformity and clarity for acoustic dosimetry, that is, the quantification of energy deposition from acoustic waves. Simple expressions developed using plane-wave assumptions may be used widely, but not universally. In the focal zone, and for much of an unfocussed near field, the dissipation function (and hence the acoustic dose-rate) is essentially the same whether bulk or shear viscosity is the primary mechanism [8]. Close to the transducer this is not so, but information may, in principle, be obtained by measuring Q distributions in known acoustic fields [8].

Accepting the plane-wave approximation, acoustic dose-rate may be calculated from the product of exposure, specified in terms of intensity, and energy transfer, specified by the absorption coefficient. The measurement of acoustic dose, therefore, requires the measurement of both. Each quantity may vary spatially, and this will be reflected in general by a 3D variation of acoustic dose

Table 1. Calculated values of acoustic dose-rate, Q_m, and acoustic dose per pulse, Φ, associated with several conditions of medical uses of ultrasound at 3 MHz. Standard values of absorption coefficient and mass density have been used [9]. An intensity transmission coefficient of 0.5 from tissue to bone is assumed. A pulse duration of 1s is assumed used for HITU exposure, and 1μs for a diagnostic pulse.

	Tissue	Intensity (W cm^{-2})	Acoustic Dose-rate, Q_m (W g^{-1})	Acoustic Dose per pulse, Φ (J g^{-1})
FDA *in-situ* I(spta)	Soft tissue	0.72	0.216	
	Bone	0.72	2.25	
FDA *in-situ* I(sppa)	Soft tissue	190	57	5.7×10^{-5} (1 μs)
	Bone	190	500	5.0×10^{-4}(1 μs)
HITU	Soft tissue	1000	300	300 (1 s)
Physiotherapy	Soft tissue	1.0	0.3	

(and hence acoustic dose-rate) throughout any region of interest. The spatial distribution of acoustic dose-rate, Q_m, may be predicted from the local product of the *in-situ* intensity (using any standardized or more realistic model) and the absorption coefficient. Such 3D distributions of dose-rate form the appropriate input for computations of 3D temperature increase under real situations. Accurate prediction of the acoustic dose-rate therefore depends on the ability to accurately predict (or measure) *in-situ* intensity, including not only the transmission loss from overlying tissues but also the contributions from scatter and reflection from adjacent tissues. Generalized models of tissue properties are limited in value (the simplest, and therefore the least useful, being the standard soft-tissue model for which an attenuation coefficient of 0.3 dB cm^{-1} MHz^{-1}). Reliable experimental data of measured absorption coefficients of body tissues are still relatively rare, and the need to carry out further measurements is therefore emphasized. Examples of calculated values for Q_m (W g^{-1}) and Φ (J g^{-1}) are given in table 1 for a few relevant conditions, including the limits given by the FDA for marketing of diagnostic ultrasound equipment in the USA.

In principle it would be possible to develop an experimental acoustic dose-rate meter to measure directly the rate of energy deposition in a standard material by either thermal or mechanical means. A thermal test object [10] might form the basis for a calorimetric method. Radiation force balances offer the basis for a mechanical approach. Critical to the success of such devices would be the need to ensure that the device captures all the acoustic power, and that the absorber is very well characterized. "Standardized" acoustic dose-rate, measured by such an instrument, could then be used for equipment inter-comparison, calibration of therapeutic and surgical devices, and to provide input to estimates of *in-vivo* acoustic dose and dose-rate.

Each of the quantities could have been defined in terms of energy deposited per unit volume, rather than unit mass. By using the definition given here, however, the quantity is identically analogous to Specific Absorption Rate, used as the primary dosimetric quantity for non-ionizing electromagnetic radiation in the RF and microwave frequency range, for which heating is the major, and perhaps the only important, bio-effects mechanism. For acoustics, the dose-rate relates to two bio-effects mechanisms, heating and radiation stress. Thus acoustic dose-rate, and not acoustic dose, is the more meaningful dosimetric quantity for ultrasound studies. Indeed, considerable benefit may accrue to those considering biological outcomes from exposure to low-level ultrasound from a careful review of the very large literature on the biological effects of RF and microwave radiation. If the assumption is made that, in each case, it is heating that dominates the biological processes, then SAR and acoustic dose-rate values are equivalent. It is recognized that there are conditions for which whole-body absorption is studied for RF radiation, conditions not generally relevant in an ultrasonic context. But equally many studies of localized microwave heating, for mobile phone dosimetry for example, concern highly localized patterns of energy deposition. The geometric conditions are not very different from those used during ultrasound physiotherapy, and approach those arising during some diagnostic ultrasound exposures.

Comparison with absorbed radiation dose for ionizing radiation is perhaps less immediately relevant, because radiation dose relates directly to the biological outcome of exposure to ionizing radiation, whilst there is no evidence of similar integration of the effects of repeated ultrasonic exposures. Nevertheless, and in spite of the widespread use of radiation absorbed dose for ionizing radiation dosimetry, it is insufficient to completely characterize the associated biological responses. In order to allow for the fact that different ionizing radiations (X-rays, α particles, π mesons) cause different biological outcomes for the same energy deposition, a weighting factor, the relative biological effectiveness (RBE), has been introduced. Consideration may be necessary to introduce into ultrasound dosimetry some equivalent means to characterize, for example, dosimetry in the presence of bubbles, for which the energy transfer from the wave to the tissue follows very different routes.

8. Conclusion

Simple universal definitions for acoustic dose and acoustic dose-rate are presented. Using as precedent the definition of dose used for ionizing radiation, acoustic dose is defined as the energy deposited by an acoustic wave per unit mass of the medium supporting it. Under plane-wave assumptions, simple expressions for acoustic dose and acoustic dose-rate may be derived. Using well-established expressions, simple relationships between acoustic dose-rate and both rate of temperature rise and radiation force are demonstrated. The acceptance of well-defined measures of acoustic dose and acoustic dose-rate may help to clarify the usage of terminology associated with ultrasonic exposimetry, ultrasonic dosimetry and biological effects mechanisms.

References

[1] Sapareto S A and Dewey W C 1984 Thermal dose determination in cancer therapy *Int. J. Radiation Oncology Biol. Phys.* **10** 787-800

[2] Dewey W C 1994 Arrhenius relationships from the molecule and cell to the clinic *Int. J. Hyperthermia* **10** 457-83

[3] Chen W-S, Brayman A A, Matula T J and Crum L A 2003 Inertial cavitation dose and hemolysis produced in vitro with or without Optison *Ultrasound Med. Biol.* **29** 725-37

[4] Huber P E and Debus J 2001 Tumor cytotoxicity and radical formation in vitro depend on the shock wave-induced cavitation dose *Radiation Res.* **156** 301-9

[5] Duck F A 2009 Acoustic dose and acoustic dose rate *Ultrasound Med. Biol.* **35** 1679-85

[6] Advisory Group on Non-ionising Radiation 2010 *Health effects of exposure to ultrasound and infrasound* Documents of the Health Protection Agency RCE-14 pp17-8

[7] Nyborg W L 1981 Heat generation by ultrasound in a relaxing medium *J. Acoust. Soc. Am.* **70** 310-2

[8] Nyborg W L 1986 Sonically produced heat in a fluid with bulk viscocity and shear viscosity *J. Acoust. Soc. Am.* **80** 1133-9

[9] ICRU 1998 *Tissue substitutes, phantoms and computational modelling in medical ultrasound. ICRU Report 61* Bethesda USA International Commission on Radiation Units and Measurements pp 44-7

[10] Shaw A, Bond A D, Pay N M and Preston R C 1999 A proposed standard thermal test object for medical ultrasound *Ultrasound Med. Biol.* **25** 121-32

Journal of Physics: Conference Series **279** (2011) 012007 doi:10.1088/1742-6596/279/1/012007

Building and assessing anatomically relevant phantoms for neonatal transcranial ultrasound

G.Memoli[1], M.Gatto[2], N. Sadhoo[1], P. Gélat[1], R.A. Harris[2], A. Shaw[1]

[1] Acoustics and Ionising Radiation Division, National Physical Laboratory, Teddington (UK)

[2] Wolfson School of Mechanical & Manufacturing Engineering, Loughborough University, Loughborough (UK)

E-mail: gianluca.memoli@npl.co.uk

Abstract. This study describes the design and construction of a clinically relevant phantom to survey the temperature increase caused by ultrasound equipment, as currently used in neonatal head-scanning in the UK. The phantom is an ellipsoid of bone-mimic material, filled with brain-mimic; a circular hole in the external surface mimicks the fontanel, through which most clinically relevant scans are made. Finite-element simulations were used to identify possible hot spots and decide the most effective thermocouple positions within the phantom to investigate temperature rise during a typical scan. Novel materials were purposively designed to simulate key acoustic and thermal properties. Three Dimensional Printing (3DP) was employed for the fabrication of the skull phantom, and a specific strategy was successfully pursued to embed a thermocouple within the 3DP skull phantom during the manufacturing process. An in-process Non-Destructive Analysis (NDA) was used to assess the correct position of the deposited thermocouple inside the fabricated skull phantom. The temperature increase in the phantom for a typical trans-fontanellar scan is also presented here. The current phantom will be used in a hospital survey in the UK and, in its final design, will allow for a more reliable evaluation of ultrasound heating than is currently possible.

1. Introduction

Clinical practice is seeing an increasingly widespread use of ultrasound in neonatal units, together with the growing sophistication of ultrasound scanners and the associated potential for increased ultrasound exposure to neonates. In addition to intrauterine fetal imaging, there are other cases where ultrasonography is considered an essential, non-invasive, imaging tool. A key example is trans-cranial imaging, commonly used to diagnose congenital and acquired anomalies of the neonatal brain, particularly in premature babies and in the first weeks after birth [1]. In this context, the quasi-total absence of proper epidemiological studies, evidencing the effects of prolonged exposure to ultrasound, makes it nearly impossible to quantify the risks connected to the use of ultrasonography. This is why medical physicists and practitioners refer to guidelines on thermal indexes (related to the local increase of temperature) to reduce the risks [2]; it is also why a survey of the equipment and its use is needed.

This study is part of a larger project, funded by the UK Department of Health (DH), aimed at investigating the safe use of ultrasound for neonatal head scanning. The project proposes to design a validated, clinically relevant phantom, to be used in a hospital survey across the UK. The scope is to measure the temperature increase caused by ultrasound equipment currently used in neonatal head scanning, using a repeatable and traceable protocol [3]. Other activities include a paper-based survey of neonatal trans-cranial ultrasound practices, two training workshops, as well as an investigation of how to properly measure the temperature rise.

The present work marks a particular stage of the process, conducted in collaboration with Loughborough University. Following feedback from clinicians, the first generation of phantoms (PH1) – cylindrical in shape, tailored for quality assurance and characterised elsewhere [4] – was upgraded to a more anthropomorphic shape, based an appropriate bone-mimicking material and 3D printing technology (PH2). The resulting phantom constitutes a significant step forward from the typical phantoms for testing ultrasound equipment, usually limited to standard and simple geometries.

2. Designing the phantom

While there are specific guidelines for building validated phantoms to measure exposure to ionising radiation [5], nothing similar exists for ultrasound. Moreover, for the particular ages of interest, very little data could be found in the literature [6] for the acoustical parameters of interest (i.e. speed of sound and absorption). This information was the benchmark, on which specific materials were designed and manufactured, to replicate as faithfully as possible the real neonatal head.

At the start of the process it was decided that PH2 (i.e. the phantom described in this study) would:

- model the skull of a 33 weeks pre-term neonate (i.e. a premature baby);
- be constituted by a prolate ellipsoid (i.e. the skull) filled with brain-mimic material (i.e. the brain) and covered with silicon rubber (1 mm thick);
- have a hole, with the same area and position as that of a typical fontanel at 33 weeks.

These choices fixed the dimensions of the prolate ellipsoid (figure 1), which were obtained by intersecting the reference data commonly used in UK neonatal units [7] (i.e. in the tables to assess the normal growth of a child, the 50th percentile line was used-) with those reported in [5].

2.1. From thermal modelling to phantom requirements

Finite element modelling [3] and preliminary results with PH1 phantoms [4] showed that, while the highest temperature increase is to be expected near the surface of the transducer (in correspondence to the highest change of acoustical and thermal impedance), there are other locations within the phantom that may be interesting to monitor. Figure 2 shows the temperature dependence, obtained by finite element modelling (PAFEC), for a semi-spherical skull filled with brain-mimic material (see section 2.2.), and subjected to the transmission from a cylindrical transducer through a hole in the skull (the fontanel, on top in figure 2). Even if the values of the temperature rise in figure 2 are probably much higher than the ones that can observed in reality (due to the particular, "worst-case" model used for the transducer), a careful analysis shows four interesting positions to monitor: (1) just below the transducer, under the skin-mimic; (2) in the bulk of the brain; (3) between the skin and the bulk, to check the gradient; (4) in the bone opposite the fontanel, where the exciting transducer is.

It was then decided to provide the phantom with at least 4 thermocouples, one of which is to be embedded in the portion of skull opposite the fontanel. The challenging task of embedding the thermocouple in the skull will be described in section 3.2.

2.2. Brain mimic material

According to [6], at 1 MHz the brain of a 28 weeks old neonate has a speed of sound of 1528 ± 5 m/s and an absorption coefficient of 0.16 – 0.3 dB/cm. Considering the large range of frequencies used in neonatal trans-cranial ultrasound (contemporary scanners offer frequency as an adjustable parameter), the value at one single frequency is not sufficient to characterise the material. Assuming a power dependence on frequency, an exponent of 1.05 can be obtained by the data in [6].

The materials were acoustically characterized at NPL in the range 2-20 MHz and the values at 1 MHz were obtained by interpolation (i.e. a power fit). Comparing the results with the target data from literature (table 1) the TMM with 0.91 g/l was chosen to represent the brain in our phantom.

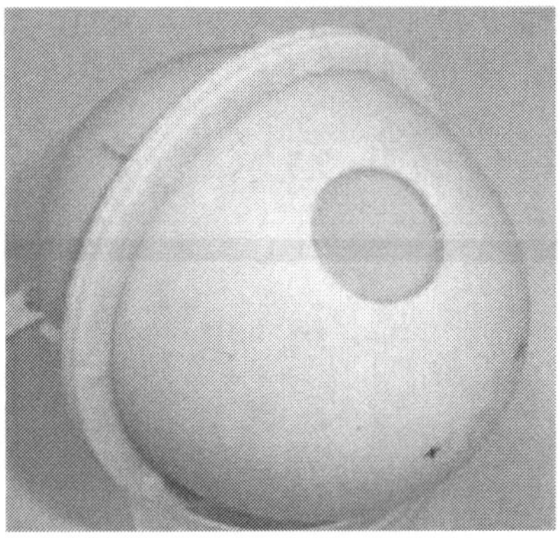

Figure 1. Appearance of the PH2, once built.

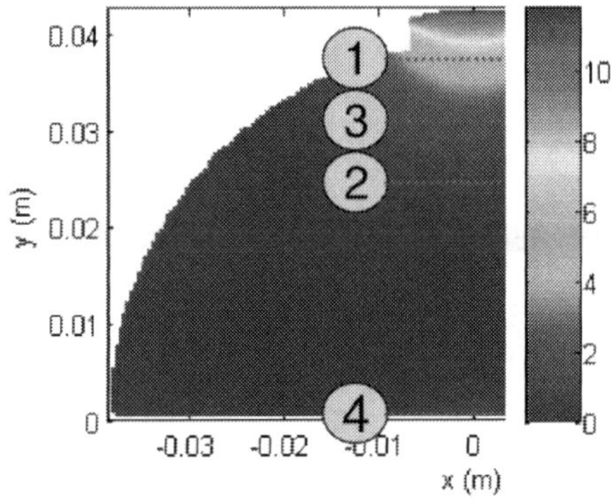

Figure 2. Ideal positions of the thermocouples, according to FEM modeling.

2.3. Skull mimic material

A procedure similar to that in section 2.2. was used to select the best candidate for the skull phantom, starting from data in [6]. The base material was the ceramic powder normally used in 3D printing (see section 3.), and infiltration with different fluids was used to change its acoustic properties.

Since it is particularly important to fill the porous ceramic uniformly and to avoid undesired air bubbles. For this reason, the infiltration was performed by immersion, under vacuum conditions (-95 kPa), in a dedicated chamber. The test samples were left to cure for 72 hours at 25 °C.

In this study, three fluids were chosen: epoxy resin (Clear Coat™ by System Three Resins Inc.), polyester resin (Crystic® 2-414PA by Scott Bader Ltd.) and wax (Paraplast™ X-TRA by McCormick Scientific™). Preliminary tests for these materials showed a penetration of the infiltrant inside the ceramic up to a few millimeters (perfect for a thickness of the skull of 2 mm), so their acoustical characteristics in the range 2-20 MHz were measured using disks of 2, 3 and 5 mm. The use of disks with different thicknesses reduced the uncertainty on the measured parameters.

A comparison with the target value from the literature (table 2) identified the ceramic infiltrated with epoxy resin as the best candidate to mimic neonatal cranial bone.

Table 1. Acoustical characteristics of different candidate brain mimics and target values from the literature [6]. The brain mimics share a similar formulation, but differ in the concentration of Al_2O_3.

Al_2O_3 (g/l)	Speed @ 1 MHz (m/s)	Absorption @ 1 MHz (dB/cm)	Power dependence of absorption with frequency
0.91	1528	0.264	1.06
1.82	1531	0.437	1.02
2.73	1537	0.571	1.00
Target	1528 @28 weeks	0.16 to 0.3	1.05

3. Building the skull

The efficiency of rapid prototyping (RP) technologies for medical applications has been widely documented and different authors have already pioneered the adoption of RP for the manufacturing of phantoms for ultrasound tests [9]. In particular, 3DP (Three-Dimensional Printing) developed by ZCorporation® (Burlington, USA) is an increasingly popular additive manufacturing process. Some key attributes of 3DP, like the relative low cost of building material, the speed of manufacturing and its level of accuracy [10], make this technique suitable for several precision applications, including the building of anatomical models for surgical planning, education or training [11].

Particularly important for this study is the fact that the porous structure of the ceramic powder used in 3DP allows post-processing treatments, such as infiltration with liquid substances. With this technique, the physical properties of the final material can be adjusted over a wide range [12].

A further reason to choose 3DP as prototyping technology for this work comes from the great accessibility to the building chamber during the printing process. This peculiarity of 3DP allowed a Fine-Wire Thermocouple (FWT) to be embedded in the phantom skull during printing, in a position opposite to the fontanel (i.e. position 4 of figure 2), as required in section 2.1.

Table 2. Comparison between the target values from [6] and the candidate materials for skull bone.

Infiltrant	Absorption @ 1 MHz (dB/cm)	Power dependence of absorption with frequency
Epoxy	6.8	1.5
Polyester	64.9	1.37
Wax	23.6	0.97
Target value	5 to 7 for neonates	1.2 to 2.1

3.1. Production process

A Z510 Spectrum 3D printer (24-bit color models at a resolution of 650 × 540 dpi) was employed for the manufacturing process and set so that, with each printing layer in the XY plane, the phantom grew in thickness steps of 0.1 mm in the z direction. During the printing process, a FWT was embedded at the top of each semi-ellipsoid composing the skull (figure 2), positioned in the XY plane at 0.5 mm from the inner surface of the skull. After completion, the parts were then placed into an air-circulating oven at 80°C for 2 hours, and at 120°C for 1 hour. This guaranteed the full curing of the water-based binder within the parts and the elimination of excess moisture, which could impede infiltration. Infiltration with epoxy resin followed soon afterwards, as described in section 2.3.

3.2. Thermocouple embedding on 3DP parts

Preliminary trials showed that only very fine objects could be embedded in a 3DP prototype. The strength of the prototype, in fact, mainly relies on the quality of the bond created between the deposited layers. The presence of an external body (e.g. a thin-film thermocouple) within the jetted layers might compromise the bonding between the printed planes, and, consequently, it could affect the integrity of the final structure. Fine-wire thermocouples (FWT) with a diameter of 0.005 in. (0.127 mm) proved to be the most effective, while still offering a reasonable resistance to handling.

The printing process could be monitored thanks to a novel Non-Destructive Analysis (NDA) technique, specially developed for 3D printing at Loughborough [12]. In particular, the image of each jetted layer, constituting the final object, was acquired in real time using a modified CMOS camera (Trust® WB-6250, 1240 x 1024 pixels), with a resolution of 876 ppi. An automatic acquisition program (designed using RoboRealm®-) allowed the operator to follow the embedding of the selected FWT on a monitor, greatly improving the accuracy in positioning the thermocouple (figure 3).

After reaching the desired height, the 3D printer was paused and the FWT was positioned over the un-bound powder layer (layer A in figure 3). The FWT was deposited along the roller movement direction, thus avoiding the obstruction of the powder distribution on the next spreading phase. After the working conditions (i.e. 38 °C) were restored in the printing area, the job was resumed. The next layers were additionally built over the deposited FWT, which was embedded inside the phantom.

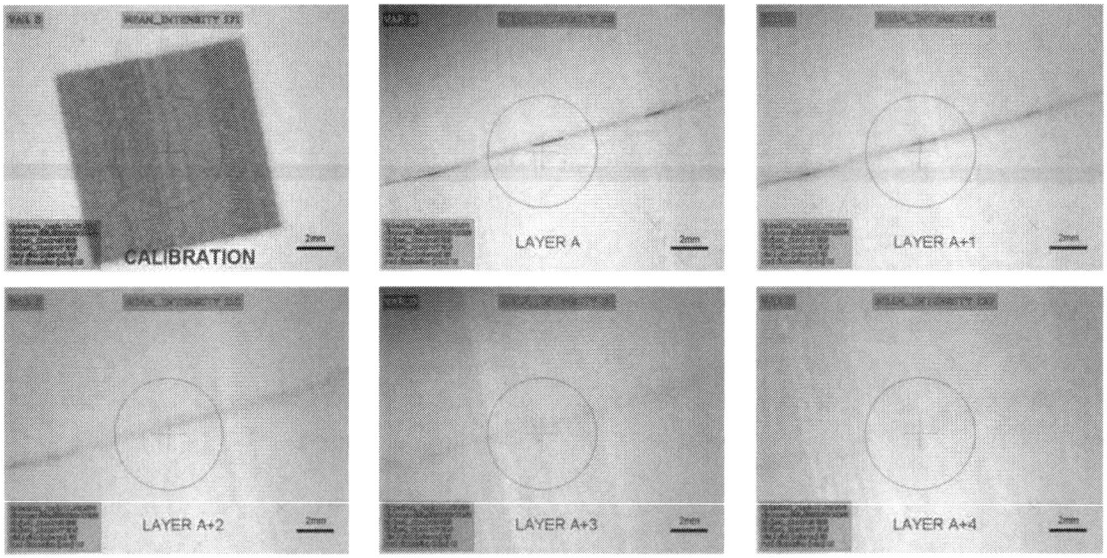

Figure 3. A series of subsequent printed layers, showing how the FWT is embedded. Calibration was needed to fix the coordinate system in the following frames.

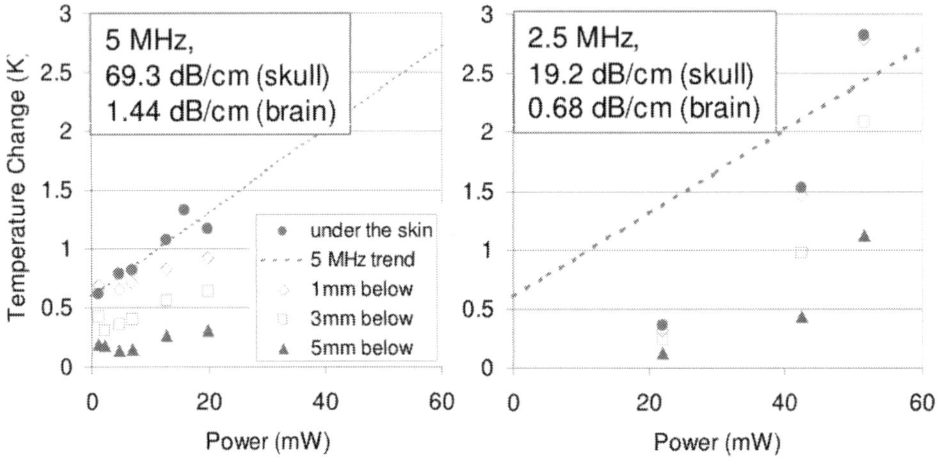

Figure 4. Temperature rise at fixed distances for transfontanellar 15 minutes scans with PH2, using two transducers (different for shape and working frequency). Values of the absorption coefficient in the two materials have been reported, for comparison.

4. Temperature increase in the PH2 phantom

The phantom was tested using a HP 77020A scanner and two transducers, different in shape, power and operating frequency: HP21200B (2.5 MHz) and HP21211B (5 MHz). Emitted powers were measured using an acoustic balance before the tests, while the temperature rise was measured using a profile thermocouple by Omega (containing 3 thermocouples spaced 2 mm), which was aligned along the transducer axis (figure 4). The scanning conditions chosen represent a "worst case scenario" for a trans-fontanellar scan: the transducer is placed on the fontanel for 15 minutes (the average duration of a scan in hospital practice [3]) and aligned along the thermocouple probe, in absence of a thermal dissipation mechanism mimicking perfusion. This considered, it might not be surprising that values as large as 2.5 °C were measured. On the other hand, it is worth noticing that powers as high as 150 mW might be currently in use in UK hospitals. Figure 4 also highlights a possible effect of the shape of the transducer; a more comprehensive picture will be obtained during the final part of the project.

5. Conclusions

This paper described the process of building an anatomically relevant phantom (PH2), designed to assess the temperature increase in the neonatal head during transfontanellar scans, and the first measurements with it. A careful choice of tissue mimicking materials was crucial for the reliability of the final phantom and, to the authors' knowledge, this is the first time that the skull of a neonate is reproduced with such an accuracy (at least from the ultrasonic point of view). This study also highlighted the advantages offered by rapid prototyping (and in particular by 3D printing) in the development of phantoms for medical ultrasound. The innovative techniques developed at Loughborough to embed thermocouples in 3DP prototypes and to monitor the printing process were paramount for this achievement. Further developments will include an even more anatomically relevant phantom, based on the feedback received so far from clinicians. Last but not least, the (high) temperature increases measured with PH2 confirm the need for a hospital survey, to test equipment as it is used in neonatal units. Such a study will represent the completion of the DH funded project.

Acknowledgements

The authors acknowledge the funding by the UK Department of Health, through contract RRX 113, and the help of Mr T. Korrison in the temperature measurements.

References

[1] Couture A., Veyrac C. 2001 *Transfontanellar doppler imaging in neonates* (Berlin: Springer).
[2] British Medical Ultrasound Society 2009 Guidelines for the safe use of diagnostic ultrasound equipment, available online on www.bmus.org .
[3] Memoli G., Shaw A., Gélat P., Sadhoo N., Osborne J., Robinson H., Bartle D., Duck F. 2009 A survey on clinical practice in neonatal ultrasound: preliminary results, poster presented at *IPEM meeting 2009*, York.
[4] Memoli G., Shaw A., Gélat P., Sadhoo N., Osborne J., Robinson H., Bartle D., Duck F. 2009 Towards a clinically relevant phantom for neonatal transcranial ultrasound: preliminary results, poster presented at *Euroson 2009*, Edinburgh.
[5] The International Commission on Radiological Protection 2002 ICRP PUBLICATION 89 – Basic Anatomical and Physiological Data for Use in Radiological Protection: Reference Values, *Annals of the ICRP* **32**.
[6] Duck F. A. 1990 *Physical properties of tissue* (Academic Press: London).
Hill C. R. 1987 *Physical principles of Medical ultrasound* (Wiley WCH: Weinheim)
[7] Child Growth Foundation 1996/1 *Duodecimal growth charts* (Harlow: Tyne & Wear).
[8] International Electrotechnical Commission 2007 *IEC 60601-2-37 Medical electrical equipment – Part 2-37: Particular requirements for the basic safety and essential performance of ultrasonic medical diagnostic and monitoring equipment.*
[9] Demitri C., Sannino A., Conversano F., Casciaro S., Distante A., Maffezzoli A. 2008 Hydrogel based tissue mimicking phantom for in-vitro ultrasound contrast agents studies. *J. Biomed. Mater. Res. Part B: Appl. Biomater.* **87**, 338-345.
[10] Radley G.J., Sama A., Watson J. and Harris R.A. 2009 Characterisation, Quantification and Replication of Human Sinus Bone for Surgery Simulation Phantoms, *Proceedings of the Institution of Mechanical Engineers: Part H Journal of Engineering in Med.* **223** 875-888.
[11] Suwantreeb J. 2006 Comparative study of 3DP material systems, *Rapid Prot. Journal* **13**, 48-52.
[12] Gatto M., Harris R.A. Non-Destructive Analysis (NDA) of Internal and External Structures in 3DP, *Rapid Prototyping Journal* In press

Advanced Metrology for Ultrasound in Medicine (AMUM 2010) IOP Publishing
Journal of Physics: Conference Series **279** (2011) 012008 doi:10.1088/1742-6596/279/1/012008

The Edinburgh Pipe Phantom: characterising ultrasound scanners beyond 50 MHz

C M Moran[1], W Ellis[2], A Janeczko[2], D Bell[3] and S D Pye[2]

[1] Medical Physics, University of Edinburgh, Edinburgh, EH16 4TJ, UK
[2] Medical Physics Department, NHS Lothian University Hospitals Division, Royal Infirmary, Edinburgh EH16 4SA, UK
[3] Precision Acoustics Ltd, Hampton Farm Business Park, Dorset, DT2 8QH, UK

E-mail: carmel.moran@ed.ac.uk

Abstract. The ability to measure the imaging performance of pre-clinical and clinical ultrasound scanners is important but difficult to achieve objectively. The Edinburgh Pipe Phantom was originally developed to assess the technical performance of clinical scanners up to 15MHz. It comprises a series of anechoic cylinders with diameters 0.4 – 8mm embedded in agar-based tissue mimic. This design enables measurement of the characteristics (Resolution Integral *R*, Depth of Field L_R, Characteristic Resolution D_R) of grey-scale images with transducer centre frequencies from about 2.5 to 15MHz. We describe further development of the Edinburgh Pipe Phantom as a tool for characterising ultrasound scanners with centre frequencies up to at least 50MHz. This was achieved by moulding a series of anechoic pipe structures (diameters 0.045 – 1.5mm) into a block of agar-based tissue mimic. We report measurements of *R*, L_R and D_R for a series of 10 transducers (5 single element and 5 array transducers) designed for pre-clinical scanning, with centre frequencies in the range 15-55 MHz. Values of *R* ranged from 18-72 for single element transducers and 49-58 for linear array transducers. In conclusion, the pre-clinical pipe phantom was able to successfully determine the imaging characteristics of ultrasound probes up to 55MHz.

1. Introduction

The ability to measure the imaging performance of both pre-clinical and clinical scanners is important but difficult to achieve objectively. We have previously shown the versatility of using the Resolution Integral as a figure-of-merit for characterising clinical scanners [1,2]. More recently we have extended this work to include the quantitative assessment of scanners and transducers for pre-clinical applications [3]. In this paper we employ the Resolution Integral to compare 10 transducers designed for pre-clinical applications: 5 single element transducers ranging in centre frequency from 25-55MHz and 5 linear array transducers ranging in centre frequency from 15-40MHz.

The Resolution Integral, *R*, has been defined elsewhere [4] and will only briefly be described here. Essentially *R* is the ratio of the penetration of an ultrasound beam in soft tissue to the ultrasound beam width. Transducers that perform well combine good penetration with narrow beam width, and thus *R* will be large. Two other parameters are also defined in [4], Depth of Field L_R, and Characteristic Resolution D_R. The Depth of Field defines the axial extent of a region of optimum imaging and the Characteristic Resolution is representative of the resolution within the Depth of Field, taking into consideration both the lateral and elevation planes. The three quantities are related by the expression $R=L_R/D_R$.

2. Methods

Measurements of R, L_R and D_R were made on 10 transducers from two different pre-clinical ultrasound scanners. Five linear array transducers on a Vevo 2100 scanner were tested, and also 5 single element transducers on a Vevo 770 scanner (Visualsonics Inc, Toronto, Canada). The manufacturer's data is summarised in Table 1. Further details of the measurement method described here are contained in [2] and [4].

Measurements were made using two Edinburgh Pipe Phantoms [1] both composed of agar-based tissue mimicking material (TMM) with attenuation coefficient 0.5dBcm^{-1} MHz^{-1} and speed of sound of 1540ms^{-1} determined at frequencies up to 20MHz by [5]. A series of anechoic wall-less pipes were moulded within the TMM at an angle of 40^0 to the vertical. The first ("clinical") pipe phantom was constructed from a block of TMM (250 x 250 x 100mm) contained in a perspex casing and with pipe diameters from 0.35mm to 7.9mm. The second ("pre-clinical") pipe phantom comprised a series of cylindrically-shaped blocks of TMM with diameter 60mm and height 40mm. Embedded within these cylinders were pipes of diameter 1470, 550, 330, 193, 139, 92, 68 and 45 microns. The TMM of both phantoms was immersed in a water and glycerol mixture (speed of sound 1540ms^{-1}) which filled the pipes and provided coupling to the upper surface of the phantoms.

The lowest frequency transducers (15 and 21MHz) were evaluated using both the clinical and pre-clinical pipe phantoms. The other transducers were all evaluated using the pre-clinical pipe phantom. To carry out measurements, each transducer was placed on the surface of the phantom so that the axis of the largest pipe lay in the scan plane. The image was then optimised to ensure that the upper section of the pipe was visualised as superficially as possible. For the array transducers, a minimum of two focal positions were chosen around the location of the upper section of the pipe. The image was frozen and the minimum depth at which the pipe could be visualised was measured using on-screen callipers. The transducer was then re-positioned to enable the lower section of the pipe to be visualised as deep as possible within the phantom. Again, a minimum of two focal positions were chosen and located around this depth. The image was frozen and the maximum depth at which the pipe could be visualised was measured using on-screen callipers. To determine the locations of the most superficial and deepest imaged extent of the pipe, a mask was manufactured containing a slot of approximate width 15 times the wavelength of the centre frequency of the transducer. This was placed over the frozen image and the slot moved along the extent of the pipe to establish the positions where the pipe could be confidently visualised with respect to surrounding speckle at the same depth. The difference between the most superficial and the deepest imaged depths is L (figure 1). Three sets of measurements were taken from three pipes of the same diameter, and the mean of these was calculated. The procedure was repeated for each pipe and each mean value of L was plotted against the appropriate value of α. α is the inverse of the effective pipe diameter, calculated as the geometric mean of the pipe diameter in the image and elevation planes, and given by $d/\sqrt{(\cos 40^0)}$ where d is pipe diameter. The low contrast penetration depth in the image was identified as the depth at which speckle could just be differentiated from system noise. The dead zone was identified as the most superficial depth at which speckle could be imaged. The axial distance between the dead zone and the low contrast penetration is given by the intercept of the L-α curve with the ordinate in figure 1. The area under the curve is equal to the Resolution Integral [4].

The same procedure was followed for the single element transducers except that the deepest and most superficial positions where each pipe could be visualised were both measured using the same optimised image. In addition, an alternative method of measuring the Resolution Integral was employed for the RMV704 single element transducer, whereby the distance between the transducer and surface of the phantom was varied to enable the extent of the pipe to be tracked through the depth of the phantom. This technique of tracking the pipes throughout the phantom is consistent with the recommended use of the single element transducers for pre-clinical scanning, where coupling gel is often used as a stand-off to ensure that superficial regions of interest can be located within the focal region.

Transducer	Single element (SE) or linear array (LA)	Centre Frequency (MHz)	Focal-Length (mm)	Field of View (mm)
RMV710	SE	25	15	20
RMV707B	SE	30	12.7	20
RMV704	SE	40	6	14.6
RMV708	SE	55	4.5	10.7
RMV711	SE	55	6	8.5
MS200	LA	15	18	36
MS250	LA	21	15	30
MS400	LA	30	9	20
MS550S	LA	40	6	15
MS550D	LA	40	7	15

Table 1. Manufacturer's data for the 10 pre-clinical transducers.

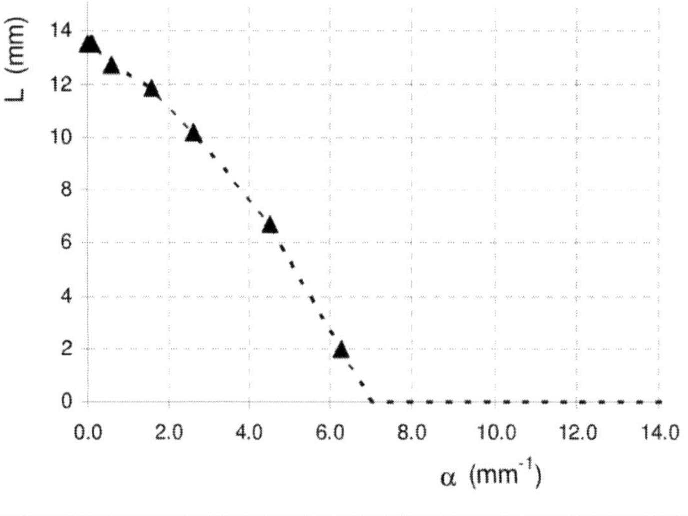

Figure 1. Plot of L (depth range over which pipe can be visualised) against α (1/effective pipe diameter) for a Vevo 2100 MS550S linear array centred at 40MHz.

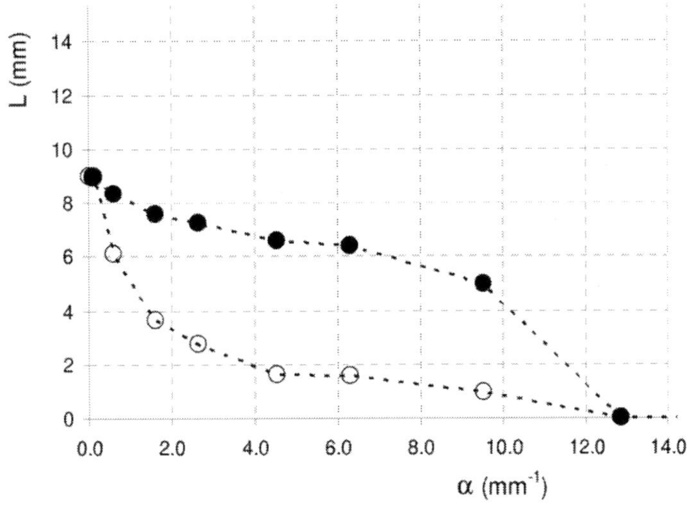

Figure 2. Plots of L (depth range over which pipe can be visualised) against α (1/effective pipe diameter) for a Vevo 770 RMV704 single element probe centred at 40MHz. Measurements were made with the probe on the surface of the pipe phantom and the focal region at a fixed depth (○), and also by tracking the focal region down through the phantom (●).

3. Results

The L-α curve for each transducer typically comprised 5 to 7 data points. Figure 1 shows an example obtained using the MS550S 40MHz linear array. The area under the curve is equal to the Resolution Integral. Figure 2 illustrates two sets of results obtained from the RMV704 40MHz single element probe: one with the probe resting on the surface of the phantom, and one obtained using the tracking technique. The smallest diameter pipes (45μm) were not imaged by any of the pre-clinical transducers.

Values of L_R and D_R for each of the 10 transducers were calculated from their L-α curves and plotted on a diagram of L_R vs D_R as described in [4]. The results are shown in figure 3. The value of R for each transducer is equal to the gradient of the line joining the point (D_R, L_R) to the origin, since $R=L_R/D_R$. The values of L_R and D_R calculated from the three L-α curves shown in figures 1 and 2 are labelled in figure 3 as "40MHz (S)", "40MHz(a)" and "40MHz(b)".

4. Discussion

It can be seen in figure 3 that the 5 single element transducers all had values of R close to 25 (range 18 - 25), with the exception of the tracked measurements for the RMV704 where R=72. For the linear array transducers, values of R were close to 50 (range 49 - 58). The smallest Characteristic Resolution (93μm) was observed for the RMV704 tracked measurements. The points representing the two linear array transducers with centre frequencies of 40MHz (MS550D and MS550S) lie close together, with R values of 55 and 56 respectively. The MS550D had a slightly longer Depth of Field (10.9mm) than the MS550S (10.5mm) which is consistent with the manufacturer's recommendations that the MS550D transducer should be used for imaging slightly deeper structures. The RMV704 single element probe with the same nominal centre frequency of 40MHz had a significantly different performance, with a smaller Depth of Field (3.3mm) and Characteristic Resolution (135μm) compared to 188μm and

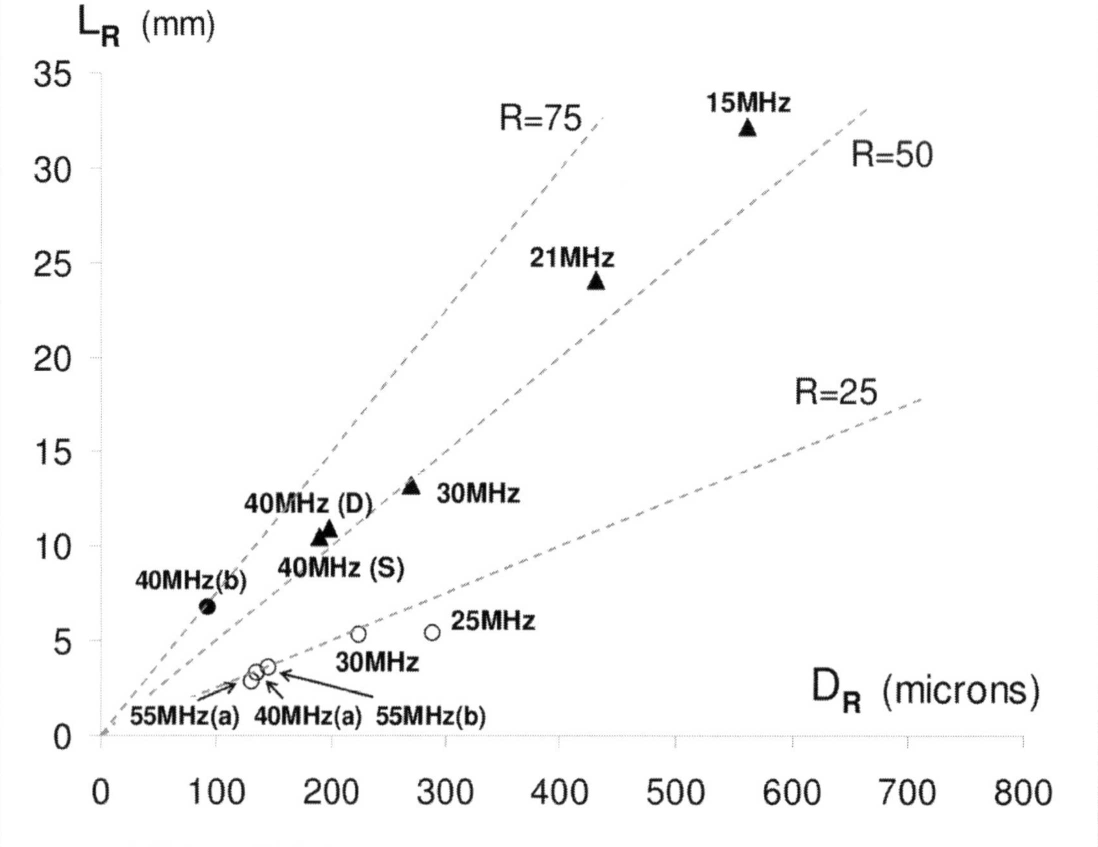

Figure 3. Measured values of L_R and D_R for 10 pre-clinical transducers. ▲ denotes Vevo 2100 linear array transducers [40MHz (S) = MS550S, 40MHz (D) = MS550D]. ○ denotes Vevo 770 single element transducers [55MHz(a) = RMV708, 55MHz(b) = RMV711, 40MHz(a) = RMV704, • 40MHz(b) = RMV704 tracked through the phantom]. Lines of constant R are also shown (R=25, 50, 75). Expanded uncertainty in L_R and D_R (95% confidence) is approximately ±8%.

197μm for the MS550S and MS550D respectively. This is due to the greater degree of focussing of the single element transducer, yielding smaller values of Characteristic Resolution and Depth of Field for the same nominal centre frequency. The same comparison can be made between the single element and linear array transducers with a nominal centre frequencies of 30MHz: the RMV707B with D_R=225μm, L_R=5.3mm, and the MS400 with D_R=269μm, L_R=13.2mm.

Using the tracking technique with the RMV704 ensured that all the superficial aspects of the pipe could be visualised within the focal region, thus effectively lengthening the Depth of Field to 6.7mm, compared to 3.3mm when the tracking technique was not applied. In addition, since each superficial aspect of the pipe was visualised at optimal resolution in the focal region, the Characteristic Resolution was reduced from 135μm to 93 μm. The Resolution Integral increased from 25 to 72, which is the highest of all the pre-clinical transducers and comparable with values achieved by premium range clinical scanners [4].

The shapes of the L-α curves for the pre-clinical linear array transducers (e.g. the MS550S shown in figure 1) were all close to linear. This is very similar to the shape of the L-α curves reported for clinical scanners in [4], and indicates a weakly focussed ultrasound beam. In contrast, the results for the RMV704 in figure 2 show two different curve shapes. The lower curve, obtained with the probe resting on the surface of the phantom, is distinctly concave. This is indicative of the relatively strong focussing of the single element transducer. The upper curve, produced by tracking the focal region through the phantom, is convex and approaches the rectangular "box" shape indicative of a collimated beam [2].

5. Conclusions

Measurements of R, L_R and D_R have been carried out on 10 pre-clinical ultrasound transducers, and the characteristics of single element and linear array technologies have been compared. The results demonstrate that the clinical and pre-clinical pipe phantoms, incorporating anechoic targets from 45μm to 8mm diameter, are able to quantify and differentiate the characteristics of transducers with centre frequencies up to at least 55MHz. The characteristics of the pre-clinical transducers reported here can also be compared with a range of clinical transducers reported in [4]. It may be possible to extend Resolution Integral measurements to transducers above 55MHz, given that the smallest pipes were not imaged by any of the pre-clinical transducers.

References

[1] Pye S D and Ellis W 2002 UK Patent GB2396213

[2] Pye S D, Ellis W and MacGillivray 2004 *Journal of Physics: Conf. Series* **1** 187-92.

[3] Moran C M, Smart S, Ellis W and Pye S D 2008 *Proc 2008 IEEE Ultrasonics Symposium* pp1724-27.

[4] Pye S D and Ellis W 2010 *Journal of Physics: Conf. Series (this issue)*

[5] Brewin M P, Pike L C, Rowland D E and Birch M J 2008 *Ultrasound in Med and Biol* **34** 1292-1306

Journal of Physics: Conference Series **279** (2011) 012009 doi:10.1088/1742-6596/279/1/012009

The Resolution Integral as a metric of performance for diagnostic grey-scale imaging

S D Pye and W Ellis

Medical Physics Department, NHS Lothian University Hospitals Division, Royal Infirmary, Edinburgh EH16 4SA, UK

E-Mail: stephen.pye@luht.scot.nhs.uk

Abstract. There is a strong clinical need for objective measurements of grey-scale ultrasound imaging performance. We have previously proposed the Resolution Integral R as a suitable metric, because it defines the ratio of the penetration of an ultrasound beam in soft tissue to the ultrasound beam width. Transducers with good performance combine deep penetration with high lateral resolution, giving a large ratio of penetration to beam-width. Depth of Field L_R and Characteristic Resolution D_R are defined such that $R= L_R/D_R$. L_R defines a region of optimum resolution, and D_R is representative of the lateral and elevation resolution within L_R. We report the results of a survey of the imaging performance of 79 models of ultrasound scanner made using an Edinburgh Pipe Phantom to measure values of R, L_R and D_R. The scanners were manufactured between 1986 and 2009 and were from 23 manufacturers. A total of 171 different transducers were tested, including linear and convex arrays, and mechanical sector probes. The results demonstrate that R successfully distinguishes transducers with differing levels of performance, and that L_R and D_R characterise probes suitable for different clinical applications. The characteristics of individual probes are concisely quantified and displayed on a plot of L_R vs D_R.

1. Introduction

The objective assessment of diagnostic ultrasound imaging performance has several roles: to ensure that equipment performs to a defined standard and is clinically fit-for-purpose; to assess new imaging modalities and signal processing techniques; to underpin decision making in the procurement and replacement of equipment. Technical assessments of grey-scale ultrasound have been limited by the fact that the resolution of ultrasound images is a strong function of depth and varies in three orthogonal directions [1]. As a consequence, measurements of spatial and contrast resolution made with test objects often produce large volumes of data which may be useful to monitor equipment during its life cycle [2], but are restricted in their ability to compare the performance of different ultrasound systems. International standards are limited to describing procedures for testing the accuracy of measurement callipers and for measuring resolution and point spread function at discrete depths [3, 4]. The limitations of current methods for performance evaluation are discussed by Shaw and Hekkenberg [5].

Pye and Ellis [6] introduced the concept of a Resolution Integral and published preliminary work [7]. This was the first description of a fundamental physical approach to combining measurements of resolution made at different depths in the ultrasound image. Moran et al [8,11] describe the use of the Resolution Integral to characterise pre-clinical ultrasound scanners, and Rowland et al [9] used a test phantom containing a step-change in backscatter to make independent measurements of Resolution Integral in the scan and elevation planes. In this paper we report the results of a survey of ultrasound equipment carried out in Lothian Region, Scotland between 2002 and 2010. Measurements of Resolution Integral were made using a number of Edinburgh Pipe Phantoms [6, 7]. The aim of the survey was to determine whether measurements on individual transducers could characterise their performance and suitability for their specified clinical application. In particular, whether R could discriminate improvements in transducer technology, and whether L_R and D_R could characterise clinical application

2. Resolution Integral

The Resolution Integral R for a collimated beam is defined as the ratio of penetration to beam width [7]. For real ultrasound beams, the variation of beam characteristics with depth is accounted for using an integral approach. Figure 1a illustrates this concept. The definition of R has been amended slightly here, so that it is defined in terms of beam width rather than beam radius as in [7]. Two other parameters are defined in figure 1b: Depth of Field L_R and Characteristic Resolution D_R. The Depth of Field defines the axial extent of a region of optimum imaging, analogous to the focal region. The Characteristic Resolution is representative of the beam width within the Depth of Field. Medical ultrasound transducers with good performance combine deep penetration with high resolution, so that the ratio of penetration to beam width is large. A transducer suited to a particular clinical application will have a small Characteristic Resolution together with sufficient Depth of Field for the particular application.

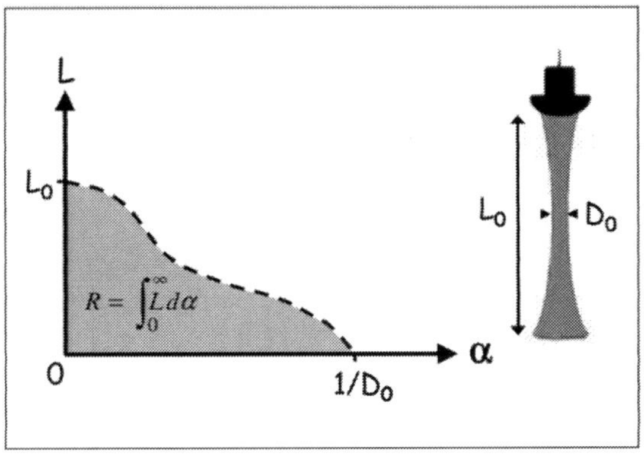

Figure 1a. The graph (left) shows a plot of L vs α for an ultrasound beam (right) with low contrast penetration L_0 and minimum beam width D_0. α is the reciprocal of beam width, and $L(\alpha')$ corresponds to the depth range in the image over which the beam width is less than $1/\alpha'$. The Resolution Integral R is equal to the area under the curve.

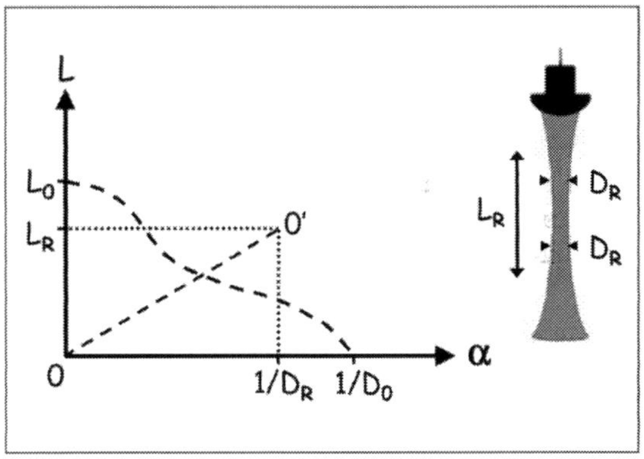

Figure 1b. The graph (left) shows the same plot of L vs α as Fig 1a, and illustrates Depth of Field L_R and Characteristic Resolution D_R. The area under the dashed curve 0-L_0-1/D_0 is equal to the Resolution Integral R. The rectangle 0-L_R-0'-1/D_R also has an area equal to R, and line 00' bisects both the rectangle and the area under the dashed curve, giving $R = L_R / D_R$

3. Materials and Method

Measurements were carried out on 79 different models of ultrasound scanner manufactured between 1986 and 2009, and sited at various hospitals and clinics in Lothian Region. These included top- and mid-range scanners and a smaller number of portable and hand-held machines, employing linear arrays, convex arrays and mechanical sector probes. A total of 171 different scanner/transducer combinations from 23 manufacturers were assessed: 78 abdominal transducers, 57 small parts transducers, 23 transvaginal (TV) endoprobes, 9 cardiac transducers, 3 vascular access imagers, and 1 bladder imager. Scanners were either in clinical service when tested, withdrawn from service and maintained for laboratory use, or undergoing pre-purchase evaluation. They were all in good working order with no known faults.

The measurements of *R* were made by three operators using six Edinburgh Pipe Phantoms, manufactured in-house as described in [7]. Each pipe phantom was constructed from a 250 x 250 x 100 mm block of agar based tissue mimicking material (TMM), prepared using the method described in [10]. The TMM had an attenuation coefficient of 0.5dB cm^{-1} MHz^{-1} and speed of sound of 1540ms^{-1}. Eleven wall-less pipes ranging in diameter from 0.35mm to 7.9mm were moulded into the TMM at an angle of 40^0 to the vertical and filled with water-glycerol mixture (c=1540ms^{-1}) [7]. The six phantoms were cross-compared using small parts and abdominal probes. Measurements of *R* made using the same probes on different pipe phantoms were within a 5% range. The stability of the pipe phantoms over time was tested by making periodic measurements of *R* during the course of the survey using the same scanner/transducer combination. Results were within the repeatability of the measurement (3%).

Scanning took place under a controlled level of ambient lighting (<20 lux at the surface of the display monitor). Each transducer was placed on the surface of the pipe phantom and coupled to it with water/glycerol mixture. An image of one of the pipes was obtained, and the upper portion of the pipe was imaged with the pipe axis positioned in the scan plane of the transducer. The image was optimised to visualise the pipe as superficially as possible, with the transducer held vertically, and the most superficial region of the pipe that could be detected positioned directly below the transducer (within $\pm 10^0$ of the vertical). All available controls were used to optimised the image e.g. field of view, TGC, transmit focus, transmit frequency, harmonic imaging, speckle reduction, compound imaging. The image was then frozen and a visual assessment was made of the minimum depth at which the pipe could be visualised. To aid this assessment, a mask of buff-coloured paper with a slot of width approximately 15 times the wavelength of the centre frequency of the probe was used to mask out echoes from neighbouring regions of the pipe and TMM. The mask was placed onto the frozen ultrasound image and the slot was moved up and down along the length of the pipe image to allow the observer to visualise each short section of the pipe in turn, compare it with adjacent speckle, and identify and measure the minimum depth below the scanning surface at which the pipe could confidently be detected. The transducer was then re-positioned, the image was optimised, and the same procedure was carried out to identify the maximum depth below the scanning surface at which the pipe could confidently be detected. The difference between the maximum and minimum depths at which the pipe could be detected gave one value of *L* (figure 1). This procedure was repeated for each pipe and each value of L was plotted against α, the reciprocal of the effective pipe diameter. The effective pipe diameter was calculated as the geometric mean of the pipe diameter in the image and elevation planes and is equal to $d/\sqrt{(\cos 40^0)}$ for a pipe of diameter d. The low contrast penetration (LCP) of each transducer was determined by imaging in real time and measuring from the scanning surface to the deepest speckle identifiable from noise. Care was taken to ensure that no pipes were visible in the field of view. The depth of the transducer dead zone (DZ) was measured vertically below the probe. The distance LCP – DZ represents the maximum range of depth over which anechoic objects of any size can be detected, and corresponds to the intercept (L_0) on the ordinate in figure 1. Typically, the measured *L-α* plot for each transducer comprised values of *L* from 6 - 7 pipes plus the distance LCP – DZ. The area under the *L-α* curve was integrated using a simple trapezium rule to calculate *R*. Values of L_R and D_R as defined in figure 1b were also calculated.

4. Results

Figure 2 shows the measured values of *R* for each transducer/scanner combination, plotted against year of manufacture and grouped in 5-year intervals. The median value of *R* is shown for each group, together with the interquartile range. The median value for 2005-09 is significantly different from 2000-04 ($P<0.04$); as is 05-09 vs 95-99 ($P<0.01$); and 00-04 vs 90-94 ($P<0.01$). As an illustration of some of the data summarised in figure 2, figure 3 shows values of *R* measured for a series of scanners

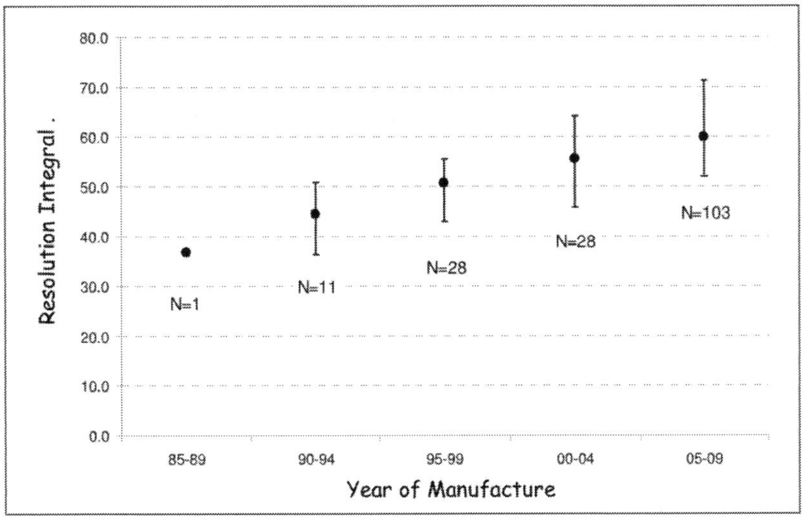

Figure 2. Median value of Resolution Integral and interquartile range, plotted against year of manufacture for 171 transducer/scanner combinations grouped in intervals of five years. N denotes the number of transducers in each group.

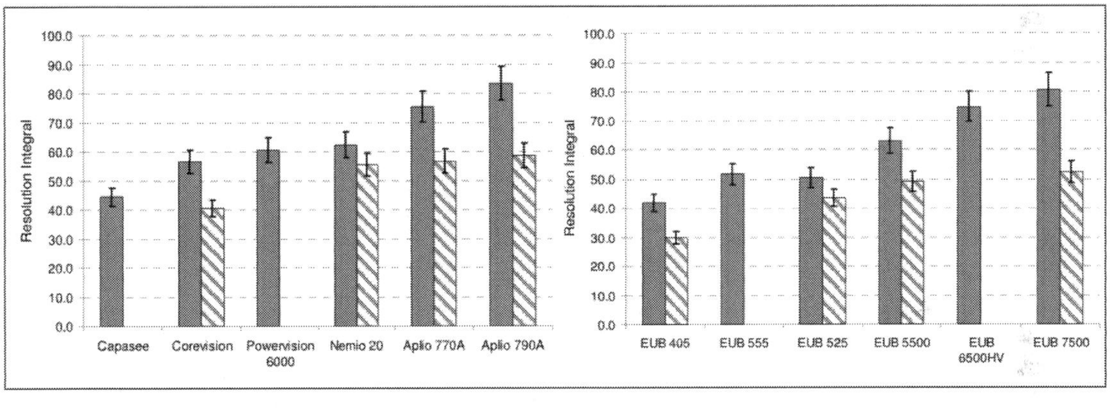

Figure 3. Resolution Integral values for different models of scanner manufactured between 1994 & 2008 by Toshiba Medical (left) and Hitachi Medical (right). Older scanners are on the left of each chart and the newer ones are on the right. Each bar represents a single transducer. Solid bars = abdominal probes. Hatched bars = TV endoprobes. 95% confidence intervals are shown.

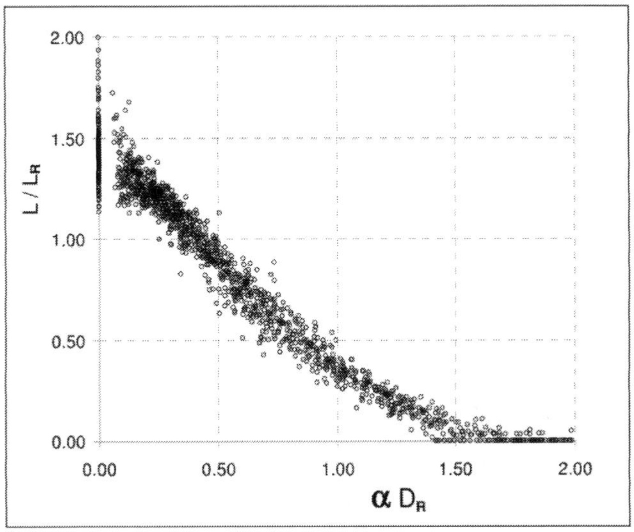

Figure 4. Plot of L normalised by L_R versus α normalised by D_R for 171 transducer/scanner combinations. All the curves lie close to each other, indicating that L_R and D_R are efficient parameters for describing beam shape. i.e. Probes with the same values of L_R and D_R have very similar beam shapes; probes with different values of L_R and D_R have different beam shapes.

produced over a period of 15 years by two major manufacturers. To test the validity of using L_R and D_R to characterise beam shape, the measured values of L for each transducer were divided by L_R for that transducer, and the values of α were divided by $1/D_R$ to produce a normalised plot of L/L_R against αD_R. The results for all 171 transducers are shown together in figure 4.

Figure 5 shows a schematic plot of L_R vs D_R, and figure 6 illustrates the results obtained for 158 probes designed for use in abdominal imaging, small parts imaging, and transvaginal scanning.

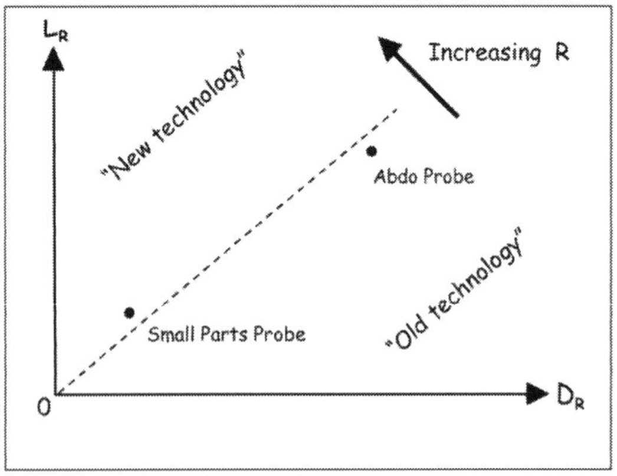

Figure 5. Schematic plot of L_R versus D_R. Two transducers are shown: a small parts probe located in the lower left quadrant, and an abdominal probe in the upper right quadrant. The value of R for each probe is equal to the gradient of the line joining it to the origin ($R = L_R/D_R$). Equipment with low values of R ("old technology") is located in the lower right quadrant. Equipment with high values of R ("new technology") is located in the upper left quadrant.

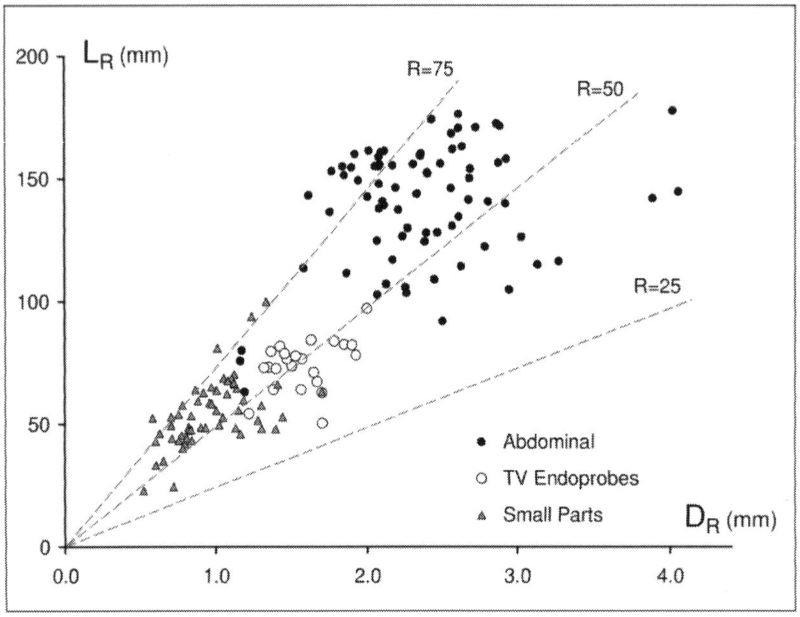

Figure 6. Measured values of L_R and D_R for 158 transducers designed for three different clinical applications. The dashed lines represent lines of constant R (25, 50 and 75). Modern premium range equipment lies close to or above the line R=75. The oldest transducers and those with the most limited performance lie close to R=25.

5. Discussion

The Resolution Integral combines a number of ultrasound image parameters (resolution at different depths, low contrast penetration, dead zone) in a physically meaningful way. The results shown in figure 2 demonstrate that R is able to characterise improvements in ultrasound imaging technology that have taken place over the last 25 years. This is supported by the data in figure 3, where the values of R for a series of scanners produced by two manufacturers increase between older models and modern high-end equipment (from 40-50 to >80 for abdominal probes; from 30-40 to 50-60 for TV endoprobes). One limitation of the survey is the uncertainty regarding any historical deterioration in performance, given that some scanners were >10 yrs old when tested. The majority of older scanners were part of a local quality assurance programme and had been assessed at regular intervals using commercial tissue mimicking test objects. Although this data did not provide useful information about resolution, measurements of low contrast penetration had been consistent over time since installation. Given that the equipment was maintained in good working order, we estimate that any change in the Resolution Integral over the lifetime of the scanners was <10%. It is also useful to compare the Resolution Integral values of five single-element transducers tested during the survey. Two non-diagnostic imagers had R values of 22 and 24 (Bardscan bladder imager, manufactured 2005; Dymax SiteRite II, manufactured 1994). Three single element diagnostic probes had R values of 36, 36.7 and 36.8 (Siemens Sonoline AC CDA3.5, 1991; Siemens Sonoline SX CDA3.5, 1986; Kretz Combison 401, 2002). The different R values of the diagnostic and non-diagnostic equipment suggest a difference in technology rather than a historical deterioration in performance.

Figure 6 shows L_R and D_R values for transducers designed for abdominal, transvaginal and small parts imaging. Transducers designed for each clinical application are grouped together in different parts of the graph, as would be expected by comparison with figure 5. Three abdominal probes shown in figure 6 with $D_R \sim 1.1$mm appear to be outliers, but these were in fact paediatric transducers operating at a higher centre frequency than probes designed for adult work. The median values of R for the three groups of transducers in figure 6 are: 62 (abdominal), 49 (TV endoprobes), 58 (small parts). The low value of R for TV endoprobes may be due to the small radius of curvature of the imaging arrays used on many of the probes. This limits the size of the acoustic aperture, and also generates rapidly diverging scan lines. These two factors may restrict the improvements that can be achieved in image quality. This hypothesis is supported by the data in figure 3, which shows a smaller increase in R for older & newer TV endoprobes, compared to older & newer abdominal transducers.

6. Conclusions

The Resolution Integral, measured using an Edinburgh Pipe Phantom, is able to discriminate between medical ultrasound transducers with differing levels of performance. The Depth of Field L_R and Characteristic Resolution D_R provide a quantitative summary of the imaging characteristics of individual probes that can be concisely displayed on a plot of L_R vs D_R.

Acknowledgements

The authors wish to thank Anna Janeczko, Carmel Moran, George Campbell and Scott Inglis for their assistance.

References

[1] Madson E L 2000 *Ultrasound in Med. & Biol.* **26** Supplement 1 ppS48-S50
[2] Thijssen J M, Weijers G and de Korte C L 2007 *Ultrasound in Med. & Biol.* **33** 460-71
[3] International Electrotechnical Commission 1996 IEC 1390
[4] International Electrotechnical Commission 2006 IEC 61391-1
[5] Shaw A and Hekkenberg R 2007 NPL Report AC 2. UK National Physical Laboratory, London
[6] Pye S D and Ellis W 2002 UK Patent GB2396213
[7] Pye S D, Ellis W and MacGillivray 2004 *Journal of Physics: Conf. Series* **1** 187-92
[8] Moran C M, Smart S, Ellis W and Pye S D 2008 *Proc 2008 IEEE Ultrasonics Symposium* pp1724-27
[9] Rowland D E, Newey V R, Turner D P and Nassiri D K 2009 *Ultrasound in Med. & Biol.* **35** 1525-34
[10] Inglis S, Ramnarine K V, Plevris J N and McDicken W N 2006 *Ultrasound in Med. & Biol.* **32** 249-59
[11] Moran C M, Ellis W, Janeczko A, Bell D and Pye S D 2010 *Journal of Physics: Conf. Series (this issue)*

Evaluation of temperature rise in a tissue mimicking material during HIFU exposure

S. Maruvada, Y. Liu, B.A. Herman, and G.R. Harris

Food and Drug Administration, Center for Devices and Radiological Health, 10903 New Hampshire Ave., Bldg., Silver Spring, MD 20993

subha.maruvada@fda.hhs.gov

Abstract. In pre-clinical testing it is essential to characterize clinical high intensity focused ultrasound (HIFU) devices using tissue-mimicking materials (TMMs) with well known characteristics, including temperature rise and cavitation properties. The purpose of this study was to monitor cavitation behavior and correlate its effect with temperature rise in a HIFU TMM containing an embedded thermocouple. A 75-μm fine wire thermocouple was embedded in a hydrogel-based TMM previously developed for HIFU. HIFU at 1.1 and 3.3 MHz was focused at the thermocouple junction. Focal pressures from 1-11 MPa were applied and the temperature profiles were recorded. Three hydrophones were used to monitor cavitation activity during sonication. A hydrophone confocal with the HIFU transducer and a cylindrical hydrophone lateral to the HIFU beam were used as passive cavitation detectors for spectral analysis of signals, and a needle hydrophone placed beyond the HIFU focus was used to record changes in the pressure amplitude due to blockage by bubbles at or near the focus. B-mode imaging scans were employed to visualize bubble presence during sonication. In a separate measurement, schlieren imaging was used to monitor the change in field distribution behind the TMM. All hydrophone methods correlated well with cavitation in the TMM.

1. Introduction

A reusable hydrogel-based tissue mimicking material (TMM), which has both thermal and acoustic properties close to that of soft tissue, has been developed in order to characterize the output of HIFU devices [1]. During HIFU exposure, the occurrence of cavitation can either enhance or diminish the treatment. In order to further characterize the cavitation phenomena in the TMM, and correlate cavitation detection techniques with temperature rise, experiments were done to expose the TMM to HIFU pressures, record the temperature rise at the focus using embedded thermocouples (TCs) and monitor the occurrence of cavitation with hydrophones placed at several locations. B-mode imaging was employed to visually monitor the effects of cavitation on HIFU temperature measurements. The reusability of the TMM was determined by performing the same set of exposures on one TMM over several weeks.

2. Materials and Methods

2.1. System Setup

The setup of the experiment is shown in Fig. 1. The driving electronics consisted of a function generator and power amplifier. A 50 dB dual-directional coupler, two power sensors and a power meter were placed between the amplifier and transducer to monitor the forward and reverse electrical power to the transducer. A single element HIFU transducer (6-cm diameter, 6-cm focal length) was used to sonicate the TMM at its fundamental frequency of 1.1 MHz and third harmonic of 3.3 MHz. This transducer had a 2-cm central hole in which a broadband confocal passive cavitation detector (PCD) was inserted. This PCD was connected to a 500 kHz low pass filter and the output was saved to a digital signal analyzer (DSA). A needle hydrophone was placed behind the TMM to monitor the variation of the HIFU transmission signal due to bubble activity. The needle hydrophone was connected to an RMS voltmeter and its output was saved to a computer-controlled acquisition system. A second cylindrical PCD hydrophone was placed behind the TMM. This PCD signal, after amplification, was saved to a digital signal analyzer for FFT analysis below 250 kHz. During the 20-sec sonication, 200 time signals were collected from each PCD. The FFT analysis was done offline and the energy in each signal versus time was calculated. The presence of cavitation bubbles was also monitored with a diagnostic imaging system. B-mode images were recorded with the image plane containing the HIFU beam axis.

A cylindrical mold 4-cm thick and having an 8-cm inner diameter was constructed with a lateral opening for the imaging transducer. A 75-μm diameter Cr-Co bare wire TC was affixed through the center of the mold. The TMM mold with TC was then cleaned by boiling in degassed water with added surfactant for 20 minutes. Once the TMM was prepared, it was immediately poured into the mold with the TC and allowed to set within the mold. The TC recorded the temperature during the HIFU sonication via a computer-controlled acquisition system.

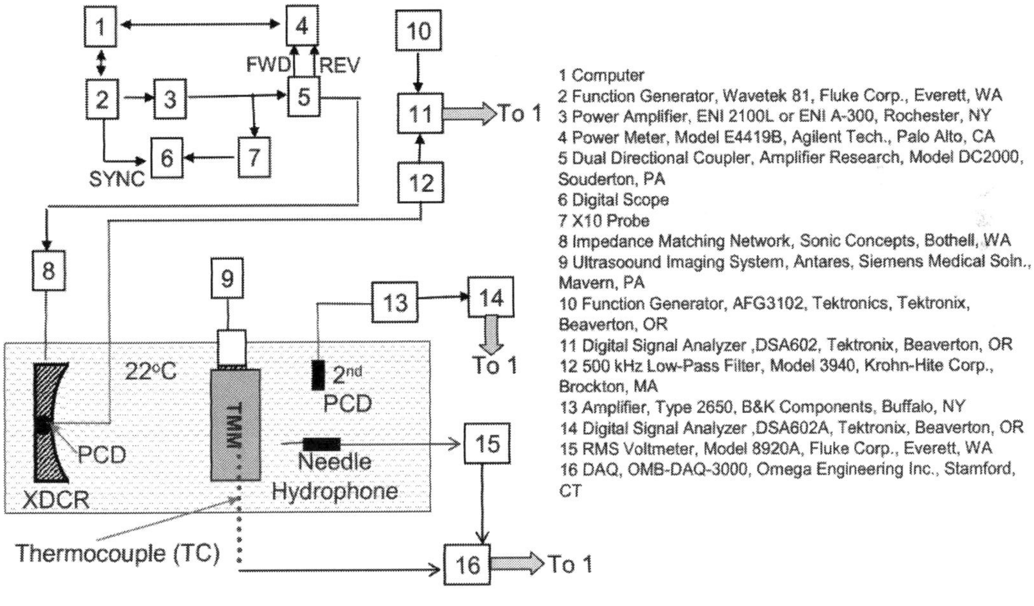

Fig. 1. Experimental setup

2.2. Experimental Protocol

The TC was positioned at the HIFU focus using low ultrasound power. The needle hydrophone was positioned for maximum output also using low ultrasound power. Each TMM was sonicated using 20 s exposures, starting at the lowest pressure level (level 1) and the temperature rise and the three hydrophone responses were recorded before, during and after sonication. B-mode images were taken during and after sonication. Each TMM underwent the complete range of sonications three times, first while the TC was positioned at the HIFU focus, next after the TC was moved 1 cm radially from the focus, and last, after the TC was moved back to the focus. To determine reusability of the TMM, one TMM was used three times, one week apart for each sonication, for a total of seven data sets. For the

1.1 MHz sonications, the focal pressure range was approximately 1 to 5 MPa. For the 3.3 MHz sonications, the focal pressures range was approximately 1 to 11 MPa.

2.3. Calculation of Threshold Pressures and Temperature Analysis

In order to obtain a more accurate assessment of the actual acoustic pressure at the focus within the TMM, simulations from a two-layer nonlinear propagation model based on the KZK equation were performed using the measured acoustic power, transducer focusing geometry, and attenuation coefficient of the TMM [2]. The simulation yielded the axial pressure and intensity waveforms, the axial harmonic distribution and the pressure waveform on the axis at the location of peak positive pressure. Directly from the simulation, we obtained the peak negative pressure and the time averaged intensity. For establishing the cavitation threshold, we used the pressure derived from the computed spatial-peak, temporal-average intensity. Both peak negative pressure values (P_-) and pressure values derived from intensity ($P_{calc} = \sqrt{2I\rho c}$) are listed in Table I.

The temperature measured by the TC at the end of sonication (EOS) may be different than the actual temperature were the TC not present due to viscous heating of the TC [3] or TC-mediated cavitation at the TC junction. To correct for these rapidly decaying measurement artifacts, the thermal decay curve, beginning 2 s after the EOS, was extrapolated back to the ultrasound "off" time [4]. An example of a normal heating curve with an overlay of the cubic fit is shown in Fig. 2a Other TC artifacts (thermal conduction due to the metal of the TC or distortion of the ultrasound beam due to the wire) are unlikely to be significant due to the small thermocouple wires.

Table I. Pressure Values obtained from Simulation for 1.1 and 3.3 MHz sonications.

Sonication level	1.1 MHz		3.3 MHz	
	P_- (MPa)	P_{calc} (MPa)	P_- (MPa)	P_{calc} (MPa)
1	0.7	1.1	1	1.4
2	1.4	2.1	2.6	4.4
3	2.0	3.1	3.9	7.4
4	2.6	4.2	4.5	9.1
5	3.2	5.2	5.1	10.9
6	3.7	6.3	5.6	12.7
7			6	14.7
8			6.4	16.6

3. Results

3.1. Thermally Significant Cavitation Threshold Results at 1.1 and 3.3 MHz

Our standard for indication of cavitation, against which we compared the different cavitation monitoring techniques, was the observation of anomalies in the temperature trace. At 1.1 MHz, 13 sets of exposures were done in five TMMs. The average cavitation threshold value was first calculated for each TMM. Then the average cavitation threshold value from the five TMMs was calculated as 4.3 ± 1.2 MPa if P_{calc} was used and 2.7 ± 0.5 MPa if P_- was used. Cavitation in the five TMMs occurred between pressure levels 3 and 6. For all cases in which the sonication level was below the cavitation threshold, the temperature traces were well behaved (Fig. 2a), the needle hydrophone signal was either constant or slowly changing, the two PCD signals were both flat and the B-mode imaging showed at the most slight streaking [5]. An example of the signals seen at a post-cavitation level is shown in Fig. 2b. In this case, anomalies in the temperature trace occurred at 9 s and continued until the end of sonication. A hyperechoic region in the B-mode images is seen at the start of the sonication and at 5 s, but then seems to elongate at 9 s. At the end of sonication, the elongated bubble pattern persists. Both PCDs showed low frequency broadband noise indicative of inertial cavitation, with strong spikes at and after 9s. The needle hydrophone output decreased significantly after 9 s.

At 3.3 MHz, eight sets of exposures were done in three TMMs. The final average cavitation threshold value calculated as 12.5 ± 3.0 MPa if P_{calc} was used and 5.4 ± 0.8 MPa if P_- was used.

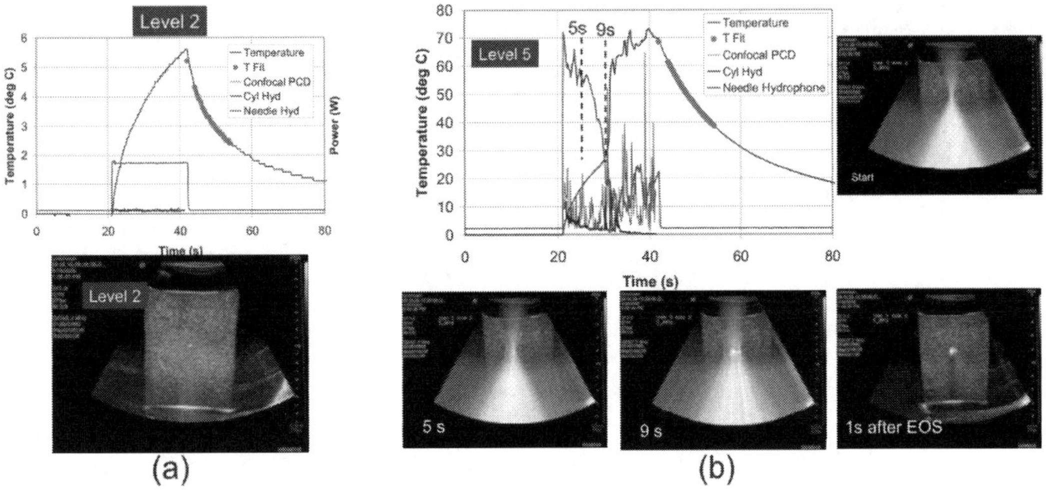

Figure 2. (a) Example of pre-cavitation detection in TMM; (b) Example of cavitation in TMM. The HIFU transducer is located to the right of the B-mode images.

3.2. Reusibility of the TMM

One TMM was used to assess reusability. A full set of exposures was done on this TMM at three different times. The average temperature rise and standard deviation of the EOS temperature results were calculated at each level when cavitation did not occur and are given in Table II

4. Discussion and Conclusions

For all cases where the sonication level was below the cavitation threshold, the temperature traces were normal, the hydrophone signals contained no sudden transitions and the B-mode images contained very little, if any, acoustic interference. When anomalies in the temperature trace indicative of cavitation were observed, there was correlation between the output of the three hydrophones and strong acoustic interference in the B-mode images, often followed by a persistence of bubble clouds in the region surrounding the focus. The B-mode images showed multiple interference patterns centered at the focus, often times culminating in a bubble cloud elongated in the direction of the HIFU beam axis.

	Level 1	Level 2	Level 3	Level 4	Level 5
T rise (^{o}C)	1.7	6.6	14.3	25.3	38.3
STDev	0.2	0.2	0.3	1.2	1.2

Table II. Average temperature rise and standard deviation of the EOS temperature results in one TMM.

Anomalies also were seen in the needle hydrophone signal such as noisy jumps and overall decrease of signal. However, at 1.1 MHz, 10% of cases produced uncertain results in that the temperature traces were well behaved but a decrease in the needle hydrophone signal was seen (Fig. 3a), indicating a change of transmitted energy that was not due to blocking by thermally significant cavitation. This phenomenon was seen in nearly half the cases at 3.3 MHz and was much more pronounced (Fig. 3b). In order to understand the reason for the decrease in signal, exposures were performed in a schlieren system (Optison, Onda Corp., Sunnyvale, CA), so that the transmitted ultrasound field could be imaged (Fig. 4). It was found that while the total transmitted energy distal to the TMM may not have decreased with time, the spatial distribution changed, which affected the needle hydrophone output. Since the levels used were below cavitation thresholds, we hypothesized that temperature changes could be causing acoustic property changes and hence altering distal intensity profiles via refraction. Therefore, another test was performed in which a small volume of the TMM around the focus was heated by passing a DC current through the thermocouple (Peltier heating). The TMM was then sonicated with very low pressure ultrasound (< 1 MPa, <<1° C temperature rise at focus). As shown in Fig. 5, the needle hydrophone signal was affected. We therefore believe that changes in temperature-dependent quantities such as acoustic impedance are occurring in the TMM that affect the distal beam. Therefore, one should be aware that slowly varying changes in the needle hydrophone output may not indicate cavitation phenomena. This affect was much more pronounced at the higher frequency and happened over a larger range of mid-level pressures.

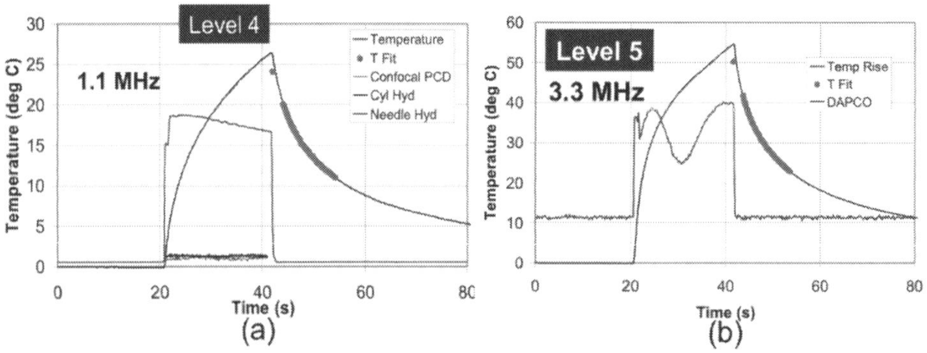

Fig, 3. Needle hydrophone anomaly during normal temperature rise at (a) 1.1 MHz and (b) 3.3 MHz.

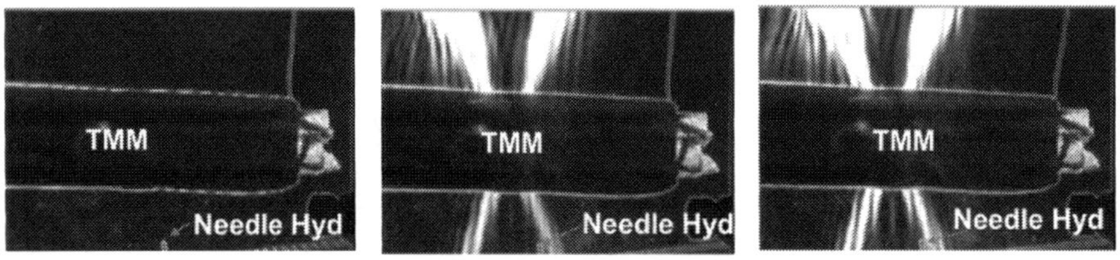

Fig. 4. Example of schlieren image of HIFU beam passing through TMM. As the TMM was heated at sub-cavitation threshold levels, the distal (lower) beam pattern was seen to distort, possibly due to refraction caused by temperature-induced impedance changes in the TMM in the perifocal region.

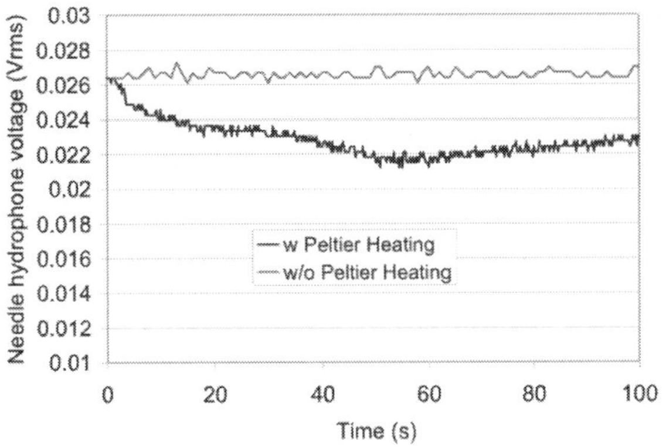

Fig. 5. Needle hydrophone voltages with and without Peltier heating by the thermocouple.

In summary, when anomalies in the temperature trace were observed, the various cavitation monitoring techniques produced corresponding results. Of the several methods studied for detecting cavitation, the needle hydrophone was a convenient adjunct to spectral analysis for evaluating cavitation activity in the TMM at 1.1 MHz but not at 3.3 MHz. The TMM proved to be reusable for cavitation experiments since non-cavitation temperature results varied very little each time the TMM was used.

Note: The mention of commercial products, their sources, or their use in connection with material reported herein is not to be construed as either an actual or implied endorsement of such products by the Department of Health and Human Services. This research was supported by the Defense Advanced Research Projects Agency (DARPA) through IAG # 224-05-6016.

References

[1] R. L. King, B. A. Herman, S. Maruvada, K. A. Wear, G. R. Harris, "Development of a HIFU Phantom," Proceedings for 6th International Society on Therapeutic Ultrasound, Oxford, UK, 2006, pp 351-356.

[2] J. Soneson, "A User-Friendly Software Package for HIFU Simulation," Proceedings for 8th International Society on Therapeutic Ultrasound, Minneapolis, MN, 2008, pp. 165-169.

[3] H. Morris, I. Rivers, A. Shaw, and G. ter Haar, "Investigation of the viscous heating artifact arising from the use of thermocouples in a focused ultrasound field", Phys. Med. Biol, (53), 2008, pp. 4759-4776.

[4] S. Maruvada, Y. Liu, B.A. Herman and G.R. Harris, "Temperature Measurements in Tissue-Mimicking Material during HIFU Exposure," Proceedings for 8th International Society on Therapeutic Ultrasound, Minneapolis, MN, 2008, pp. 286-290.

[5] C-C Wu, C-N Chen, M-C Ho, W-S Chen, and P-H Lee, "Using the acoustic interference pattern to locate the focus of a high-intensity focused ultrasound (HIFU) transducer," Ultrasound Med. Biol., 34 (1), 2008, pp. 137 – 146

Journal of Physics: Conference Series **279** (2011) 012011 doi:10.1088/1742-6596/279/1/012011

Instantaneous input electrical power measurements of HITU transducer

B Karaböce[1], Y Gülmez[1], S Rajagapol[2] and A Shaw[2]
[1] Tübitak Ulusal Metroloji Enstitüsü (UME), P.K. 54 41470 Gebze-Kocaeli, Turkey
[2] National Physical Laboratory (NPL), Hampton Road, Teddington TW11 0LW, UK

E-mail: baki.karaboce@ume.tubitak.gov.tr

Abstract. HITU (High Intensity Theraupetic Ultrasound) transducers are widely used in therapeutic ultrasound in medicine. The output ultrasonic power of HITU transducer can be measured in number of methods described in IEC 61161 standard [1]. New IEC standards specifically for measurement of HITU equipment are under development. The ultrasound power radiated from a transducer is dependent on applied input electrical voltage and current and consequently power. But, up to now, no standardised method has been developed and adopted for the input electrical power measurements. Hence, a workpackage was carried out for the establishment of such method in the frequency range of 1 to 3 MHz as a part of EURAMET EMRP Era-net plus "External Beam Cancer Therapy" project. Several current shunts were developed and evaluated. Current measurements were also realized with Philips current probe and preamplifier at NPL and Agilent current probe at UME. In this paper, a method for the measurement of instantaneous electrical power delivered to a reactive ultrasound transducer in the required frequency range is explored.

1. Introduction

The work described in this paper is a deliverable in a workpackage of EURAMET EMRP Era-net plus "External Beam Cancer Therapy" project. The aim is to develop and validate a method for the measurement of instantaneous electrical power delivered to a reactive ultrasound transducer in the frequency range 0.5 to 3 MHz with a target accuracy of ±5%.

New IEC standards specifically for measurement of HITU equipment are under development. The ultrasound power radiated from a source (transducer) is directly dependent on applied electrical voltage and current and consequently power. No method up to now has been developed and adopted for the input electrical power measurements of HITU transducers. HITU transducers are produced to work at discrete resonance frequencies e.g. at 1 MHz or 3 MHz etc. They are typically PZT type transducers a bowl shape and a single focal point. They produce ultrasonic power in water when an oscillating electrical voltage is applied at the correct frequency.

Accurate electrical power measurement is well established in low frequencies applications. When the frequency of interest is higher than 100 kHz, measurement of voltage, current and impedance are easily affected by the input impedance of the measurement equipments. Input impedance of the measurement equipment and the impedance of the transducer contain resistive, capacitive and inductive components and these components change significantly with frequency. Input voltage can be measured in number of ways such as calibrated scope (at the end of cable), rf voltmeter and TVC (thermal voltage converter) methods [2]. Input impedance of the measurement equipment is usually close to 50 Ω or 1 MΩ. If the transducer is purely resistive, only straightforward numerical corrections are necessary for between the input impedance of the measurement device and output impedance of the transducer. When the input impedance of the measurement equipment contains reactive components, it should be calculated and whether it should be taken account or not in the phase calculation.

In order to measure the current, two approaches were explored:

i) to design a current shunt

ii) to use a current probe.

2. Current shunts

Four terminal and two terminal current shunts were developed for current measurements at 1 MHz and 3 MHz frequencies (Figure 1). In order to avoid phase error observed in the shunts, compensation circuits were developed. Current, voltage and phase between them were determined. To minimise the effect of inserting a current shunt into the circuit between a power amplifier and transducer, the target impedance of the shunt was 0.1 ohm: this being one percent of the impedance of a transducer with a 10 ohm input impedance. At this level, changes in contact resistance (for instance using a BNC connector) becomes significant.

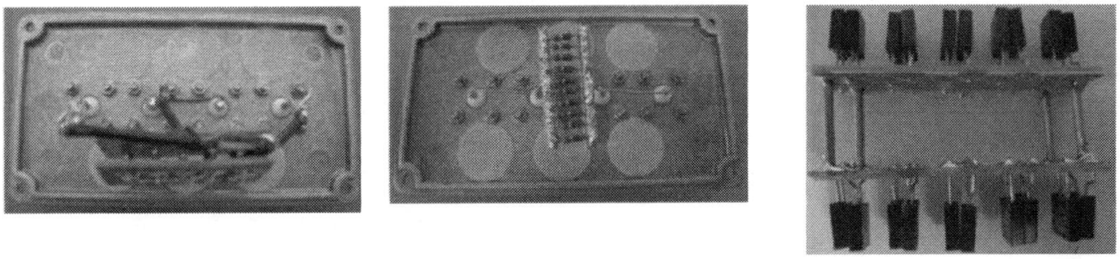

0,1 Ohm Shunt **0,1 Ohm Shunt** **0,25 Ohm Shunt**

Figure 1. Three different shunt designs.

For three different shunts, time delay between current and voltages were observed. Within the restrictions of this particular project, we were not able to achieve adequate performance of a 0.1 ohm current shunt. There were more promising indications from a 0.25 ohm shunt, however this would introduce a 2.5% voltage drop when measuring a 10 ohm transducer, so we have not pursued this option further at this stage. However, some important observations were made relating to the significance of harmonic distortion in the drive signal. Harmonics produce power half cycles of different amplitude and shape, and a reactive load impedance results in each power half cycle itself being asymmetric. In practice, the output of many HITU amplifiers will be distorted: whilst this usually only has a small effect on the acoustic output power, it can introduce large errors when measuring electrical power if the amplitude of the voltage or current is used to calculate power.

2. Current probes

Current measurements were realized at NPL with Philips current probe and preamplifier and at UME with Agilent 1147A current probe. The results presented here relate to the work at UME. The power measurement set-up is shown in Figure 2. The output of the power amplifier is directly connected to the magnetic coupled current probe. The ultrasonic transducer with cable is connected to the one port of Tee BNC which is the definition point for all electrical measurements, the other port of the Tee BNC is employed for the voltage measurements. The voltage measurement feature of the scope was calibrated by an RF Voltmeter and the correction factor of the scope voltage measurements determined. Agilent 4294A is used for the two terminal impedance measurements. The reference point for the two terminal impedance measurement is one port of the Tee BNC. The calibration of the impedance analyzer was realized at this point.

Devices used in the set up;

Generator	: Agilent 33250A
Power Amplifier	: ENI (150 W) and E&I (500 W)
Infinium Scope	: Agilent 500 MHz, 2GSa/s
Current Probe	: Agilent 1147A compatible with Agilent scope
Impedance Analyzer	: Agilent 4294A
HITU transducer	: Precision Acoustics and Sonic Concepts H-102
RF Voltmeter	: Rohde & Schwarz URE3
Balance	: Mettler Toledo PR2004

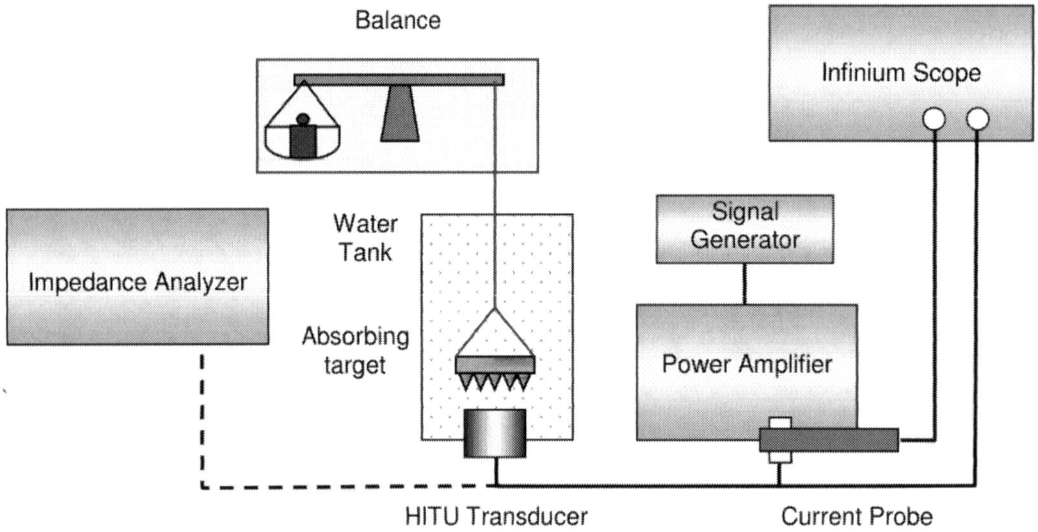

Figure 2. Ultrasound power and input electrical power measurement set-up

All the devices used in the set up were traceable to electrical standards. A 50 Ohm resistive load that was measured with the impedance analyzer used to verify the current probe readings. The current probe was kept heated, demagnetized and levelled before the measurements according to its manual and checked during measurements. The calibration of the impedance analyzer was done at the end of cable. Voltage measurements were realized with both RF voltmeter and the scope which are calibrated by a voltage laboratory.

Current measurements were realized with a calibrated current probe. Impedance and phase were also measured with impedance analyzer. Phase measured between voltage and current are in agreement with the phase measured with impedance analyzer in few degrees (Figure 3 and 4). However, there are sections of the curves which show systematic differences between the methods.

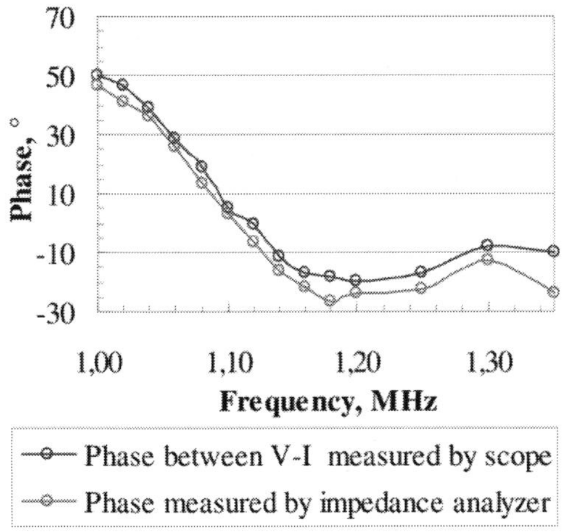

Figure 3. Difference between phases measured by scope and impedance analyzer for Sonic Concepts

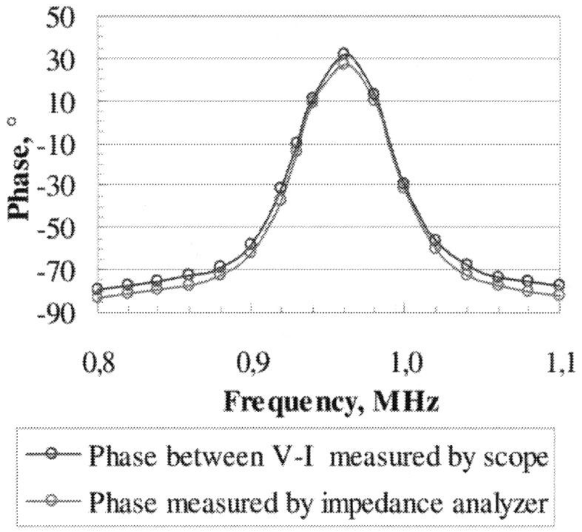

Figure 4. Difference between phases measured by scope and impedance analyzer for Precision Acoustics

3. Measurements

Ultrasound power was measured with absorbing target by conventional methods described in IEC 61161 standard [1]. Input electrical voltage, current and phase between them were measured at the same time using an scope. Impedance and phase had been previously measured with an impedance analyzer in a separate set-up. Input electrical power was calculated with three ways.

Electrical power *P1*, was calculated by multiplying *rms* input voltage (*V*), current (*I*) and phase (θ) between them as;

$$P1 = V \cdot I \cdot cos(\theta) \tag{1}$$

Electrical power *P2*; was calculated by multiplying rms electrical input current and resistance measured with Impedance Analyzer

$$P2 = I^2 \cdot R \tag{2}$$

Electrical power *P3*; was calculated from the *rms* electrical input voltage and the phase and resistance measured with Impedance Analyzer.

$$P3 = V \cdot I \cdot cos(\theta) = V \cdot \frac{V}{Z} \cdot cos(\theta) = V \cdot \frac{V}{R/cos(\theta)} \cdot cos(\theta) = \frac{V^2}{R} \cdot cos(\theta)^2 \tag{3}$$

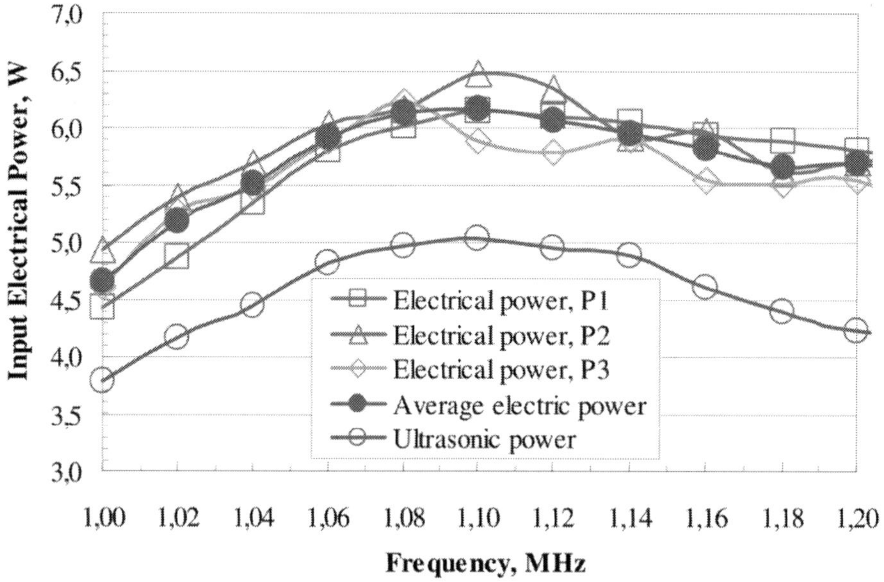

Figure 5. Electrical powers and ultrasound power for Sonic Concepts transducer

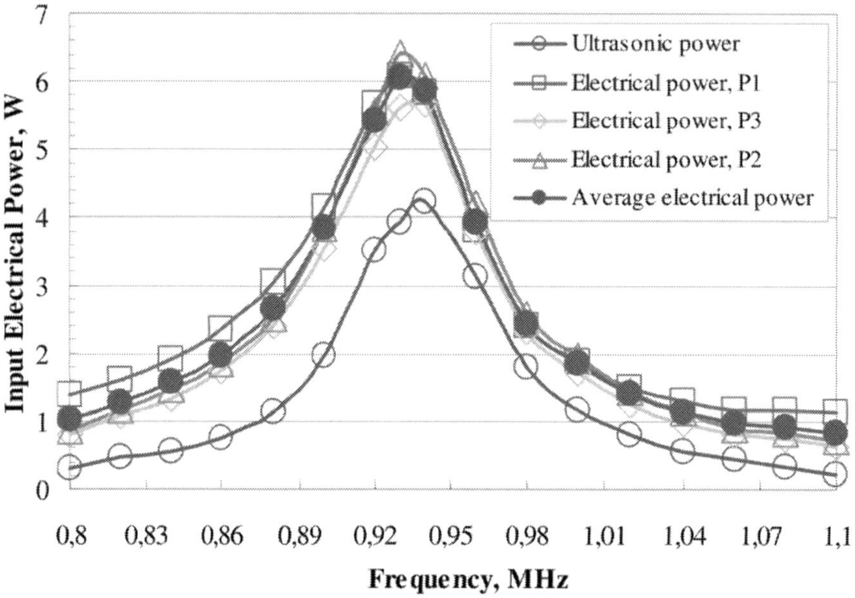

Figure 6. Electrical powers and ultrasound power for Precision Acoustics transducer.

4. Results

Measurement results are shown in Figure 5 and 6. This work shows good correlation between electrical power measurements and direct measurements of the power of the ultrasonic transducers which are immersed into the water. During the measurements, no harmonic distortion was observed in the amplifier output at the defined output power level.

Efficiencies of both HITU transducers at resonance frequency are between 70-80 % that is consistent with manufacturer specifications declared in data sheets of transducers. The three calculations agree within approximately ±5% at frequencies close to resonance where the phase angle of the transducer impedance is within about ±45°. In this range, *P1*, determined using the voltage and current probes, lies between *P2* and *P3*.

5. Discussion and conclusions

Parts of this work have highlighted the significance of harmonic distortion in the drive signal. Harmonics produce power half cycles of different amplitude and shape, and that a reactive load impedance results in each power half cycle itself being asymmetric. In practice, the output of many HITU amplifiers will be distorted: whilst this usually only has a small effect on the acoustic output power, it can introduce large errors when measuring electrical power if the amplitude of the voltage or current is used to calculate power. It also becomes difficult to determine the phase between voltage and current in the time domain for a reactive load as the two waveforms have different shapes. Even when simply determining an applied voltage (for instance to calculate radiation conductance), voltage amplitude or peak-to-peak value should not be used when there is the possibility of distortion. The *rms* voltage or, perhaps better, the magnitude of the spectral component at the fundamental frequency should be determined.

Attempts to create a low impedance current shunt to be placed between the amplifier and transducer were not successful. Instead, the method which seems likely to offer lower absolute uncertainty is to measure simultaneously voltage, current and relative phase using commercial voltage and current probe. Input electrical powers were calculated in three different measurement methods and compared their measurement results with ultrasonic power for two HITU transducers. Measurements were realised at around 1 MHz for low power levels. An analysis of the uncertainty in using simultaneous voltage and current probes to determine power has not yet been carried out The method also needs to be tested at higher frequencies and power levels.

6. References

[1] IEC61161:2007. Ultrasonics – Power Measurement –Radiation Force Balances and Performance Requirements. International Electrotechnical Commission, Geneva, 2007

[2] Karaboce B, Sadıkoğlu E, Bilgiç E, Kırbaş C, Turan A İ, Arifoviç M, Comparison of voltage measurement techniques in radiation conductance calculations of ultrasonic power transducer, Conference proceedings, Internoise 2007 Istanbul

Acknowledgements

This work was supported in part by the European Community's Seventh Framework Programme, ERA-NET Plus, under Grant Agreement No. 217257 (EURAMET joint research project).

Calorimetric method for measuring high ultrasonic power using water as a heating material

T Kikuchi and T Uchida

National Metrology Institute of Japan / AIST, 1-1-1 Umezono, Tsukuba, Ibaraki, 305-8563 Japan

E-mail:tsuneo.kikuchi@aist.go.jp

Abstract. The present study shows the calorimetric method for measuring high ultrasonic power using water as the heating material. In recent years, at the National Metrology Institute of Japan (NMIJ), an ultrasonic power primary standard of from 1 mW to 15 W has been established by the radiation force balance (RFB) method. Conventionally, the RFB method is widely used for ultrasonic power measurement, but this method is not suitable for very high power measurement due to thermal damages to the absorbing targets. High power ultrasonic standards, however, are being required by medical HITU measurements and in the sonochemistry industry. In order to meet these requirements, we have started to develop an ultrasonic power standard between 15 W and 200 W. Our final goal is an ultrasonic power standard of up to 500 W. The calorimetric method is an alternative ultrasonic power measurement method to the RFB method. We have adopted this method and use water as the heating material. Water has excellent features as a standard material, because the physical properties of water are well known. In the present study, we present an experimental system and the results for an ultrasonic power standard of up to 100 W. The measured ultrasonic power agreed well with the NMIJ primary standard up to 25 W.

1. Introduction

At the National Metrology Institute of Japan (NMIJ), an ultrasonic power standard from 1 mW to 15 W was established by the radiation force balance (RFB) method in 2006. Recently, a high-power ultrasonic standard has been required for High Intensity Therapeutic Ultrasound (HITU) measurement methods, such as those used for medical treatment and in the sonochemistry industry. To meet these requirements, we have started to develop an ultrasonic high power standard [1][2]. Our final goal of ultrasonic power measurement will be up to 500 W, because this appears to be of common interest worldwide [3]. Generally, it is difficult to apply the RFB method to ultrasonic high power measurement because of, for example, thermal damage to the absorbing targets and acoustic streaming. Experimentally, for absorbing targets, 20 W may be considered the upper limit of the RFB method. As such, an alternative ultrasonic power measurement method must be developed. The calorimetric method is well known as one of the ultrasonic power measurements. This method requires a thermal object, namely, a heating material. There are a number of general requirements for the heating material.

1) The acoustical and thermal properties of the material should be given or easily measurable.
2) The material properties of the material should be stable during measurement.
3) Steady supply should be guaranteed as the standard materials.

Based on these conditions, water is an excellent material for use in ultrasonic power standards. At the NMIJ, we have attempted to use water as the heating material for the calorimetric method.

2. Experimental methods

2.1. Measurement conditions for the calorimetric method

There are some important requirements when applying the calorimetric method to ultrasonic power measurement in water.

1) In order to avoid varying the radiation characteristics of the transducers, ultrasound should not return to the transducer. In other words, the "free field" condition must be achieved.
2) Ultrasound must be absorbed perfectly in water, i.e., the total ultrasonic power must be used for the water temperature rise. This requires ultrasound to be perfectly reflected at the walls and bottom of the water bath.
3) Thermal loss must be minimized, In other words, water must be thermally isolated.

2.2. Experimental system

The ultrasonic power measurement system using the calorimetric method has been set up so as to satisfy the conditions described in Section 2.1. Figure 1 shows a block diagram of the measurement setup and it is composed of three parts:

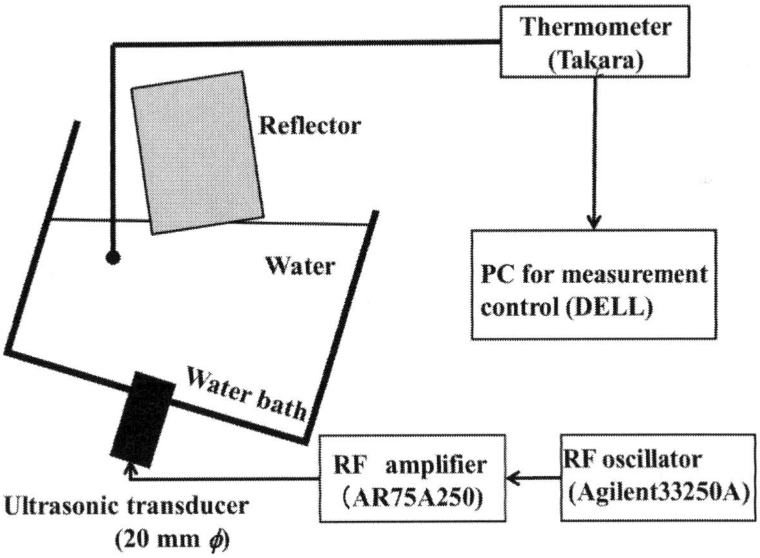

Figure 1. Block diagram of the measurement system.

1) The water bath and reflector
2) The temperature measurement system
3) The ultrasonic transducer, oscillator, and power amplifier

Figure 2(a) shows a section view of the water bath, which has an inner diameter of 15 cm and a depth of 8 cm. In order to achieve perfect reflection, the walls and bottom of the water bath are constructed of thin aluminum plates having a thickness of 0.8 mm, and the inside of walls and bottom of the water bath are hollow. The thermal loss from the water bath can also be minimized by this structure. Figure 2(b) shows a photograph of the water bath.

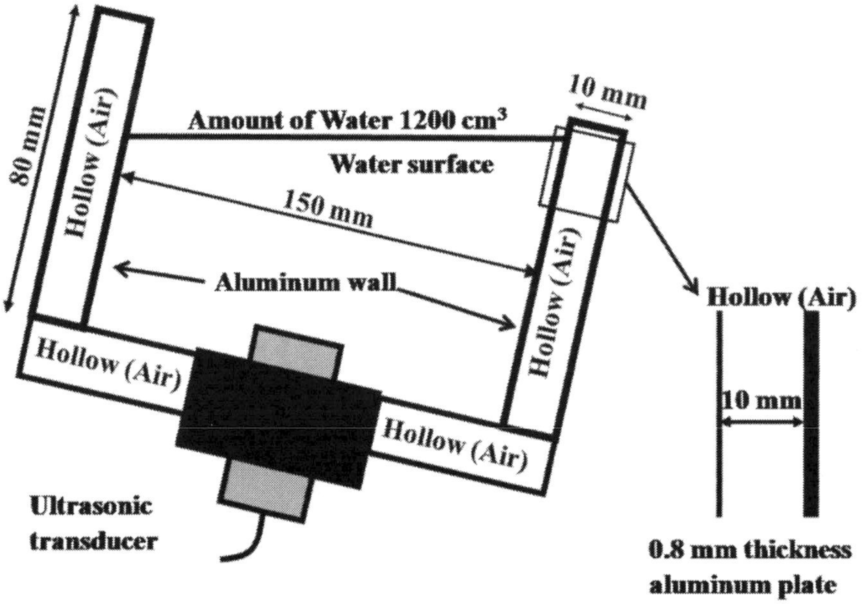

(a) Section view of the water bath and the structures of the wall and bottom plates.

(b) Photograph of the water bath

(c) Photograph of ultrasonic transducer

Figure 2. Measurement setup

Figure 2(c) shows the custom air-backing type ultrasonic transducer used in the experiments. The center frequency is 1 MHz, and the diameter is 20 mm.

Degassed water was used for the power measurements. The water temperatures were measured using a thermometer with a resolution of 0.01°C. The room temperature was maintained at 23°C ± 0.5°C during the measurements. As shown in figure 2(a), the water bath was tilted approximately 30°, and a hollow aluminum reflector was placed on the water surface. Using this setup, the free field condition is almost satisfied, namely, returning ultrasound to the transducer is minimized.

2.3. Calculation of ultrasonic power

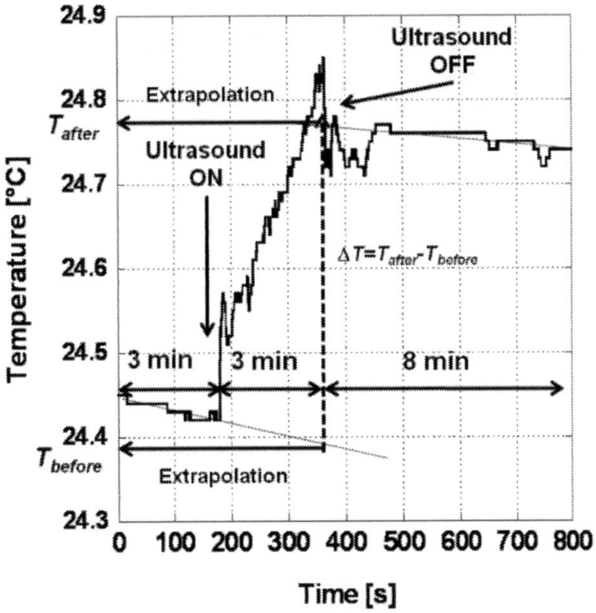

Figure 3. Example of the time dependence of water temperature during measurement.

Figure 3 shows an example of the time dependence of water temperature variation during actual ultrasonic power measurements. The water temperature measurement starts at 0 seconds, and ultrasound exposure begins three minutes after starting the measurement. The water temperature then gradually rises due to ultrasound exposure. Ultrasound exposure continues for three minutes, and then temperature measurement continues for eight minutes after exposure OFF, until the temperature stabilizes. Generally, it is difficult to perform stable measurements during ultrasound exposure, because there are several instability factors, including viscous heating, acoustic streaming, and inhomogeneous temperature distribution.

In order to avoid these instability factors, the water temperature is not measured during ultrasound exposure, but rather is measured both before and after ultrasound exposure. The temperature variations before and after the exposure are approximated by linear regression and are then extrapolated to the time that ultrasonic exposure finishes, as shown in figure 3. The true water temperature increase ΔT is then defined as the difference between the two extrapolated temperatures, namely, T_{before} and T_{after}. The ultrasonic power P can then be calculated as

$$P = \frac{\Delta T}{t_x} C \cdot M$$

where ΔT is the water temperature increase, t_x is the exposure time (three minutes in this experiment) , C is thermal constant of the water (4.18 J g^{-1} K^{-1}) [4], and M is the mass of the water (1200 cm^3).

3. Results and discussion

Figure 4 shows the powers measured by the calorimetric method and by the primary standard of the NMIJ at 1 MHz as a function of oscillator output voltage. The expanded uncertainty of the RFB

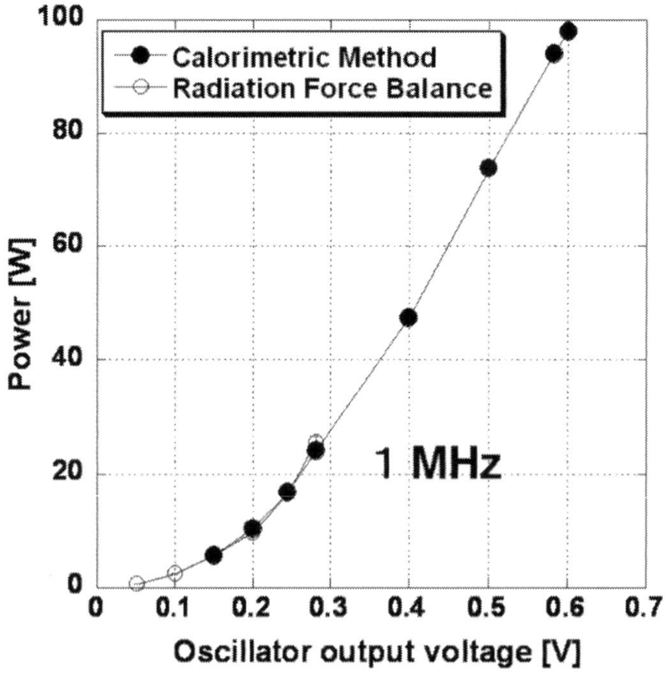

Figure 4. Ultrasonic power measured by the calorimetric method and the RFB method at 1 MHz.

method is estimated to be approximately 4%. The measurement uncertainty for the calorimetric method has not yet been estimated, but the reproducibility of the measurements (standard deviation) is estimated to be approximately 10%. The ultrasonic powers measured by the calorimetric method are validated up to 25 W through comparison with the primary standard. Under the present measurement conditions, ultrasonic power measurement up to 100 W is established, as shown in figure 4.

4. Conclusions

In the present paper, the calorimetric method using water as the heating material is proposed as an alternative ultrasonic power measurement method to the RFB method. Measured ultrasonic powers were validated by comparison with the primary standard of NMIJ (RFB) up to 25 W, and ultrasonic power measurements can be achieved up to 100 W.

Acknowledgments

The present study was supported by a Grant-in-Aid for Scientific Research B (No.21300195).

References

[1] T. Kikuchi, T. Uchida, *J. Appl. Phys. Proceeding of Symposium on Ultrasonic Electronics,* **29** 521
[2] T. Kikuchi, T. Uchida, *J. Appl. Phys .Proceeding of Symposium on Ultrasonic Electronics,* **30** 31
[3] A. Shaw and G. ter Haar, *NPL Report,* **DQL-AC 015**, p. 13
[4] National Astronomical Observatory, *Chronological Scientific Tables 2008* (Tokyo: Maruzen Co.) p 490

Journal of Physics: Conference Series **279** (2011) 012013 doi:10.1088/1742-6596/279/1/012013

HIFU Ultrasound Power Measurements at INRiM

G. Durando, C.Guglielmone, C.Musacchio

INRIM strada delle cacce 91, 10135 Torino– ITALY

E-mail: g.durando@inrim.it

Abstract. In this work the new system for the ultrasound power measurement of High Intensity Focused Ultrasound transducers realized at INRIM ultrasounds laboratory is presented. The system is based on a submersible load cell that takes the place of the balance. This solution presents essentially two advantages. The first one, of mechanical nature, is relevant to the fact that the target is directly connected to the force transducer, eliminating unwanted target motion at high power. The second, of electric nature, concerns the possibility to reduce the insonation time (the ON period of the electric driving signal to the HIFU transducer) under of 2 s, and is allowed for by the faster response of the force transducer (700 Hz bandwidth). The main components of uncertainty and the overall budget of the measurement system are presented together with the results of measures of conductance, G, carried on a HIFU transducer, at the work frequencies 2.0 MHz and 6.38 MHz, for values of power ranging from 10 W to 100 W. The results of the ultrasonic conductance, G, obtained with the new system are compared with values obtained using the traditional measuring system for low powers ($P \leq$ 20W).

1. Introduction

HITU transducers use is expanding at a high rate. The potential hazards of the use of high ultrasound power on biological tissues make the control the output of HITU devices necessary. One of the relevant parameters is the ultrasonic power itself. The measurement of high ultrasonic power poses serious challenges to conventional radiation force balances. Absorbing target are heated and pose problems of drift due to thermal expansion and, at high levels, the target may be severely damaged. The reflecting target copes well with heath problems, but the rather large radiation forces may cause significant horizontal components that provoke unwanted movements of the target and disturbance to the force readout. Although a commercial balance provides a direct traceability path, other solutions may prove more efficient for measuring the radiation force, at the cost of some additional passes in traceability to mass unit and more complicated analysis of the force readout.

2. A new technique

The problems outlined above required the construction of a device with the following characteristics:

1. Faster dynamic response, in order to be able to use shorter sinusoidal wave trains, for heating reduction
2. A solid fastening of the target to the sensing device, compliant only on the vertical axis and very rigid in the directions normal to this axis

3. Positioning on the bottom of the tank, so that the transducer mounting is simplified, especially for HITU it may be cumbersome to provide an adaptor for mounting the transducer on the bottom of the tank.

At INRiM we choose to use a submersible load cell, with full scale sensitivity of 50 g, to which the target is fastened by use of a screw. This device fulfills requirements 1 to 3, but there are drawbacks:

1. The output is fast but noisy
2. The cell must be calibrated in situ by standard calibrated weights to provide traceability to mass quantity. A balance instead may be more conveniently calibrated in an accredited laboratory.

An additional advantage is that there are no effects due to surface tension, as all the measuring system is under water.

In order to fully exploit its possibilities, the measuring system must be treated as a wide band linear system, whose response must be characterized as a function of frequency. In practice, a traceable calibration is carried out at frequency 0, and a relative frequency response can be obtained by using wave trains at different rates with a known transducer. This calibration procedure over a wide frequency range allows measurement with short signals. In the present paper however, the duration of the insonation is such that no correction for the frequency response is needed. The fast response, with its associated relatively large pass band (about 700 Hz), allows the use of time series processing for obtaining a transfer function between the radiation force and the voltage driving the transducer, a quantity that at the first harmonics of the driving signal is proportional to the conductance G.

3. Measurement Set-Up

The device used for the ultrasonic power measurement is based on the radiation force balance method. The ultrasonic power is determined from the measurement of the force exerted on a target by the sound field generated by an ultrasonic source.

The target is connected to a load cell (Honeywell mod. 31) which measures the apparent target mass variation due to ultrasonic field when the source is alternately switched on and off. The absorbing target is rigidly connected to the sensing element of the load cell with a screw (figure 1).

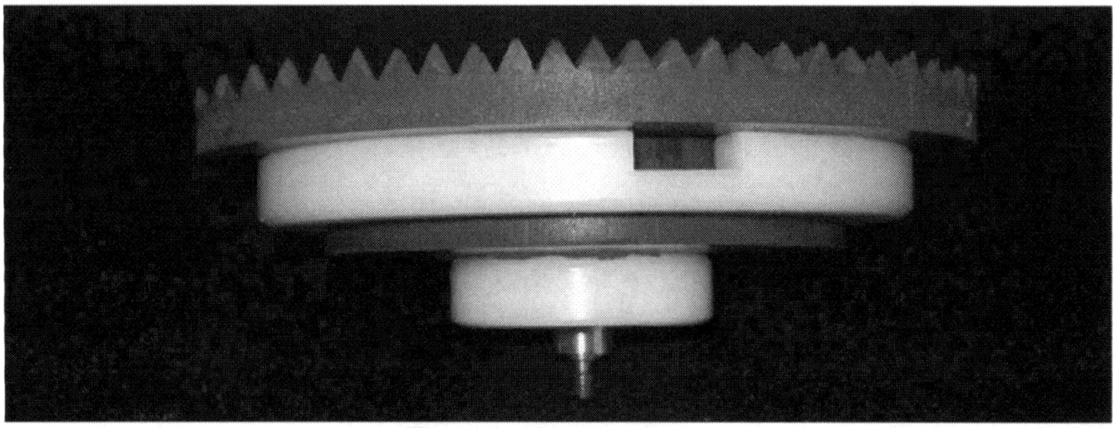

Figure 1: Absorbing target

The signal produced by the load cell is conditioned by a strain gauge amplifier (Sensotec mod. UV-10) and then measured by a nano-voltmeter (Agilent 34420 A). The load cell is screwed on the bottom of the tank, with a volume of approximately 15 litres. The short duration of the tone burst ($t \leq 1$s) used for the measurement of the ultrasonic power, P_{out}, allows the use of the absorbing target even for high values ($P \leq 200$ W) of ultrasonic power. In figure 2 the measurement apparatus is shown, with a detail of load cell and target in the upper right box.

Figure 2: System set-up

The RMS value, U_{in}, of the transducer driving signal is red, simultaneously with the mass variation, by means of a Rohde&Schwarz mod. URE-3 voltmeter. The electro-acoustic radiation conductance G is calculated according to the relation:

$$G = \frac{P_{out}}{U_{in}^2} = \frac{gu(T)\Delta M}{U_{in}^2} \tag{1}$$

where g is the gravity acceleration evaluated in the INRiM laboratory, $u(T)$ speed of sound in water and ΔM is the apparent mass variation induced by the ultrasound field.

4. Measurement Procedure/Technique

In analogy with the RFB system, for the SLC system a calibration (in measurements conditions) is necessary before the ultrasound power measurements.
The figure 4 shows the calibration (using a set of weight) curve of the acquisition chain, consisting of submergible load cell, in-line amplifier and nanovoltmeter.

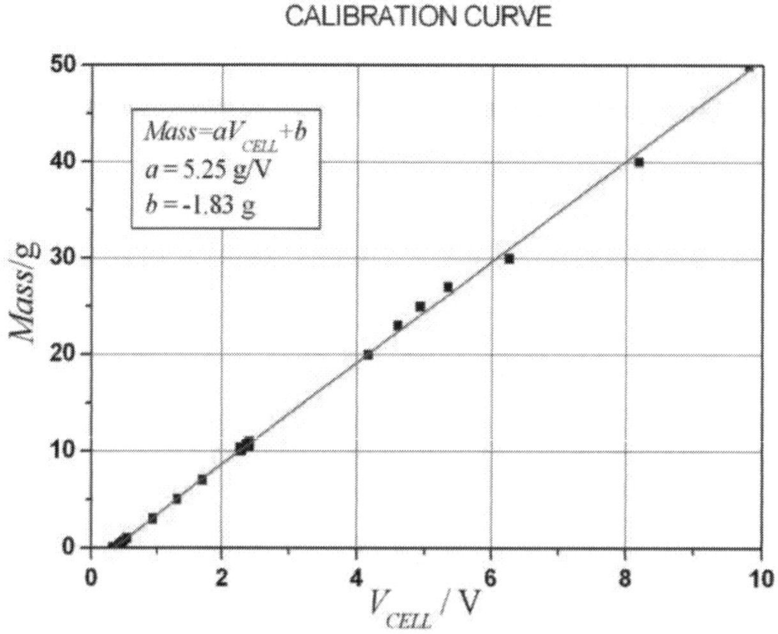

Figure 3: Calibration of the cell

Using this procedure it is easy to obtain the coefficients a and b with which it is possible to calculate the variation of mass induced by the ultrasonic field, measuring a variation of the output voltage of the cell during the insonation time.

The program for the management and the control of the many parameters influencing the measurement, water temperature, displacements of the transducer, value of power amplifier, etc... is basically the same of the program used for the radiation force balance system, the difference is related only to the use of the cell instead of the commercial balance. One of the main advantages induced to the cell is connected to a strong reduction of the insonation time necessary to perform an ultrasound power measurement.

The figure 4 shows the typical mass variation in the time domain in two cases, a) for the radiation force balance system and b) for the submergible load cell system. The integration time of the balance does not allow an insonation time lower than 10 s, with the submergible load cell it is possible to reduce the time to less than one second; one of the most important spin-off is the possibility to change the analysis method from time domain to frequency domain.

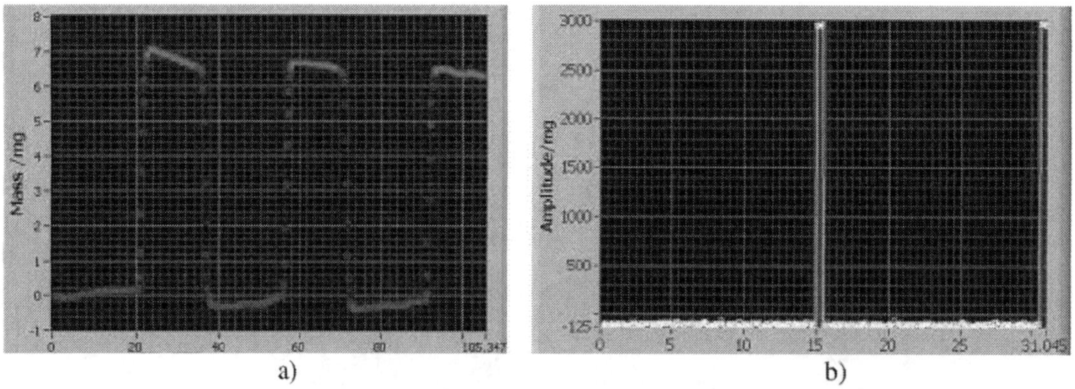

Figure 4: Variation of mass vs. time a) Radiation force balance b) Submergible load cell system (please note different timescale)

5. Results

This work shows the results of the measurements performed on HIFU transducer, its main characteristics are shown in table 1.

Table 1: Photo and main characteristics of the HIFU transducer

Manufacturer		SONIC CONCEPTS
Model		H-106-MR
Frequencies	f	2.00 MHz 6.38 MHz
Active diameter	Φ_{EXT}	64.00 mm
Radius of curvature	r_c	63.20 mm
Focal depth	z	51.74 mm
No. of elements		one
Electrical impedence	Z	50 Ω

In figure 6 the results obtained with the radiation force balance system (up to 20 W) in black and with the submergible load cell in blue for the 2.0 MHz frequency are presented.

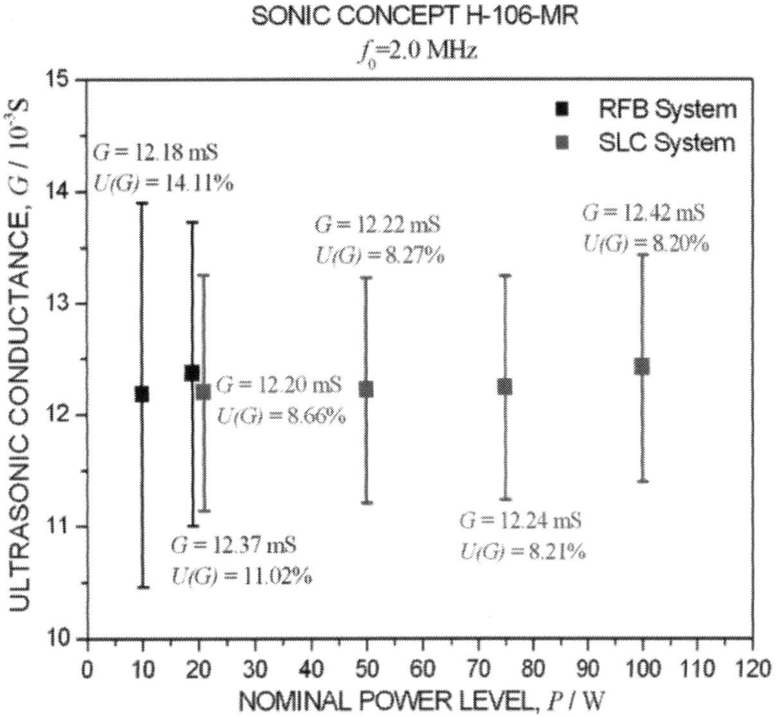

Figure 6: Results of the measurements at 2.0 MHz

Figure 7 shows the same comparison made at 6.38 MHz. The graphs indicates that the measurements obtained with the different system are not significantly different.

Figure 7 shows an higher dispersion of measurements obtained at 6.38 MHz when compared to data obtained at 2.0 MHz. One explanation for the difference between measurements at different frequencies may be related to the occurrence of parasitic capacitances that lower the signal, and hence the signal to noise ratio at higher frequencies.

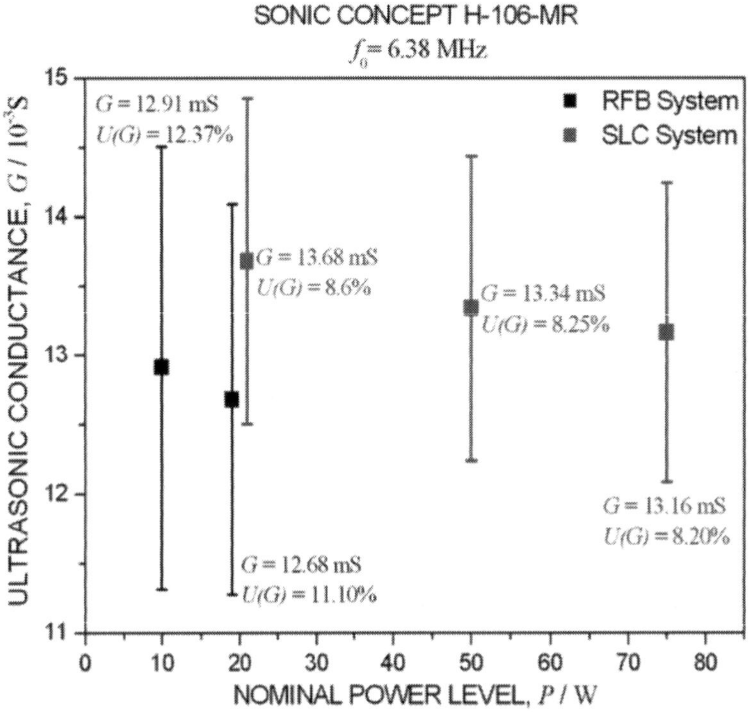

Figure 7: Results of the measurements at 6.38 MHz

6. Summary

The agreement between the traditional RFB system and the new system based on a submergible load cell is satisfactory and well within the calculated uncertainty budgets. At present the system is mostly suited for high values in ultrasound power, the effect of noise being too high for low apparent mass variations. Signal processing techniques may allow the extensions of this measurement apparatus to lower values of ultrasonic power.

7. References

[1] URL: http://www.ptb.de/EURAMET-JRP7
[2] IEC61161:2007. Ultrasonics – Power Measurement –Radiation Force Balances and Performance Requirements. International Electrotechnical Commission, Geneva, 2007
[3] Guide to the expression of uncertainty in measurement. ISO/IEC GUIDE 98-3: 2008

Acknowledgements

This work was supported by the European Community's Seventh Framework Programme, ERA-NET Plus, under Grant Agreement No. 217257 (EURAMET joint research project).

Journal of Physics: Conference Series **279** (2011) 012014 doi:10.1088/1742-6596/279/1/012014

Ultrasound power measurements of HITU transducer with a more stable radiation force balance

B Karaböce[1], E Sadıkoğlu[1], E Bilgiç[1]
[1] Tübitak Ulusal Metroloji Enstitüsü (UME), P.K. 54 41470 Gebze-Kocaeli, Turkey

E-mail: baki.karaboce@ume.tubitak.gov.tr

Abstract. A new radiation force balance (RFB) system was established at Turkish National Metrology Institute (UME) Ultrasonics Laboratory for High intensity therapeutic ultrasound (HITU) power measurements. The new system is highly stable at high power levels up to 500 Watts. The measurement system consists of a Plexiglas cylindrical balance arm, target mounting scale disks, conical reflecting and absorbing targets, adjustment nuts, and a hanging wire. Both of the two sides of balance were mounted similar size and weight targets. The equilibrium of the balance arm can be adjusted with nuts on screws located at both sides of the balance arm. Transducer was mounted to bottom of water tank. Absorbers in the bottom and the near walls of the tank were used for reflecting target case. Ultrasound power was applied to one scale of the balance where the reflecting/absorbing target was mounted and corresponding force was measured on the other scale of balance where was connected to a balance with a thin wire while the thin rest standing on a support. Ultrasound power of two HITU transducers at frequencies 0.93 MHz, 1.1 MHz and 3.3 MHz were measured with conventional and new system, the values were compared and uncertainty components were assessed in this paper.

1. Introduction

Ultrasound power was measured with absorbing or reflecting target by conventional RFB described in IEC 61161 standard [1]. Instability of the target may cause errors in alignment of the target and mathematical calculations (Figure 1). For high powers above the 50 Watts, the rolling movement of the target under the effect of the transducer may be dominant error in ultrasonic power measurements (Figure 2).

Ultrasound power directly depends on the change in apparent weight of target, m exposed to ultrasound radiation, between the on and off times of input electrical power due to formula, $P=mgc$. The other two parameters, g and c are acceleration of gravity and speed of sound correspondingly. g depends on geographical location and c depends on water temperature.

Therefore the change of apparent weight must be calculated as correct as possible.

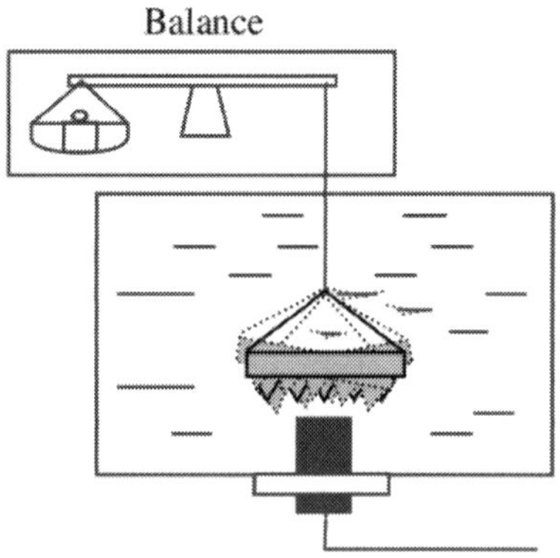

Figure 1. Schematic view of the instability of the absorbing target exposed to high power ultrasound. Similar rolling is seen for the reflecting target as well.

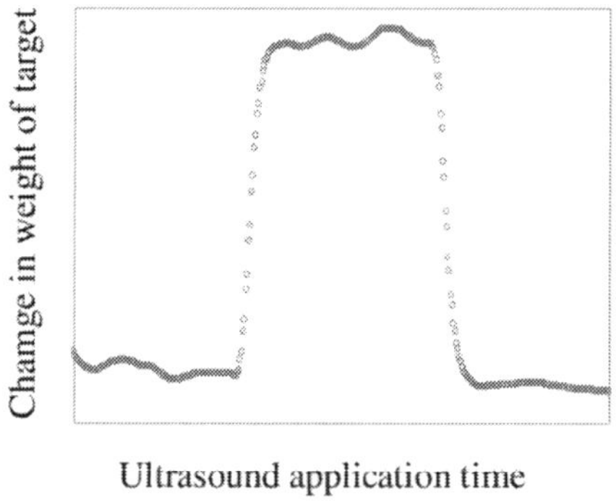

Figure 2. Errors may occur at mathematical calculations due to instability of the curve.

In order to overcome this drawback, one method is to maintain the target at the bottom of tank. In this case a stable positioning system to mount the transducer is required. But as the HITU transducer is a bowl type transducer, air bubbles can easily be occur during montage (Figure 3 and 4). Removal of the bubbles is hard and time consuming task. The air bubbles also easily could be born due to the cavitation effect at high power applications.

New set up was designed and built at UME for stable measurements of HITU power. The set up ebnables to eliminate efficiently problems mentioned above.

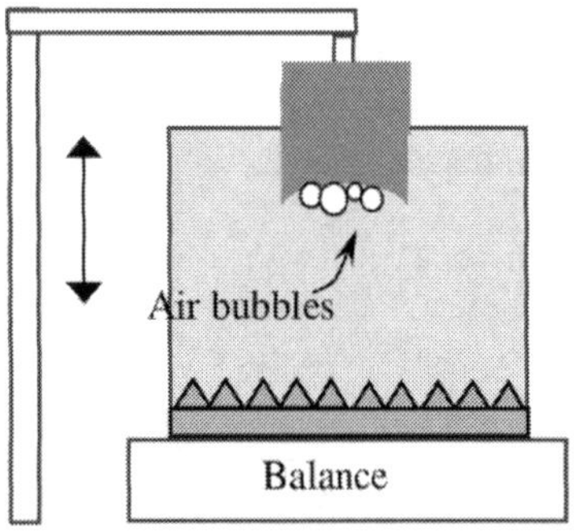

Figure 3. HITU application is downward in RFB set up with absorbing target. Bubbles are occurred.

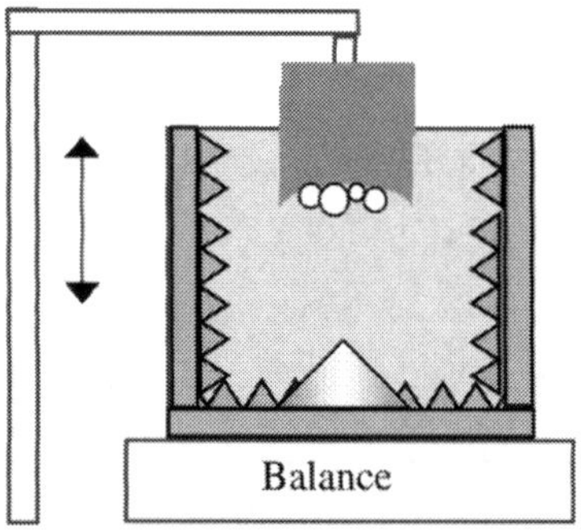

Figure 4. HITU application is downward in RFB set up with reflecting target. Bubbles are occurred.

2. Modified RFB set up

The stable RFB set up has been constructed at TÜBİTAK UME. New modified radiation force balance was used with reflecting and absorbing targets (Figure 5 and 6). Ultrasound power was applied to one scale of the balance where the reflecting target was mounted and corresponding force was measured on the other scale of balance where is connected to a balance with a thin wire, while the thin rest standing on a support. Transducer was mounted to a bottom of a water tank. Ultrasound radiation was applied upward to target. Absorbers in the bottom and the near walls of the tank were used for the reflecting target in use.

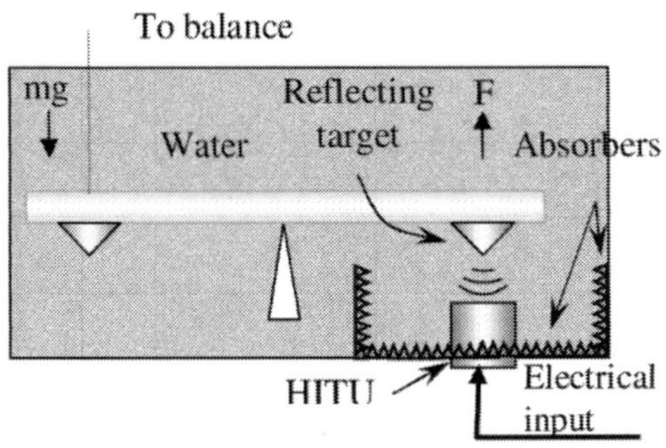

Figure 5. Schematic of the modified RFB set up. Note that an absorbing target can be mounted at the same set up.

Figure 6. View of the modified RFB set up with absorbing target

3. Measurement results

The determination of the apparent weight change in the new modified and stable set-up is better and it can be calculated with higher accuracy. This fact could be clearly seen from Figure 7. There is no fluctuation in power measurements in the modified RFB set up since it stands on a support. This gives opportunity to perform more stable power measurements.

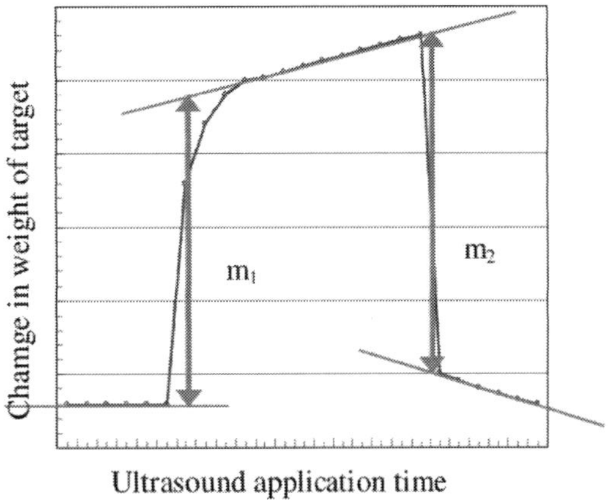

Figure 7. Qualitative presentation of apparent weight change with the modified RFB set-up.

Three uncertainty components that are influence of mathematical procedures to calculate radiation force from balance read-our curve, transducer and target alignment and repeatibility decreased as seen in Table 1. Total decrease in combined uncertainty is aproximately 1.5%.

Table 1. Comparison of three uncertainty components. Note that the combined uncertainty covers all other uncertainty components.

No	Uncertainty Component	Conventional RFB	Modified RFB
1	Influence of mathematical procedures to calculate radiation force	0.58	0.14
2	Transducer and target alignment	2.22	1.73
3	Repeatability	4.0	1.0
	Combined Uncertainty	6.95	5.50

4. Conclusion

In order to overcome the instability of power measurements of RFB set up for high power measurements, new modified set up were constructed at TÜBİTAK UME. In spite of the fact that expanded uncertainty with new set-up is a little bit higher than with conventional RFB, the overall performance of the set-up satisfies general requirements for high power ultrasound measurements. Approximately 1.5% total decrease in combined uncertainty was achieved.

However the ultrasound power measurements with reflecting target at the modified RFB, has good stability, it gives 10-15% higher radiaiton conductance values compared to RFB set ups with absorbing target and other set ups i.e. thermal methods. But ultrasound power measurements at the modified RFB set up with absorbing target gives comparable radiation conductance values copared to other methods [2]. For example, Sonic Concepts transducer whose radiaition conductance value is measured at the comparison (European Metrology Research Programme "External Beam Cancer Therapy" project comparison), measured again and same value was reached at the modified RFB set up with absorbing target.

Acknowledgements

This work was supported by the European Community's Seventh Framework Programme, ERA-NET Plus, under Grant Agreement No. 217257 (EURAMET joint research project)

5. References

[1] IEC61161:2007. Ultrasonics – Power Measurement –Radiation Force Balances and Performance Requirements. International Electrotechnical Commission, Geneva, 2007
[2] www.euramet.org, EURAMET poject no:1108 – Comparison of measurement techniques for the determination of ultrasonic power of high intensity ultrasound transducers

Journal of Physics: Conference Series **279** (2011) 012015 doi:10.1088/1742-6596/279/1/012015

Inter-laboratory comparison of HITU power measurement methods and capabilities

K V Jenderka[1], G Durando[2], B Karaböce[3], S Rajagopal[4] and A Shaw[4]
[1] Physikalisch-Technische Bundesanstalt (PTB), Bundesallee 100, 38116 Braunschweig, Germany
[2] Istituto Nazionale di Ricerca Metrologica (INRIM), Strada delle Cacce 91, 10135 Torino, Italy
[3] Tübitak Ulusal Metroloji Enstitüsü (UME), P.K. 54 41470 Gebze-Kocaeli, Turkey
[4] National Physical Laboratory (NPL), Hampton Road, Teddington, TW11 0LW, UK

E-mail: kvjend@ieee.org

Abstract. High Intensity Therapeutic Ultrasound (HITU) is gaining in importance among the spectrum of therapeutic options to combat cancer. HITU has already been approved and is in clinical use for the treatment of organs like the prostate, the liver and the uterus. Nevertheless, the metrology of the applied high power ultrasound fields, and in consequence, reliable treatment planning and monitoring, is still a challenge. As part of a European Metrology Research Programme project, the four National Metrology Institutes from the UK, Germany, Italy and Turkey conducted an inter-laboratory comparison of their power measurement capabilities at power levels of 5, 25, 75 and 150 W each at frequencies of 1.1, 1.5 and 3.3 MHz. The task was to measure the total, time-averaged ultrasonic output power, emitted by the circulated transducers under specified electrical excitation conditions into an anechoic water load, and the actual rms transducer input voltage. The output value to be reported was the electro-acoustic radiation conductance including the associated standard and expanded uncertainties. Several different measurement techniques were applied to gain further insight into HITU power measurement. The deviations from the calculated comparison reference value found for the different techniques are discussed and conclusions for the further improvement of measuring procedures are drawn.

1. Introduction

High Intensity Therapeutic Ultrasound (HITU) procedures compromise, amongst other things, the non-invasive treatment of tumourous tissue. The procedure with the highest intensities applied to the tissue (> 10,000 W/cm^2) [1] is HIFU (High Intensity Focused Ultrasound, also referred to as Ultrasound Ablation or Focused Ultrasound Surgery) [2], used for the treatment of organs like the prostate, the liver and the uterus, by inducing localized necrosis of the tissue. Even though HITU has already been clinically approved, the metrology of the applied high power ultrasound fields, and in consequence, reliable treatment planning and monitoring, is still a challenge. A crucial step for the prediction of the desired temperature rise in an HITU-treated region is the measurement of the total ultrasound output power and the sound field distribution.

This paper describes the measurements and results of an inter-laboratory comparison of HITU power measurement methods and procedures. One goal of this comparison was the investigation of different approaches to the determination of the acoustic output power. The participants, thus, offered to use different methods available in their laboratories. The comparison was part of a European Metrology Research Programme project "External Beam Cancer Therapy (EBCT)" [3] and compromised the measurement of the time-averaged ultrasonic output power of two HITU ultrasonic transducers in the nominal frequency range from 1.1 MHz to 3.3 MHz and the measurement of the applied excitation voltage. The pilot laboratory was the PTB (Germany) and the other participants were UME (Turkey), INRIM (Italy) and NPL (UK). The measurements of this inter-laboratory comparison, including the re-measurements at the pilot laboratory, were carried out between June and December 2009.

2. The measurement task

The task was to measure the total, time-averaged ultrasonic output power P_{out}, emitted by the provided transducers under specified conditions of electrical excitation into an anechoic (i.e., free-field) water load, and the rms value of the input voltage U_{in} applied to the transducer socket.

Finally, the electro-acoustic radiation conductance G was to be calculated according to

$$G = \frac{P_{out}}{U_{in}^2} \ . \tag{1}$$

Three acoustic frequencies (1.1 MHz, 1.5 MHz and 3.3 MHz) and four power levels (very low - 5 W, low - 25 W, medium - 75 W and high - 150 W) were selected for the comparison, resulting in eleven different combinations of transducer type, frequency and power level. Two types of HITU transducers (Sonic Concepts H-102 and Imasonic 5706A), capable of providing the necessary frequencies and power levels, were used as travelling standards.

The participants applied different ultrasonic power measurement methods available in their laboratories. The reported output power was to relate to the transducer surface, which corresponds to the rim of the focusing bowl. For the measurement of the rms transducer input voltage U_{in} the participant was to use his own methods and instruments.

3. Measurement setup of the participating laboratories

3.1. PTB

The PTB primary standard radiation force balance setup was used with absorbing targets according to IEC61161 (2007) [4]. The HITU transducers were mounted at the bottom of water tanks, radiating upwards to the targets (two layers with diameters of 136 mm and 110 mm) suspended from the balance (Mettler Toledo A250) by a thin wire.

The excitation voltage of the transducers was generated by a signal generator (R&S SMX) and amplified by an rf power amplifier (amplifier research 500A100A). Additionally, the supplied electrical power was monitored by a power reflection meter (R&S NRT-Z8). The input voltage on the transducer connecting port U_{in} was measured with a calibrated 100:1 sensor head connected to an rms level voltmeter (RD 9303).

The applied measurement cycles consisted of two on-periods embedded in off-periods. The on-time lasted from 16 s to 12 s and the off-time from 48 s to 118 s depending on the applied power. The transducer target distances varied between 3 – 9 mm, including sub-wavelength variations of $\lambda/4$. For the estimation of the power at zero distance, the measured power was corrected for the attenuation in the water path (half of the absorption coefficient in the water path due to the recovering of the acoustical streaming momentum). Additionally, correction factors based on the curved geometry of the focusing transducers were applied to the measured power values.

3.2. INRIM

The device used for the ultrasonic power measurements is based on the radiation force balance method [4]. The absorbing target was made of two polyurethane rubber discs, superimposed on each other and connected by means of a mechanical coupler. The lower disc is 14 mm thick and has a diameter of 80 mm, the upper one is 10 mm thick and its diameter is 120 mm. The target was connected to a load cell (Honeywell mod. 31) which measured the apparent target mass variation due to ultrasonic field on-off-cycles. The signal produced by the load cell was conditioned by a strain gauge amplifier (Sensotec mod. UV-10) and then measured by a nano-voltmeter (Agilent 34420A). The load cell was screwed onto the bottom of the tank (volume approx. 15 l) and the transducers were mounted above, radiating downwards.

The excitation voltage of the transducers was generated by a signal generator (Agilent 33220A) and amplified by an rf power amplifier (amplifier research 500A100A). The rms values of the transducer driving signals were acquired by means of an rms voltmeter (R&S URE-3).

The applied measurement schema consisted of two on-cycles embedded in off-cycles. The on-time was from 16 s to 12 s and the off-time from 48 s to 118 s. For the calculation of the power at zero distance (rim of the transducer), the attenuation coefficient of water was applied. Correction factors were applied to the measured power values incorporating the curved geometry of the focusing transducers.

3.3. UME

A modified radiation force balance equipped with a reflecting target [4] was used for the measurements. The ultrasound power was applied to one scale beam where the reflecting target (stainless steel conic reflecting target with 100 mm in diameter) was mounted and the corresponding force was measured on the other scale beam which was connected to the balance (Mettler Toledo PR 2004) with a thin wire. The transducers were mounted at the bottom of the water tank, emitting upwards in the direction of the reflecting target. Absorbers prevented multiple reflections inside the tank.

The driving voltage of the transducers was supplied from a combination of a signal generator and an rf power amplifier (Agilent 33250A and ENI 3100L for low power levels, Agilent 33220A and E&I for high power levels). A calibrated digital scope (Agilent 54820A Infiniium) with a 1:10 scope probe was used for measurements of the input voltages.

Balance read outs were acquired during power on/off time intervals between 15/30 seconds (5 W) and 12/120 seconds (150 W) and for different target-transducer distances between near zero and the focal point. The obtained power values were extrapolated to zero distance.

3.4. NPL

For the measurements, a modification of the NPL secondary standard setup was used. A castor oil target with its entry membrane facing vertically upwards was suspended beneath the balance (Sartorius CC1200) in a tank of water (volume approx. 20 l). The transducer was mounted on a retort stand with the radiating surface facing vertically downwards on to the castor oil target. The castor oil target was used to determine the total output power simultaneously from the radiation force and the thermal expansion of the target (buoyancy method) [5].

A programmable signal generator (HP3336C) provided the transducer drive through an rf power amplifier (amplifier research 150A100B). The measurement of the driving voltage was carried out with a digital storage oscilloscope (Tektronix DPO7254) fitted with a high impedance 100:1 probe. Since non-linearity of the drive signal was observed for medium and high power levels, the rms value of the fundamental frequency component of the rf voltage was taken into account for further calculations.

The ultrasound power was determined from the measurement of the radiation force [4] and from the change in buoyancy [5] of the target. The same data sets were used for both methods so the insonation conditions were identical. In all cases, the rf drive to the transducer was turned on and off with a period of 72 seconds (i.e. ON for 12 seconds; OFF for 60 seconds). For the estimation of the power at zero distance the measured power was corrected for the attenuation in the water path. For radiation force measurements, also the recovery of streaming momentum was taken into account for the estimation of the correction factor. Additionally, correction factors were applied to the measured power values due to the geometry of the focusing transducers.

4. Results

In table 1 the final results for the radiation conductance *G* of the participants are listed. The uncertainties are generally understood as relative uncertainties in % (or in 10^{-2}). The stated expanded uncertainties $ku(G)$ are based on a coverage factor k=2. The level of confidence is 95.45 %.

Table 1. Values for the radiation conductance together with the expanded uncertainty stated by the participants (Lab 1 – PTB, Lab 2 – UME, Lab 3 – INRIM, Labs 4 and 5 – NPL) and the determined reference values.

			Lab 1		Lab 2		Lab 3		Lab 4		Lab 5		**Ref. value**	
Trans-ducer	*f*	*P*-level	*G*	*ku(G)*	*G*	*ku(G)*	*G*	*ku(G)*	*G*	*ku(G)*	*G*	*ku(G)*	*G*	*ku(G)*
No.	MHz		mS	%	mS	%	mS	%	mS	%	mS	%	mS	%
1	1.1	vl	18.69	7.2	21.04	14.0	19.89	12.3	18.60	6.1	18.90	6.7	**18.93**	**3.6**
		l	18.85	7.2	20.84	14.0	19.61	8.5	18.60	6.1	19.50	4.6	**19.22**	**3.0**
		m	19.07	7.3	20.73	15.2	19.81	8.2	18.50	5.9	19.30	4.5	**19.16**	**3.0**
		H	19.25	8.1	20.41	15.2	20.71	8.2	18.50	6.0	19.40	4.7	**19.32**	**3.1**
1	3.3	vl	13.00	7.4	15.73	14.0	14.31	13.0	13.20	6.1	14.30	6.2	**13.66**	**3.5**
		l	13.23	7.4	15.62	14.0	13.70	8.5	13.40	6.1	14.00	4.9	**13.72**	**3.1**
		m	13.61	7.6	15.55	15.2	14.18	8.2	13.60	6.0	14.50	5.0	**14.08**	**3.2**
2	1.5	vl	15.29	7.2	16.98	14.0	15.61	14.1	15.00	5.6	16.00	5.8	**15.50**	**3.4**
		l	15.33	7.2	16.64	14.0	15.69	8.5	15.00	5.5	16.40	4.7	**15.70**	**3.0**
		m	15.28	7.3	16.71	15.2	15.47	8.2	14.80	5.5	15.90	4.7	**15.43**	**3.0**
		H	15.27	8.1	16.37	15.2	15.59	8.2	14.50	5.5	15.50	4.6	**15.19**	**3.0**

The uncertainties were derived in accordance with the "Guide to the Expression of Uncertainty in Measurement" [6]. The analysis of the reported radiation conductance followed the guidelines for the evaluation of key comparison data [7]. The reference values *G* for each combination of frequency and power level were calculated as weighted means, incorporating the submitted measurements G_i and standard uncertainties $u(G_i)$ of the participating laboratories.

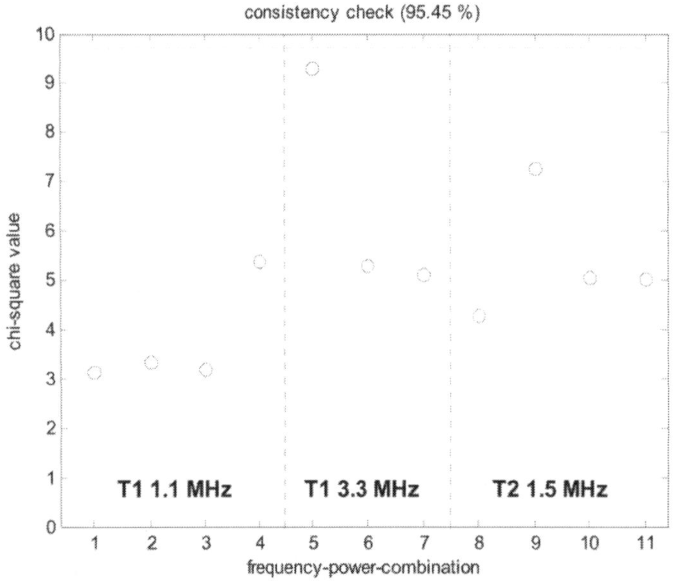

transducer 1
@ 1.1 MHz:
1 – very low
2 – low
3 – medium
4 – high

transducer 1
@ 3.3 MHz:
5 – very low
6 – low
7 – medium

transducer 2
@ 1.5 MHz:
8 – very low
9 – low
10 – medium
11 – high

Figure 1. Consistency test - result of the applied chi-squared test.

As the measurements are consistent (figure 1), the calculated reference values are valid estimates of the radiation conductance (table 1).

The degree of equivalence of each applied measurement method is given by the pair of values d_i and $ku(d_i)$ (k=2). The values were calculated using

$$d_i = G_i - G \quad \text{and} \quad u(d_i) = \sqrt{u^2(G_i) - u^2(G)} \,, \tag{2}$$

with G_i the reported values and G the reference value (figures 2 and 3).

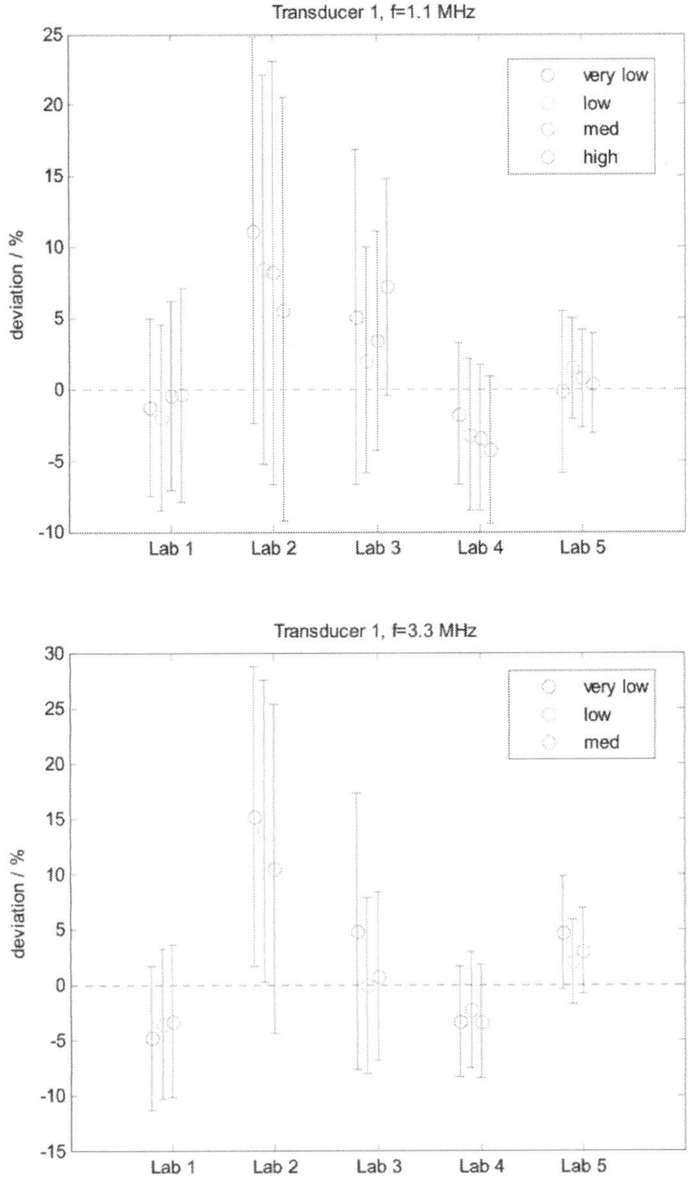

Figure 2. Degrees of equivalence d_i of the participating laboratories with the reference value for the radiation conductance G of transducer 1 (Sonic Concepts H-102) and associated expanded uncertainties $ku(d_i)$ (k=2).

5. Discussion and conclusion

With respect to the reported uncertainties of the participating laboratories, the circulated transducers appeared sufficiently stable over the measurement period from June to December 2009. For transducer 1 @ 1.1 MHz, the uncertainty contributions are much smaller than the uncertainty of the reference values. As reported by the participants, transducer 1 @ 3.3 MHz showed instabilities during measurements at higher power levels (drop in output) and transducer 2 @ 1.5 MHz experienced a leakage, which was successfully fixed. In summary, these facts induced an increase of the uncertainty contributions due to temporal instabilities, which are still smaller than the uncertainty of the reference values.

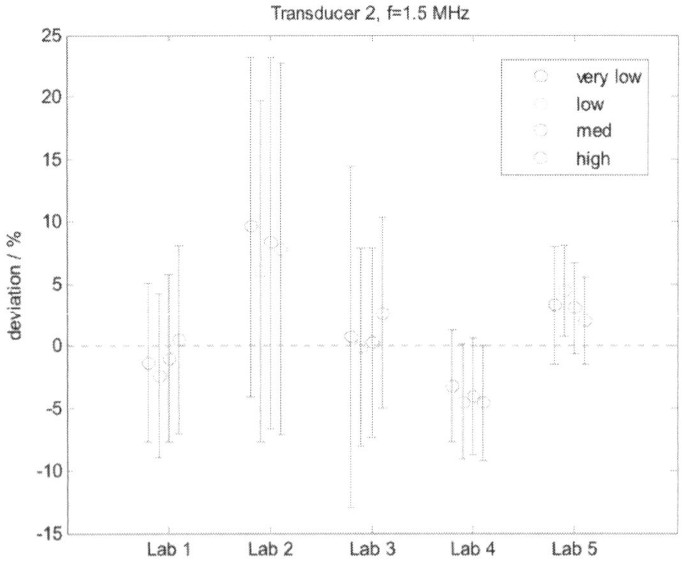

Figure 3. Degrees of equivalence d_i of the participating laboratories with the reference value for the radiation conductance G of transducer 2 (Imasonic 5706A) and associated expanded uncertainties $ku(d_i)$ (k=2).

The participants' relative expanded uncertainties for the radiation conductance at the various combinations of transducer type, frequency and excitation level range from 4.5 % to 15.2 %. The majority of deviations from the reference value were below 5 %, except the results obtained by means of a radiation force balance equipped with a reflecting target (max. deviation 15 %). This leads to the conclusion that the use of reflecting targets in strong focusing fields is not recommended.

6. Summary

The outcome of this first inter-laboratory comparison of power measurement methods provides an insight into the current state of HITU power measurement capabilities. The obtained agreement between the results of the participants can generally be considered satisfactory. Based on the experience of this comparison, a repetition with more participants is recommended.

7. References

[1] Shaw A and ter Haar G 2006 *NPL Report* DQL AC 015
[2] ter Haar G 2001 *Physics Today* **54** 29
[3] URL: http://www.ptb.de/EURAMET-JRP7
[4] IEC61161:2007 Ultrasonics – Power Measurement –Radiation Force Balances and Performance Requirements. International Electrotechnical Commission, Geneva
[5] Shaw A 2008 *Ultrasound in Medicine and Biology* **34** 1327
[6] ISO/IEC GUIDE 98-3: 2008 Guide to the expression of uncertainty in measurement.
[7] Cox A G 2002 *Metrologia* **39** 589

Acknowledgements

This work was supported by the European Community's Seventh Framework Programme, ERA-NET Plus, under Grant Agreement No. 217257.

Measurement of Angle of Ultrasound Propagation from Phase

John Civale, Ian Rivens, Gail ter Haar

Joint Department of Physics, The Royal Marsden Hospital, Institute of Cancer Research, Downs Road, Sutton, Surrey, SM2 5PT

john.civale@icr.ac.uk

Abstract. Acoustic field calibrations often use Fast Fourier Transforms (FFT) to quantify the spectral amplitude components of detected acoustic waveforms. The ability of FFTs to provide phase information is often overlooked. This phase data may be useful in determining the angle of propagation of the ultrasound beam. The angle of propagation at the focal peak (and any other point in the field) can be calculated easily and quickly without additional measurement, and may be the quickest and most accurate method of aligning the sound axis with respect to the beamplotting system's co-ordinates.
Acoustic fields have been measured experimentally using a system capable of waveform acquisition. Measurements were made using a fibre-optic hydrophone (Precision Acoustics, UK) which provided spatial resolution of <100 μm. Two operating configurations of a 10 strip array HIFU (high intensity focused ultrasound) transducer were tested, as was a single element HIFU device. Theoretical pressure and phase distributions for these transducers were predicted using a linear acoustic field model. Results show that for the single element, radially symmetric device, beam alignment measurements using phase data at the focal peak are in agreement with the more conventional method based on finding the on-axis peak positions. In the case of a transducer with a number of elements de-activated to produce an asymmetric ultrasound source, the angle of propagation at the focal peak was altered, indicating a change in performance of the transducer which otherwise might not have been detected using the "on-axis peaks" method. Simulations agreed with the experimental data.

1. Introduction

Detailed acoustic field measurements are made using beam plotting systems to quantify the pressure distributions generated by both therapeutic and imaging ultrasound transducers. Such systems usually characterize properties such as peak acoustic pressure using calibrated hydrophones. If a system allows waveform acquisition, the use of a Fast Fourier Transform (FFT) allows spectral characterization of the detected signal. The phase information also provided at each frequency is usually ignored. However this data can provide useful information regarding the nature of the ultrasound beam.

In this paper, we show how the phase data can be used to calculate the direction of propagation of the ultrasound beam. Perhaps more importantly than allowing visualization of the way the ultrasound beam may be converging or diverging across the acoustic field, it could provide rapid evaluation of the correct alignment of the sound axis with the measurement system. Alternatively, a measurement of the angle of propagation at the focal peak may provide a quantity to be routinely measured when monitoring transducer operation over time. This may be particularly useful for phased array devices with many elements, which allow electronic focal steering.

In the case of High Intensity Focused Ultrasound (HIFU) transducers the focal peak is the obvious position at which to measure beam alignment. In this paper, the measured angle of propagation at the focal peak is compared to the more standard technique of identifying, and then aligning, the on-axis pre- and post-focal peak positions to determine the 3D alignment of the sound beam with respect to the beam plotting coordinate system. Firstly, a single element HIFU transducer is tested, and secondly a 10 strip array transducer is used to illustrate how both techniques deal with an asymmetric ultrasound source. A linear ultrasound propagation model is used to predict the pressure and phase distributions for comparison with the experimental results.

2. Methods

2.1. Transducers and drive system: The HIFU transducers were driven using a pulse generator (HP33120A Agilent) set to burst mode, and an ENI A300 amplifier providing a gain of 55 dB. The tone burst length was set to between 40 and 80 cycles, depending on transducer and scan parameters, at a pulse repetition frequency of 100 Hz. The synchronisation signal from the pulse generator was used to trigger the oscilloscope used for measurement purposes. Transducers were connected to the drive system via their respective impedance matching circuits. A single element transducer and a 10 strip array transducer were tested. Details of transducer parameters are given in Table 1 along with the resolution for large area beam plots in the *X-Z* plane. Two cases were tested for the single element transducer: 1) visual (by eye) alignment of the transducer sound axis with the *Z* axis of the positioning system, 2) deliberate misalignment by a few degrees in the *X-Z* plane (azimuthal angle). The 10 strip transducer was tested using two configurations: 1) all ten segments active and working in phase, 2) segments numbers 1, 2, 3 and 4 deactivated resulting in a non axisymmetric ultrasound source (as shown in Figure 1). The propagation angle calculation based on phase measurements was performed in the *X* axis direction. The transducers were driven at relatively low powers (<30 W acoustic power).

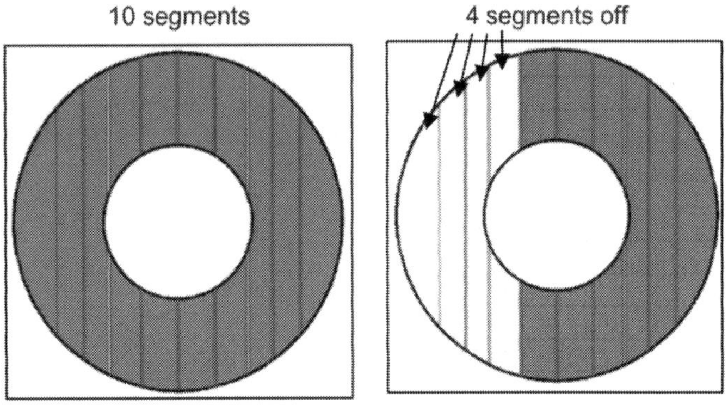

Figure 1. The 10 strip array transducer . In the first configuration (left) all 10 segments were active and driven in phase, in the second configuration (right), the first four elements indicated by the arrows were deactivated

2.2. Beam plotting system: A fibreoptic hydrophone [1] (Precision Acoustics, UK) was used to measure the acoustic field. The fibreoptic sensor is of the Fabry-Perot interferometer type and was chosen for its excellent spatial resolution (sensor diameter of 10 µm; fibre thickness ~ 100 µm) whilst maintaining sufficient sensitivity (>100 mV/MPa). The sensor was mounted on a motorised 3D positioning system under PC control (Labview). Measurements were performed in deionised, degassed water. The sensor signal output was connected to an oscilloscope (Lecroy Waverunner 64Xi) and the signal windowed to display a short number of cycles (<10) to give an apparent CW representation of the signal. The oscilloscope signal was averaged over several repeats (>100) to improve the signal to noise ratio. One average waveform per position was stored on the PC hard drive.

Type	Diameter (cm)	Hole (cm)	Focal length (cm)	Frequency (MHz)	Resolution (mm – X/Z)
(1) Single element	6.5	2	6.3	1.7	0.5 / 1
(2) Segmented	11	5.5	15	1.7	0.5 / 2

Table 1. Transducer specifications, including diameter, imaging aperture diameter, working frequency and scan resolution for the large area *X-Z* plane beamplots.

2.3. Phase measurement: Phase was measured from the stored average waveforms. The timebase and time delay of the oscilloscope were fixed during measurements so that any changes in the position of the sensor in the acoustic field were seen as apparent changes in phase of the measured waveform. To obtain a measurement of phase, the measured waveform data set was trimmed to provide an integer number of cycles to serve as input to the FFT. The real and imaginary components from the FFT output at the fundamental frequency were used to calculate the phase angle.

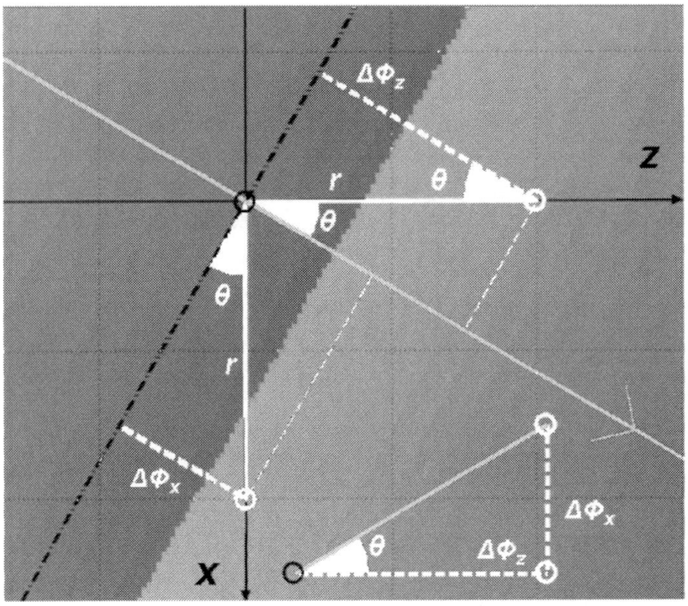

Figure 2. Propagation angle calculation. Background shading represents phase in the *X-Z* plane, the white arrowed line the direction of propagation and the black dashed line a constant phase line parallel to a planar wavefront. Measured phase shifts $\Delta\Phi_x$ and $\Delta\Phi_z$ are obtained when a sensor is moved by a distance r in the orthogonal *X* and *Z* directions, respectively. Propagation angle θ can be calculated using either phase shift between any two points with an assumed wavelength λ (1D approach, equations 1 and 2), or by using both phase shifts together without the need to know λ (2D approach, equation 3).

$$\sin(\theta)=\frac{\lambda\Delta\phi_x}{2\pi r} \quad (1) \qquad \cos(\theta)=\frac{\lambda\Delta\phi_z}{2\pi r} \quad (2) \qquad \tan(\theta)=\frac{\Delta\phi_x}{\Delta\phi_z} \quad (3)$$

2.4. Ultrasound propagation direction calculation: The direction of ultrasound propagation with respect to the beam plotting system *Z* axis (nominal sound axis direction) can be computed by moving the sensor small distances r (typically less than a wavelength) in either the *X* or *Z* directions. Plane wave propagation was assumed over these small distances. The phase difference measured between the two points are used to estimate the angle of propagation θ (see Figure 2). There are two possibilities: a one dimensional approach in which the calculation is performed using phase differences in either *X* or *Z* directions only (equations 1 and 2), or a two dimensional approach in which phase differences in two orthogonal directions are used (equation 3). The advantage of the two dimensional approach lies in the fact that it can be used to measure angles over the entire 360° range accurately, and the wavelength does not have to be known precisely. However, the distance r must be kept smaller than half the wavelength for the calculation to produce the correct result. The one dimensional approach can only yield measurements over a 180° range and it is preferable to perform the calculation in the direction lying most orthogonally to the expected principal propagation direction component, in order to avoid ambiguity when solving the inverse sine and cosine functions in equations 1 and 2. The one dimensional approach also requires the wavelength to be accurately known, however for 2D mapping of the acoustic field the condition imposed on the step size r need only apply in the measurement direction (*e.g.* X) meaning that the step size in the orthogonal direction (*ie.* Z) can be increased, leading to faster scan times. The results shown in this paper for large area beamplots were calculated using the one dimensional approach in X, with the wavelength based on an assumed speed of sound of 1500 m/s. For more detailed orthogonal linear scans across the focal peak, the above two dimensional description was extended to three dimensions. The phase differences in the three orthogonal directions were used to describe a unit vector, in a spherical coordinate system these produced estimates of the elevational (θ) and azimuthal (φ) angles describing the direction of ultrasound propagation.

2.5. Beam alignment: It was assumed that the angle of propagation at the focal peak would lie parallel with the sound axis for radially symmetric transducers. Plane wave conditions were assumed to be valid in the volume close to the focal peak. Measurements were performed in the three orthogonal directions across the focal peak using an r value of 100 µm. A total of 6 repeat measurements were made in each direction for each transducer, alternating between scanning in the positive and negative directions to reduce any bias or systematic error due to temperature (sound speed) instability of the water, or backlash in the positioning system. Sound axis alignment was computed by identifying the 3D co-ordinates of 6 on-axis peaks (including the main focal peak). A linear regression was performed on X-Z and Y-Z data points, allowing computation of the elevational and azimuthal angles by determination of the gradients of the linear fits, together with their uncertainties.

2.6. Ultrasound field model. A linear ultrasound acoustic field model based on Raleigh-Sommerfeld diffraction theory [2] was used to predict acoustic field distributions, including amplitude and phase. The program was used to model acoustic fields at the fundamental frequency, assuming a speed of sound in water of 1500 m/s.

3. Results

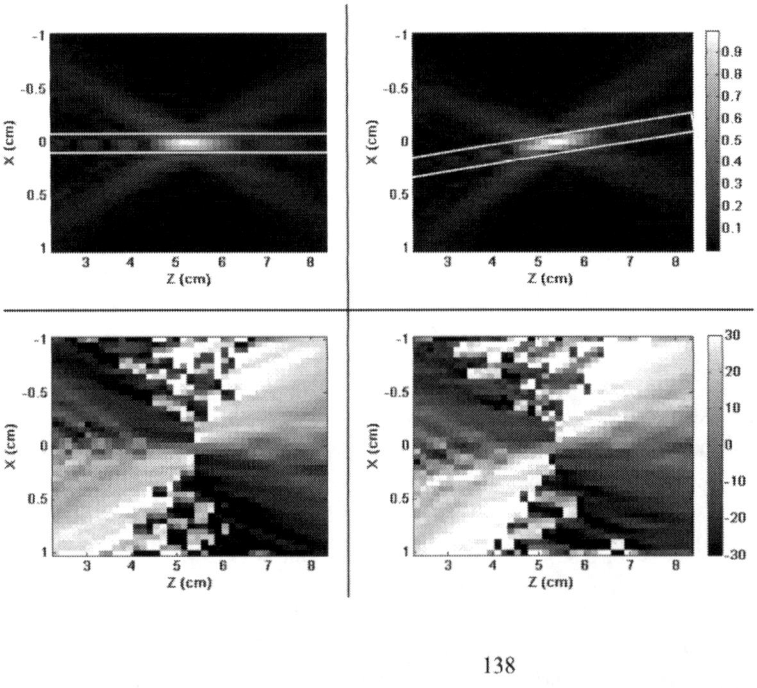

Figure 3. Normalised pressure distributions at the focal peak (top) and angle of propagation plots (bottom) in the X-Z plane for HIFU single element transducer (#1). The plots represent the parallel alignment "by eye" (left) and deliberately misaligned (right) arrangements. The white rectangles on the pressure plots indicate the alignment of the on-axis peaks.

Figure 3 shows the pressure and propagation angle plots in the *X-Z* plane for the second single element HIFU transducer which had been initially aligned by eye with the *Z* axis, and subsequently for the condition in which the transducer was rotated azimuthally in the *X-Z* plane. In the resulting angular plot there is an overall increase in all the angular values including at the focal peak position. In the pressure plots, the on-axis peak positions can also be identified, illustrating the ability to measure the amount of rotation using this method. Table 2 shows the measurement of the elevational and azimuthal angles as measured using both the 3D linear scan phase data and the on-axis peaks method. The measurement of the angle of the sound axis using the on-axis peaks method agrees, within errors, with the mean angle of propagation measured at the focal peak for both the aligned and mis-aligned arrangements.

	Elevational angle θ		Azimuthal angle Φ	
Method	Phase	Peaks	Phase	Peaks
"Straight"	-0.24±0.32	-0.07±0.05	0.23±0.16	0.38±0.03
"Angled"	0.95±0.33	1.18±0.40	6.25±0.24	6.06±0.30

Table 2. Sound axis elevational and azimuthal angles (in degrees) as calculated using the propagation angle at the focal peak and the on-axis-peaks method for single element HIFU transducer. The uncertainties represent the standard deviation for the phase method, and the standard error of the linear fit for the on-axis-peaks method respectively.

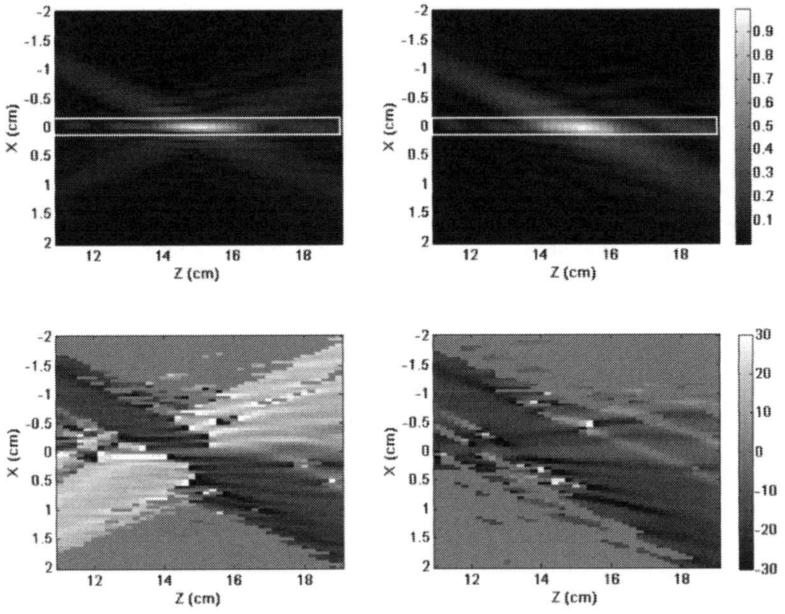

Figure 4. The normalised pressure distributions (top) and angle of propagation plots (bottom) in the *X-Z* plane of the 10 strip HIFU transducer (#2). The plots represent the configurations with all strips on (left) and with four strips off (right). The white rectangles on the pressure plots indicate the alignment of the on-axis peaks.

Figure 4 shows the pressure and angular plots obtained using the 10 strip HIFU transducer for both configurations tested. A clear pressure field asymmetry arises with the four elements switched off. The on-axis peaks appear in approximately the same positions for both configurations, giving similar results. If the user was unaware of the malfunction of some of the device's elements, the angle of propagation measurement at the focal peak would produce a result which might alert the user to the problem, whereas the on-axis peak method would not. However, it could be argued that the on-axis-peak method is a robust way of determining the geometric sound axis of the transducer, irrespective of how the transducer may be responding, while the propagation angle method gives a better indication of from where the energy in the ultrasound beam as a whole is originating. Experimental data agreed with the simulation data for both methods. Results are summarised in Table 3.

	Elevational angle θ				Azimuthal angle Φ			
Configuration	Phase	Peaks	Model phase	Model peaks	Phase	Peaks	Model phase	Model peaks
All elements on	-0.39± 0.62	-0.09± 0.04	0	0	-0.99± 1.10	-0.27± 0.03	0	0
Elements #1-4 off	0.92± 0.30	-0.03± 0.04	0	0	-8.21± 0.81	-0.17± 0.05	-8.37	0

Table 3. Sound axis elevational and azimuthal angles (in degrees) calculated using the propagation angle, and on-axis peaks method for the segmented HIFU transducer. The uncertainties represent the standard deviation for the phase method, and the standard error of the linear fit for the on-axis peaks method respectively. Results from the simulation data are included

4. Discussion and Conclusion

Phase data measured from acoustic field measurements can be used to derive estimates of the ultrasound beam propagation angle at any point in an ultrasound field. The method relies on measuring changes in phase with position as a hydrophone is scanned through the acoustic field. The data presented shows it is possible to visualise the convergence and divergence of focused ultrasound beams, but measurements around the focal peak may also help in aligning the transducer sound axis to be parallel to the beam plotting coordinate system. The advantage of the phase method lies in the relative ease and speed with which an estimate of the angle can be measured directly from the waveform data acquired during the acoustic field acquisition. The phase method has, therefore, a major advantage over the more traditional method which consists of identifying on-axis peak positions using a number of trans-axial scans. This can be time consuming when performed carefully, or lead to errors if done in an approximate manner. The principal sources of error in the phase measurement method are due to the stability and accuracy of the measurement system. If water temperature is not stable, changes in sound speed introduce apparent changes in phase over the course of a scan which will bias the result. The accuracy of the positioning gantry is also important, particularly for measurements at the focus of a HIFU field where small phase differences are measured over limited distances. Steps must therefore be taken to eliminate positioning errors (arising,for example, from backlash). The data reported in this paper shows good agreement between the phase and on-axis peaks methods in determining the relative alignment of the sound axis and beam plotting system. In the case of the 10 strip transducer, measurement of the propagation angle at the focal peak may indicate mis-functioning of the device. More rigorous analysis of the sources of errors in the phase method is now needed, together with further study of the technique when applied to the acoustic fields from a variety of transducers.

References

[1] Morris P, Hurrell A, Shaw A 2009 *J. Acoust. Soc. Am.* **125** 3611
[2] Clarke R L 1995 *Ultrasound in Medicine and Biology.* **21** 353

Hydrophone spatial directivity and the induced difference in HIFU acoustic field test

T Chen[1,2], J M Hu[2], W Huang[2], W Zhang[2] and D Zhang[1]

[1] Institute of Acoustics, Nanjing University, Nanjing 210093, China

[2] Jiangsu Province Institute for Medical Equipment Testing, Nanjing 210012, China

Chent_qxs@jsfda.gov.cn

Abstract. Hydrophone has been widely used in testing ultrasonic field distribution. But the hydrophone has a certain directivity, which may bring error in the actual acoustic field measurements. So it is necessary to meet some requirements about the directivity of hydrophone. In the measurement, the spatial directivities of needle hydrophone, membrane hydrophone and robust hydrophone is measured and a comparison about measured data of HIFU acoustic field with them is given. Experimental results indicate that the asymmetric special directivity of hydrophone is due to manufacturing processes and other reasons. As a result, it is not sufficient to test the directivity of hydrophone along one axis. It is necessary to adjust the direction of the hydrophone when characterizing the HIFU field.

1. Introduction

High Intensity Focused Ultrasound (HIFU) is an acoustic ablation technique that utilizes the power of ultrasound to destroy deep-seated tissue with pinpoint accuracy for treatment of cancers. The principle of this technique is to raise the tissue temperature to relatively high values and cause thermal coagulation and ablation of cells by using an ultrasonic focusing transducer. As we know, it is highly desirable in HIFU therapy to produce lesions with well-controlled spatial characteristics. To achieve this purpose, it is important of well understanding of the acoustic field generated by the HIFU transmitter.

Now, HIFU is popular in China. More than 10 companies are developing HIFU instruments, more than 200 hospitals using HIFU machine, and tens of thousands of patients have been treated by HIFU. For industry development and the safety of using HIFU, it is important to measure HIFU device, as excessive power or lost control of the sound field could give the patient harm.

Hydrophone is an important method of measuring the sound field, but its spatial directivity may affect the accuracy of the measurement. The purpose of this study is to examine the spatial directivity of the hydrophone and the induced difference in HIFU acoustic field test.

2. Methods

In order to test the spatial directivities of hydrophone, we set up an experimental system with two degree of freedom as shown in Fig. 1. The first is to determine the hydrophone's sensitivity surface by sliding the hydrophone from -90 to 90 degree. The second is to determine the hydrophone's sensitivity surface by rotating the hydrophone from 0 to 360 degree. In the measurement, the emitted signal is a 670 kHz sinusoidal signal with voltage of 70V to drive the HIFU transmitter.

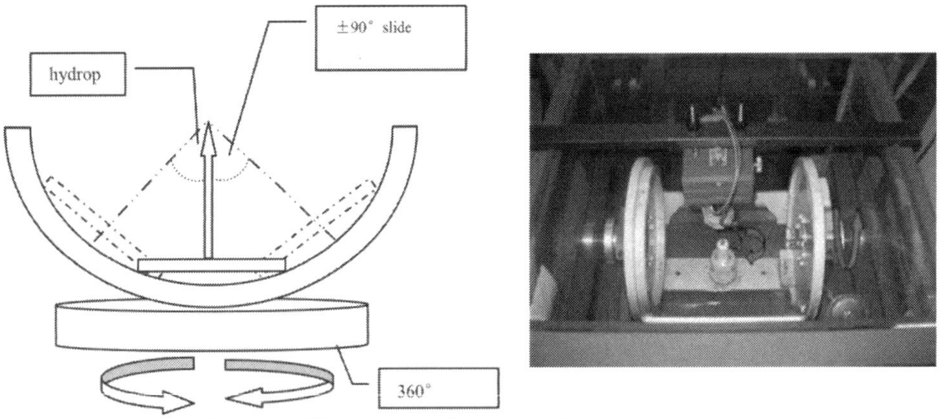

Figure 1. Illustration of the experimental system.

Figure 2 shows the steps of measurement.

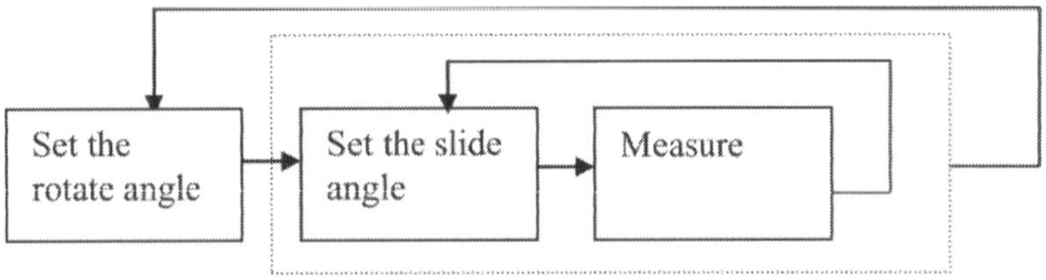

Figure 2. Steps of measurement

3. Results and discussions

In this study, total 4 tests are performed to test the influence of the hydrophone spatial directivity. Test 1 is to check spatial property of the measurement system. In the measurement, the sliding angle is from -90 to 90 degree, while the rotating angles are 0, 180, 90, and 270 degrees, respectively.

Figure 3 shows the measured results at 0 and 180 degrees, while Fig. 4 gives the measured results at 90 and 270 degrees. The data between 0 and 180 degrees are symmetrical, and the data between 90 and 270 degrees are also symmetrical It indicates that the repeatability and stability of the measuring instrument meet our requirement.

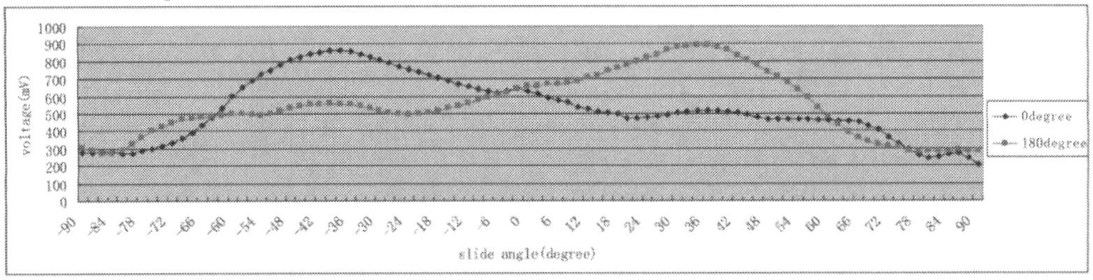

Figure 3. Measured results at 0 and 180 degrees

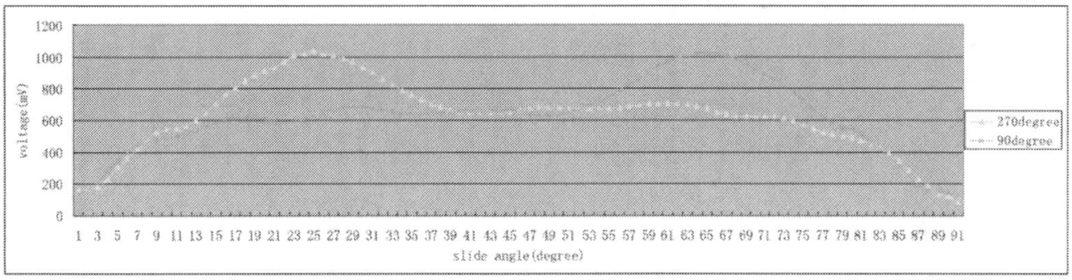

Figure 4. Measured results at 90 and 270 degrees

Test 2 is to examine the spatial difference of the hydrophone. In the measurement, the sliding angles are from -90 to 90 degrees, and the rotating angles are set to be 0, 45, 90, 135 and 180 degrees.

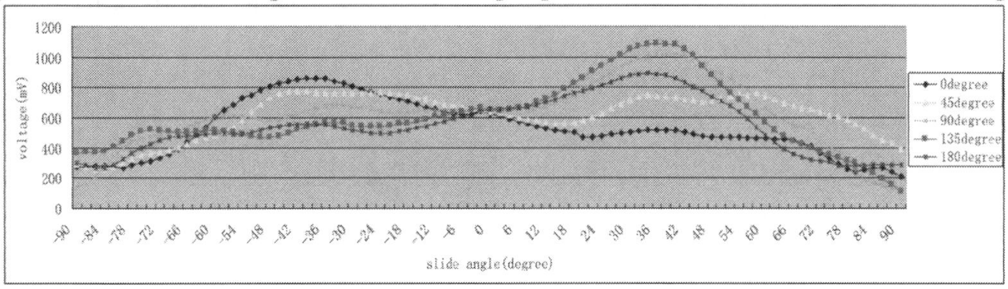

Figure 5. Spatial difference of the hydrophone.

Figure 5 compares the spatial difference of the hydrophone at different rotating angles. We can find that bigger differences are observed except for the slide angle close to 0 degree.

Test 3 is to examine the individual difference among three kinds of hydrophones (A/B/C). The type A is a needle hydrophone protected by copper, the type B is a robust needle hydrophone, and the type C is a simple needle hydrophone. In the measurement, the sliding angles are from -90 to 90 degrees; while the rotating angles are set to be 0, 45, 90, 135 and 180 degrees.

The measured results for the three kinds of hydrophones are shown in Fig.6. The difference in directivity and sensitivity for various of hydrophones is obvious. We need test the directivity of hydrophone before using in the HIFU test.

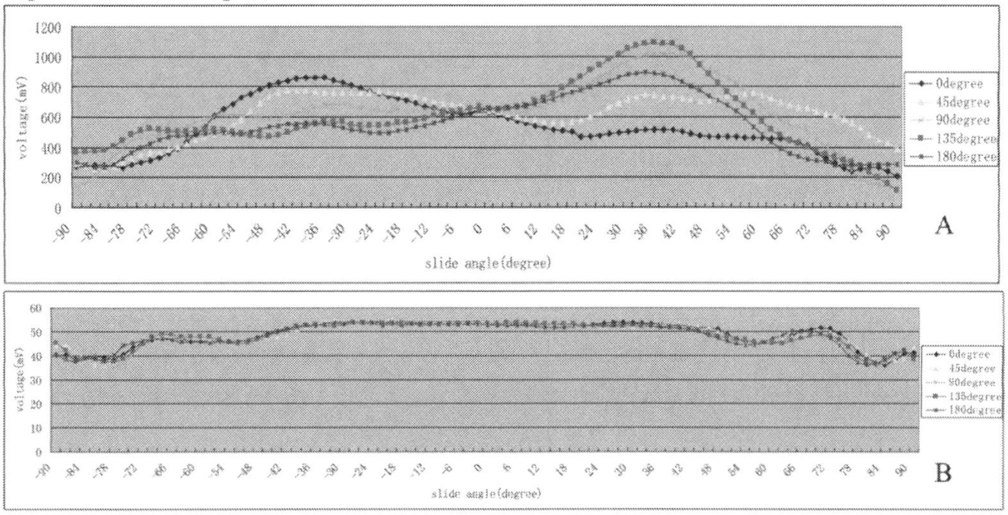

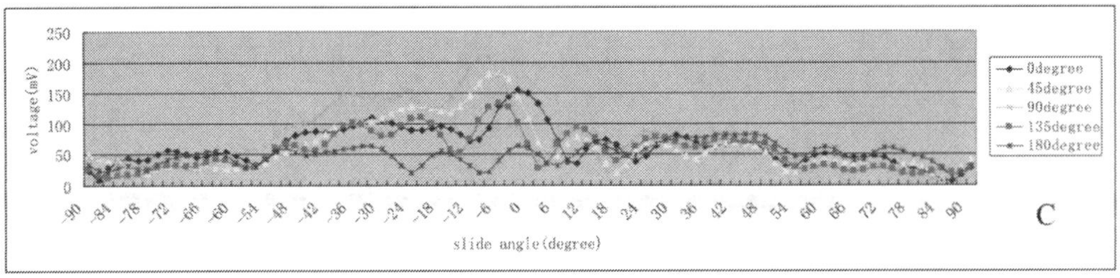

Figure 6. The spatial differences for three kinds of hydrophones.

Test 4 is to examine the frequency dependence of the hydrophone. In the measurement, the frequency is changed from 600 to 900 kHz. Figure 7 shows the frequency dependence of the hydrophone B. It is concluded that there exists frequency difference in the measurement using the hydrophone.

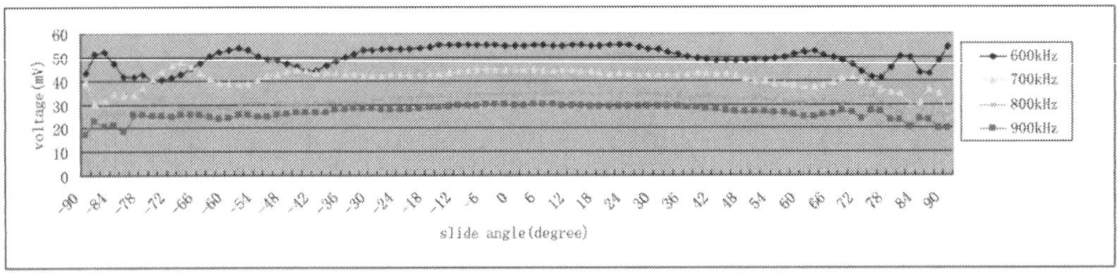

Figure 7. Frequency dependence of the hydrophone B

4. Conclusions

In the measurement of HIFU acoustic field using hydrophone, the requirement for directivity (5.2.1.3, GB/T19890-2005) and positioning system for the fixture and orientation (5.2.2 , GB/T19890-2005) are important. In this study, we performed four tests to examine the spatial directivity of the hydrophone and the induced difference in measurements. It is concluded that it is necessary to test the directivity of the hydrophone before our HIFU acoustic field testing at working frequency. In addition, we also should adjust the hydrophone positioning system in at least two degree freedoms.

References

[1] GB/T19890:2005 Acoustics- High intensity focused ultrasound (HIFU) --Measurements of Acoustic Power and Field Characteristics

[2] Adam Shaw and Gail ter Haar .NPL REPORT DQL AC 015

[3] Mark E. Schafe, James Gessert and Wayne Moore. Development of a High Intensity Focused Ultrasound (HIFU) Hydrophone System

Beam profile measurement on HITU transducers using a thermal intensity sensor technique

V Wilkens, S Sonntag, K-V Jenderka
Physikalisch-Technische Bundesanstalt, Bundesallee 100, 38116 Braunschweig, Germany

E-mail: volker.wilkens@ptb.de

Abstract. Thermal intensity sensors based on the transformation of the incident ultrasonic energy into heat inside a small cylindrical absorber have been developed at PTB in the past, in particular to determine the acoustic output of medical diagnostic ultrasound equipment. Currently, this sensor technique is being expanded to match the measurement challenges of high intensity therapeutic ultrasound (HITU) fields. At the high acoustic power levels as utilized in the clinical application of HITU transducers, beam characterization using hydrophones is critical due to the possible damage of the sensitive and expensive measurement devices. Therefore, the low-cost and robust thermal sensors developed offer a promising alternative for the determination of high intensity output beam profiles. A sensor prototype with a spatial resolution of 0.5 mm was applied to the beam characterization of an HITU transducer operated at several driving amplitude levels. Axial beam plots and lateral profiles at focus were acquired. The absolute continuous wave output power was, in addition, determined using a radiation force balance.

1. Introduction

The determination of the acoustic output of medical ultrasound equipment is important as regards aspects of quality assurance and the safety of the patient. The procedures to measure and characterize the ultrasound beams are described in several international standards for diagnostic and physiotherapy devices. These standards are continuously developed to comply with the technical progress and increasing fields of application of ultrasound equipment. More recent developments of using high intensity therapeutic ultrasound (HITU) for the ablation of tumors require new methods and standards for characterizing the high power transducers [1]. The measurements are usually performed using radiation force balances to determine the output power and using hydrophones to determine local pressure and intensity parameters. However, hydrophone measurements are difficult in the case of HITU fields at high clinical power levels since these sensors may easily be damaged when scanning the focus region. Damage and measurement distortion may occur due to heating, cavitation, and radiation force. As an alternative to the hydrophone-based technique, the local temporal-average ultrasound intensity and beam dimensions can be measured using thermal sensors. This technique – based on the transformation of the incident ultrasonic energy into heat inside a small cylindrical absorber – was developed at PTB in the past, in particular to determine the acoustic output of medical diagnostic ultrasound equipment [2,3,4]. Currently, this sensor technique is being expanded to match the measurement challenges of HITU fields. This paper describes the application of a novel sensor prototype with a spatial resolution of 0.5 mm to the beam characterization of an HITU transducer operated at several driving amplitude levels. Axial beam plots and lateral profiles at focus were acquired. The absolute temporal-average output power was, in addition, determined using a radiation force balance.

2. Sensor principle and design considerations

The thermal measurement technique used is based on the transformation of the incident ultrasonic energy into heat within a small-sized cylindrical absorber (figure 1). The front face of the absorber is in contact with the surrounding water to allow the ultrasound wave to pass into the absorber. Due to this contact, the temperature at the absorber front face is the same as in the water bath. Since the absorber is in part thermally insulated by an air-filled housing, the temperature at the rear increases during insonation. Part of the heat produced inside the absorber permanently flows through the front face to the water. After a period of insonation with constant intensity, thermal equilibrium will appear between the heat produced by absorption and the heat given off to the surrounding water. From the temperature enhancement ΔT at the rear of the absorber at this equilibrium measured by a small resistor temperature probe, the ultrasound intensity averaged over the absorber cross section can be determined [5]. Since all thermal processes are slow in comparison with the ultrasound frequencies and the repetition frequencies of bursts, the sensor inherently temporally averages over all incident waveforms. In contrast to the plexiglas absorbers formerly used for diagnostic ultrasound [3], the absorber material applied here is polymethylpentene, providing an acoustic impedance of $\rho c = 1.69 \times 10^6$ kg m^{-2} s^{-1}, being better matched to that of water. It is expected that this will reduce the frequency response variations due to lower multi-reflections within the absorber rod. However, this has to be confirmed by future calibration measurements. The absorber sizes are 0.5 mm in diameter and 2 mm in length. A steel housing is used to enhance the robustness of the sensor and to increase the cooling of the absorber at the front face rim. Finally, a conical shape for the front of the housing was chosen to reduce acoustic reflections between the sensor and the source transducer which led to distortions of measurements at short distances with the previous planar sensors [4].

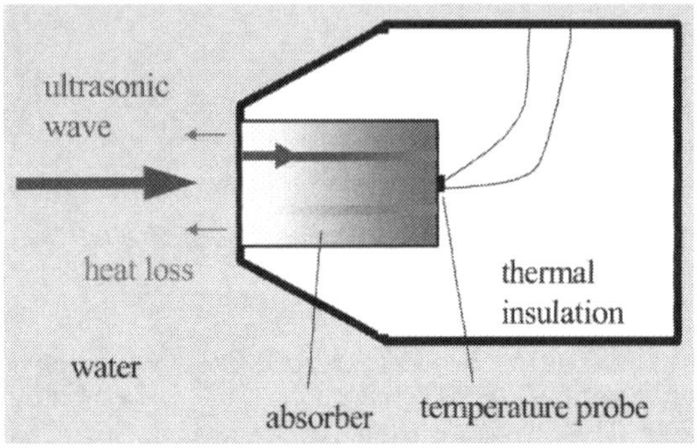

Figure 1. Left: thermal ultrasound intensity sensor. The intensity of the ultrasonic wave is determined from the temperature increase ΔT at equilibrium measured at the rear of an absorber. Right: photograph of the prototype sensor.

3. Comparison measurement of a diagnostic ultrasound field

To test the new sensor prototype manufactured, comparison measurements of a high intensity pulse Doppler field of a commercial diagnostic machine were performed first. The set-up for the intensity scans is depicted in figure 2. The 3.5 MHz convex array transducer is slightly immersed into the degassed and deionized water contained in a measurement tank and adjusted to emit the ultrasound beam vertically downwards. The thermal sensor is mounted onto a stepper motor-driven *xyz*-positioning system which is controlled by a computer. The measurement software program controls the sensor movement for axial scans (*z*-direction) and for beam profile line scans in the azimuth and elevation directions (*x*- and *y*-directions, respectively) and records the data from the temperature probe of the ultrasound sensor and from a second probe for the water bath reference temperature. The sound field axis was determined by lateral maximum search at two distances from the transducer: The nominal focal length of the elevational lens and 30 mm beyond that distance. Then a line scan was performed along this axis to obtain the axial beam profile. Afterwards, at the distance of maximum intensity, lateral beam profiles were measured in azimuthal and elevation directions. The temperature-versus-time curves recorded were automatically evaluated regarding the equilibrium temperature enhancement in comparison to the switched-off condition of the transducer. The measurement time for one line-scanning step including the time to reach equilibrium and the time to average 20 data points at equilibrium was about 60 s. A mechanical stirrer was used to achieve a homogeneous temperature distribution in the water bath and to avoid clouds of warmth in front of the thermal sensor front face. An additional acoustic absorber avoided backreflections from the bottom of the tank. The temperature data of the new HITU intensity sensor were compared with the results of the previously calibrated 0.6 mm diameter sensor as given in [4]. By equally scaling the data at the position of maximum intensity as shown in figure 3, the temperature-to-intensity transfer factor of the HITU sensor was estimated to be approximately $H(3.5\ \text{MHz}) = 0.17\ \text{mK}\ \text{W}^{-1}\ \text{m}^2$. In addition, the result of a needle hydrophone measurement is shown in figure 3. The comparison of the curves shows good agreement. The systematic overestimation of the planar thermal sensor at distances below 28 mm due to multi-reflections between the sensor and the transducer is significantly reduced with the conical HITU intensity sensor design.

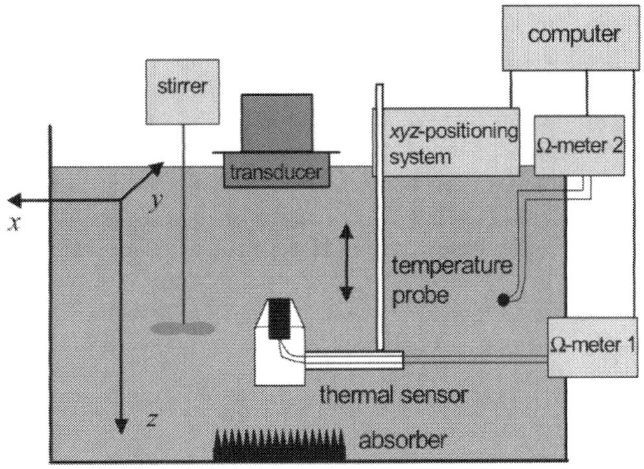

Figure 2. Set-up for intensity scans using the thermal sensor technique.

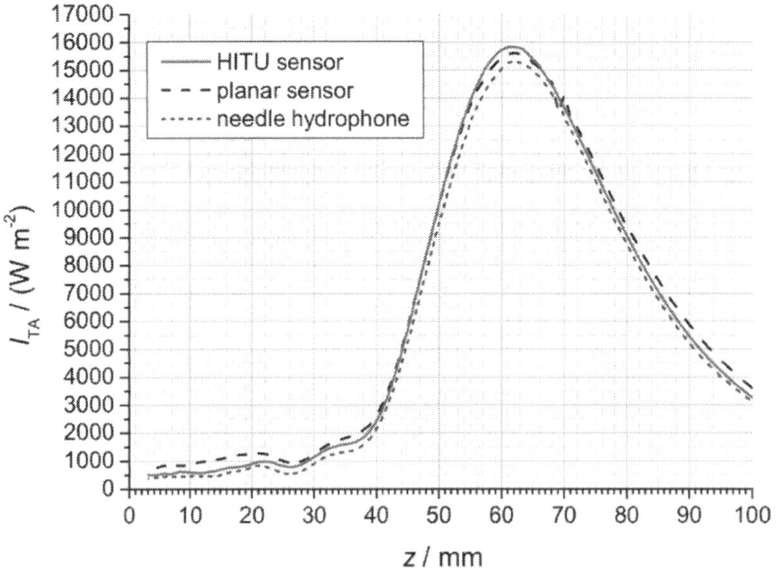

Figure 3. Beam plots in the axial direction obtained by two different thermal sensors and by needle hydrophone measurements for the high intensity pulse Doppler mode of a diagnostic 3.5 MHz convex array transducer.

4. Measurement of HITU fields

The performance of the HITU intensity sensor was tested in the field of a 64 mm diameter HITU source transducer (Sonic Concepts Inc., model H-101) driven by a signal generator (Tektronix, AFG 3251) and power amplifier (Electronics&Innovation Ltd., model 350L) at a working frequency of f = 1.06 MHz. The measurement set-up was similar to that depicted in figure 2, with the difference that the transducer was mounted at a side wall of the water tank with the sound beam axis being horizontally aligned. The measurements of the beam profiles in axial and lateral directions were performed in burst mode using 20 cycles and a pulse repetition period of prp = 1 ms for three different driving levels. The results are depicted in figure 4. The profile data are given in the form of temperature increases here, since the temperature-to-intensity transfer factor at 1 MHz has not been determined so far by an absolute calibration. The full width at half maximum beam dimensions obtained are d_x = 1.5 mm, d_y = 1.4 mm, and d_z = 9.2 mm. The beam width at focus and the position of the focus do not depend on the driving levels used. In addition to the beam plots, the continuous wave temporal-average output power P_{cw} of the transducer was determined by a radiation force balance measurement. The continuous wave power levels used were 5 W, 15 W, and 25 W. For higher driving levels, cavitation led to distortions in the thermal sensor measurements. Improvements may be possible using permanent degassing of the water and by using chemical additives. Furthermore, a smoother front face at the absorber-housing transition may shift the occurrence of cavitation towards higher driving levels.

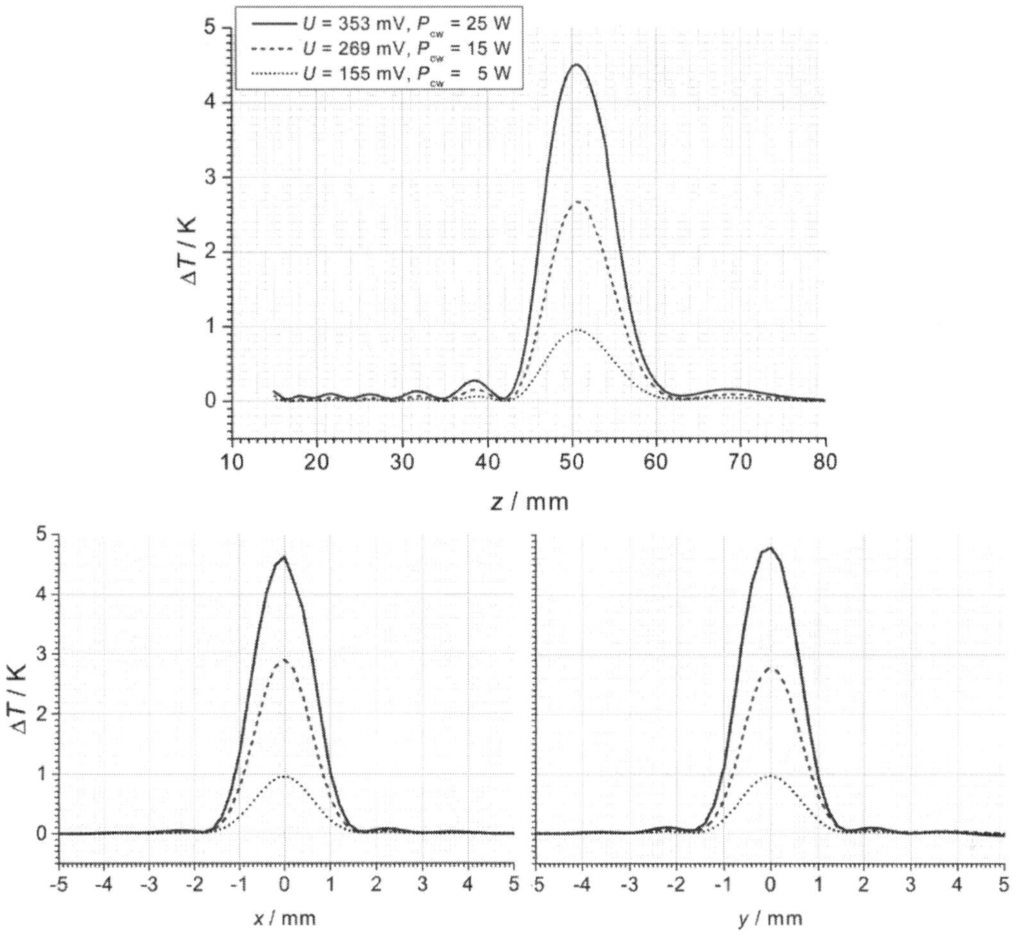

Figure 4. Beam plots in axial (upper diagram) and lateral (lower diagrams) directions obtained with the thermal intensity sensor for a 1.06 MHz HITU field at different driving levels.

5. Conclusions

A thermal sensor technique for local intensity measurements has previously been applied successfully to the acoustic output characterization of diagnostic ultrasound devices and provides an easy way of overall intensity determination. The simple principle of operation and the robustness of the sensors are principal advantages, for instance, in comparison to hydrophones, which are much more sophisticated, and the concept has been further developed here to address the specific measurement challenges in the case of high intensity therapeutic ultrasound. A new prototype with a different absorber material, reduced sensitivity, increased robustness, and less distortions formerly caused by reflections between the sensor and the transducer has been manufactured and tested. The first HITU output beam characterization measurements were performed. Currently, the measurements are limited regarding the driving level by the occurrence of cavitation leading to distortions of the beamprofiles acquired due to acoustic and thermal shielding of the sensor front face. Another limitation is given by the burst length. For duty factors above 0.3, distortions were observed. To verify the option of intensity profile measurement at reduced burst lengths compared to the clinical settings, future investigations are planned to investigate the linearity of the output power of typical HITU transducers with burst lengths up to continuos wave operation using the radiation force balance.

Acknowledgments
Financial support for the project was received from the German Bundesministerium für Wirtschaft und Technologie within the MNPQ program (Measurement, Standardisation, Testing, and Quality Assurance). Mechanical preparation of the sensor prototype was supported by the project partner Gesellschaft für Angewandte Medizinische Physik und Technik (GAMPT), Zappendorf, Germany.

References
[1] IEC 2010 *Requirements for measurement standards for high intensity therapeutic ultrasound (HITU) devices* (Geneva: International Electrotechnical Commission Publication Technical Report 62649 Ed.1)
[2] Wilkens V and Reimann H-P 2004 Output intensity measurements on a diagnostic ultrasound machine using a calibrated thermoacoustic sensor *Journal of Physics: Conference Series 1* (Advanced Metrology for Ultrasound in Medicine UK IoP Publishing) p 140-5
[3] Wilkens V 2010 A thermal technique for local ultrasound intensity measurement, part 1: Sensor concept and prototype calibration *Meas. Science Technol.* submitted for publication
[4] Wilkens V 2010 A thermal technique for local ultrasound intensity measurement, part 2: Application to exposimetry on a medical diagnostic device *Meas. Science Technol.* submitted for publication
[5] Fay B and Rinker M 1996 Determination of absolute value of ultrasonic power by means of thermoacoustic sensors *Acustica* **82** 274-9

Journal of Physics: Conference Series **279** (2011) 012019 doi:10.1088/1742-6596/279/1/012019

Calibration of HIFU intensity fields measured using an infra-red camera

A Shaw[1], V Khokhlova[2,3], S Bobkova[2], L Gavrilov[4], and J Hand[5]

[1]Acoustics and Ionising Radiation Division, National Physical Laboratory, Teddington TW11 0LW, UK
[2]Dept. of Acoustics, Physics Faculty, Moscow State University, Moscow 119991, Russia
[3]Center for Industrial and Medical Ultrasound, Univ. of Washington, Seattle 98105, USA
[4]N.N. Andreev Acoustics Institute, 117036 Moscow, Russia
[5]Imaging Sciences Dept, Imperial College London (Hammersmith Hospital Campus), London W12 0NN

E-mail: vera@acs366.phys.msu.ru

Abstract. A trend in HIFU technologies is to use 2D phased arrays that offer electronic steering of a single focus and formation of patterns of multiple foci. Conventional methods to characterize array fields using scanned hydrophone would be prohibitively slow given the potentially large number of focusing conditions. An alternative technique for rapid qualitative assessment of intensity distributions was recently developed [1, 2]. The method is based on infrared camera measurements of the temperature rises induced by low amplitude short ultrasonic bursts in a thin absorber. Here, the method is extended to estimate the absolute values of intensity in a field of a 2D 1-MHz randomized phased array. Two approaches were implemented. In the first approach it was assumed that the measured temperature rise at the surface of the absorber is proportional to the free field intensity. The second approach correlated the temperature rise measured in an absorber and calculated from the modelled acoustic field and the heat transfer equation. Corresponding correction factors between the free field intensity and temperature was obtained and introduced in the conversion of temperature images to intensity. Free field distributions in water and focusing through ribs were recorded and simulated. Good correlation between the measured and modeled results in both spatial distributions and the absolute values of intensity was demonstrated.

1. Intoduction

In recent years, there has been considerable interest in using the infrared technique to investigate thermal and acoustical fields of high-intensity focused ultrasound (HIFU) transducers [1-5]. Various other methods can be employed, such as thermocouples, MRI, and ultrasound thermometry, to measure the temperature rise and spatial temperature distribution in excised tissues. However, thermocouples are local and invasive, MRI is expensive, and acoustic methods have poor accuracy and spatial resolution. The infrared method has specific advantages in comparison with these other methods. Ultrasound field mapping in 3D is very slow, with even a single planar raster scan taking typically several hours. In addition, another trend in HIFU technologies is to use 2D phased arrays that offer electronic steering of a single focus and formation of patterns of multiple foci. Studies of numerous intensity distributions using a scanned hydrophone would be prohibitively slow in this particular situation taking into account the potentially large number of focusing conditions. Moreover, hydrophones that are used for field mapping are expensive and can be damaged in some HIFU fields. An alternative technique for rapid qualitative assessment of intensity distributions was recently developed which enable mapping of the ultrasound field for the time comparable with units or tens seconds [1, 2]. In the present paper, the method was extended to provide estimates of the absolute values of intensity from the measured spatial distributions of temperature.

2. Experimental setup

A schematic diagram of the proposed measurements of intensity distributions is shown in Fig.1. The method is based on IR camera measurements of temperature rises induced in a thin sheet of absorber with known acoustic and thermal parameters by short ultrasonic bursts of 0.1 – 0.3 s duration and relatively low acoustic power of the ultrasound radiator (a few watts acoustic power). The low output regime provided conditions of linear acoustic propagation and less than 50°C temperature rise to avoid damage of the calibrated absorber. Short duration sonication was applied to reduce diffusion effects so that the temperature rise is proportional to the intensity.

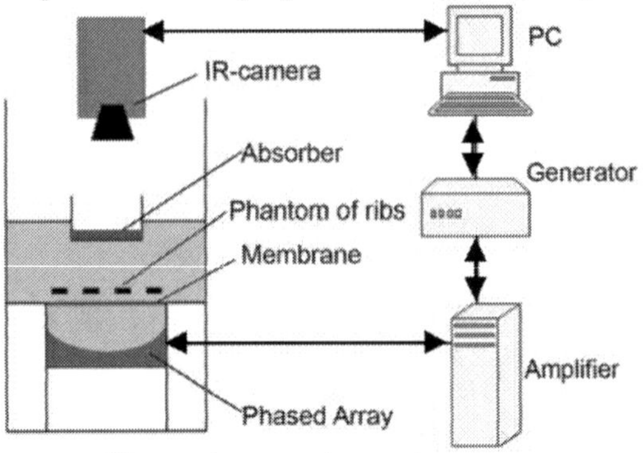

Figure 1. Experimental arrangement.

IR measurements were carried out using a modified ThermoScope® pulsed thermography system (Thermal Wave Imaging Inc., Ferndale, MI) that included a pulsed light source, a Phoenix MWIR 9705 IR camera (FLIR Systems, Boston, MA), and a PC running EchoTherm® v6.4 software [1]. The ultrasound beam was directed vertically upwards through water onto a thin (1.8 mm thickness) layer of ultrasonic absorber (Aptflex F28, Precision Acoustics, Dorchester, UK) with a single-pass attenuation coefficient of 6 dB at 1 MHz. The reflection on the water/absorber interface was -25 dB. The other side of the absorber was air-backed (fully reflective to ultrasound) and viewed from above by the IR camera. The distance between the IR camera lens and the absorber was 40 cm, the thermal resolution was 5.6 mK. Propagation was either in the free field in water or in the presence of a phantom of ribs between the array and the absorber [2].

A randomized phased array was used as the ultrasound source [1, 6]. The array consisted of 254 circular elements, each of 7 mm in diameter and 1 MHz operational frequency. The elements were distributed on a spherical surface in a quasi-random manner. The minimum centre-to-centre spacing between the elements was 7.9 mm and the largest spacing was 9.4 mm. The radius of curvature of the spherical shell and its diameter were 130 mm and 170 mm, respectively. The array was developed according to the design and parameters described in previous publications [6, 7] and fabricated from piezocomposite material (Imasonic, Voray sur l'Ognon, France).

Acoustic power of the array was measured using the radiation force balance method [1]. Generation and control of RF signals applied to the array elements were provided by a commercially available 256 channel system (Model 500-013, Advanced Surgical Systems, Inc., Tucson, AZ). The operating frequency (1 MHz), phase (0 to 360 degrees in increments of 2.25 degrees), and relative power (8 bit control) of each element of the array were downloaded to the controller from a computer. Phase and power data for each channel were generated using the theoretical data obtained in the modeling to produce and steer a single focus or create patterns of multiple simultaneous foci [1].

3. Calibration of IR measurement

The infrared method is primarily a relative method, since results are dependent on acoustic frequency and interference due to reflections from the air interface. Nevertheless, it is desirable to estimate intensity values if possible and two methods have been implemented.

In the first approach it was assumed that the initial rate of change of IR signal distribution in the measurement plane is proportional to the free-field intensity distribution with some unknown linear coefficient:

$$\frac{\Delta A_{IR}}{\Delta t} = \beta \cdot I_{water} \tag{1}$$

where A_{IR} is the IR camera signal, β is the unknown coefficient, and I_{water} is the intensity distribution in the measurement plane. The IR measurements were performed in the plane that was 5 cm below the focal plane. The acoustic power was 11 W, the duration of heating was 0.2 s, all the elements of the array were active and operated in phase to create a single focus in the center of curvature of the array. No ribs were located on the beam path. The temperature distribution was measured at the surface of the absorber facing the IR camera and the power of the IR signal, W_{IR}, which is proportional to the acoustic power, was calculated from the image:

$$W_{IR} = \sum_i \left(\Delta A_{IR}^{(i)} \cdot dS_i\right) = \beta\,\Delta t \sum_i \left(\Delta I_{water}^{(i)} \cdot dS_i\right) = \beta\,\Delta t\, W \,. \tag{2}$$

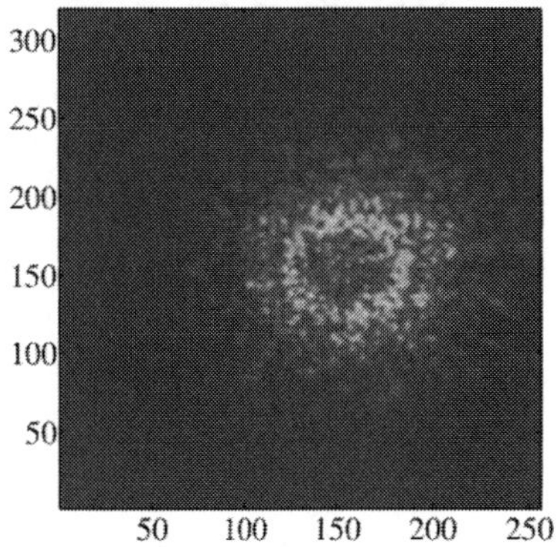

Figure 2. IR signal distribution in the measurement plane. The image size is 256x320 pixels.

Here $A_{IR}^{(i)}$ is the amplitude and dS_i is area of the i-th pixel in the IR camera image. The dimensions of the image were 256 x 320 pixels with a pixel size 0.43 x 0.43 mm (Fig.2). A value for the product $(\beta \Delta t)$ = 80 levels per W cm^{-2} was obtained from IR images (such as Figure 2) using Eq. 2 and the IR images were converted to intensity distribution using Eq. 1. This calibration method includes significant experimental uncertainty because of the noise in the image data and the small number of relatively large pixels in the image which strongly contributed to the integral in Eq.(2). Heat diffusion in the absorber was also not included in the method. Knowing that Δt = 0.2 s and that the thermal resolution was 5.6 mK/level, the coefficient β can be calculated as 400 levels/s (W/cm^2)$^{-1}$ or, more generally, the sensitivity in this configuration at 1 MHz was 2.2 K/s (W/cm^2)$^{-1}$.

To evaluate the effect of diffusion and to validate the hypothesis that the temperature rise is proportional to the intensity, the second approach applied. The approach included three main steps: modelling of temperature rise in the absorber, comparison of the IR measured temperature with modelling, and correlation of modelled and IR measured temperature with free field intensity in water.

The heat transfer equation for the temperature rise T

$$\frac{\partial T}{\partial t} = \chi_1 \Delta T + \frac{q(x,y,z)}{\rho C_p} \tag{3}$$

was solved in the absorber layer of l = 1.8 mm thickness whilst

$$\frac{\partial T}{\partial t} = \chi_2 \Delta T \tag{4}$$

was solved for the temperature rise in the water layer adjacent to the absorber. In the equation (3), $q(x,y,z)$ is ultrasound induced heat deposition in the absorber, $\chi_1 = 1.92 \times 10^{-7}$ m^2 s^{-1} is the thermal diffusivity, $\rho = 1010$ kg m^{-3} is the density and $C_p = 1800$ J kg^{-1} K^{-1} is the specific heat capacity of the absorber. In Eq. (4) $\chi_2 = 1.44 \cdot 10^{-7}$ m^2 s^{-1}. Equations 3 and 4 were solved using a finite difference method [7]; the boundary condition at the absorber/air interface was chosen as $\partial T/\partial z = 0$ assuming no heat conduction to the air. The layer of water in calculations was assumed to be thick enough so that no temperature rise, $T = 0$, occurs at the outer boundary of the layer. Heat deposition $q(x, y, z)$ in the absorber, Eq. (3), was calculated using the results of modeling the array field in water, whilst that in water (Eq. (4)) was assumed to be negligible. Reflections from the water/absorber interface were ignored but total reflection from the absorber/air interface was included in modeling the intensity field in the absorber. The acoustic field in the layer thus consisted of the primary and the reflected waves, with absolute values of intensities decreasing in the corresponding direction of propagation due to absorption in the layer, $I(x, y, z) = I_{water}(x, y, z)(\exp(-2\alpha z') - \exp(-4\alpha l + 2\alpha z'))$. Heat deposition $q(x,y,z)$ inside the layer was calculated as a combination of the absorption of the primary and the reflected waves:

$$q(x, y, z)) = -\frac{\partial I(x, y, z)}{\partial z} = 2\alpha I_{water}(x, y, z)(\exp(-2\alpha z') + \exp(-4\alpha l + 2\alpha z')). \quad (5)$$

Here $\alpha = 346$ m^{-1} is the absorption coefficient in the layer, $z' = z - F + l$ is the propagation distance inside the layer, $z' = 0$ at the water/absorber and $z' = 1.8$ mm at the absorber/air interfaces. Calculation of temperature rise was performed with and without accounting for diffusion effects during the period of heating $\Delta t = 0.2$ s. If diffusion is negligible, then the energy deposition and temperature rise in the absorber are linearly related and the distribution of temperature rise at the surface of the layer is:

$$T = \frac{4\alpha \Delta t}{\rho C_{p_v}} I_{water}(x, y, F) \exp(-2\alpha l). \quad (6)$$

If diffusion is included in the calculations, the intensity distributions are not exactly proportional to the temperature change, but simple corrections to the measured temperature distributions can be made. The results of temperature rise modeling in the focal plane accounting for diffusion were related to those that are proportional to the intensity, Eq. (6), and a corresponding correction factor between the intensity $I_{water}(x, y, F)$ and temperature was obtained.

The second method was implemented when the absorber was located in the focal plane ($z = F = 130$ mm) and the array operated so that to enable propagation of ultrasound through ribs [4]. The essence of the approach was based on switching off the array elements that are located in front of ribs. Figure 3 shows simulated spatial distributions of temperature rise in the axial plane that is perpendicular to the direction of the ribs. The upper distribution corresponds to calculations without accounting for diffusion, and the lower distribution shows the results of simulations when diffusion was taken into account. The acoustic power in the simulations was 11 W, the duration of heating was 0.2 s, and the number of active array elements was 114. The modeling was made for the case as if the ribs were located (but not physically present) at $z_0 = 45$ mm from the radiator. The spatial splitting of the focus caused by the periodic spatial structure of the active array elements is seen from this figure.

In both cases, the distribution of temperature rise inside the absorber is not uniform because of the strong absorption in the layer. The temperature is higher close to the surface that is in contact with water and lower in the plane of IR measurements, $z = F$. If diffusion is neglected, the maximum temperature rise occurs at the water/absorber interface, where intensity is the highest (Fig. 3a). If diffusion is included in the calculations, the heating moves outward from its maximum, some temperature rise in water is observed, and the maximum of temperature is shifted inside the layer (Fig. 3b).

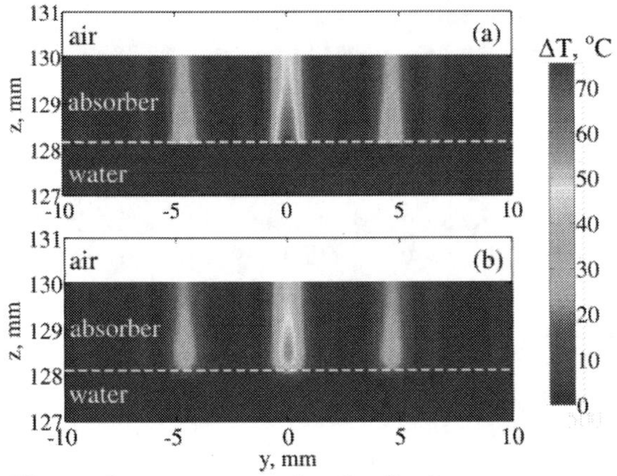

Figure 3. Temperature rise distributions in the absorber modelled without (a) and with (b) account for diffusion.

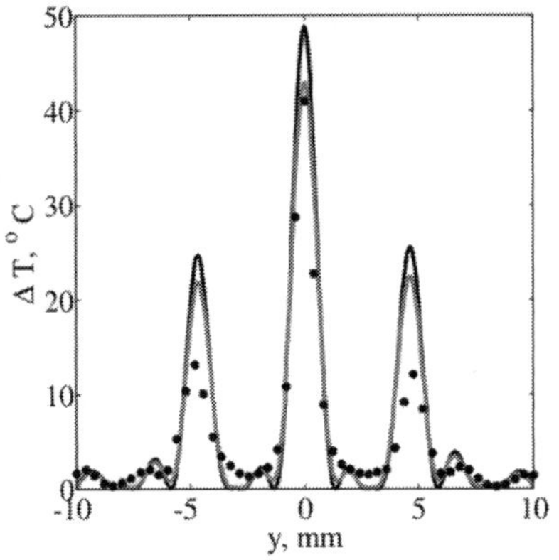

Figure 4. Temperature rise distribution in the focal plane calculated without (black) and with diffusion (red) and measured using IR camera (dots).

The simulated temperature rise along the transverse y axis at the absorber/air interface is shown in more detail in Fig. 4 and compared with measurements (dots). The results of modeling were obtained without (black curve) and with (red curve) including the diffusion effect in calculations. It is seen that for 0.2 s of heating the peak temperature in the main focal lobe modeled without diffusion (i.e. proportional to intensity) is 13 % higher than the peak temperature modeled with the real diffusion taking place under experimental conditions. The broadening of the main focal lobe due to diffusion is very small for such a short period of time: from 1.40 to 1.47 mm, i.e. only 5%. Here, the width of the lobe is defined at the $1/e$ level of its maximum. In converting from temperature to intensity this broadening of the beam was neglected and intensity was considered to be proportional to the temperature. However, the correction factor that accounts for 13% lowering of the peak intensity due to diffusion was applied in processing the data. The results of modeling taking the effect of diffusion into account agree well with the measured data within the main focal lobe. Some differences between measured and predicted temperature data can be seen at the side foci (Fig. 4).

The correction factor to convert temperature rise to intensity was defined as the ratio of peak temperatures obtained in simulations without and with accounting for diffusion. If the diffusion is negligible, then free field intensity is related to the temperature rise as in the Eq. (6) (black curve Fig. 4):

$$I_{water} = \frac{\rho C_v}{4\alpha} \cdot \exp(2\alpha l) \cdot \frac{T}{\Delta t} \,. \tag{7}$$

If the 13% difference of peak temperatures referred to above, Fig. 4, is taken into account, then the Eq. (6) is corrected as:

$$I_{water} = K \frac{\rho C_v}{4\alpha} \cdot \exp(2\alpha l) \cdot \frac{T}{\Delta t} \,, \tag{8}$$

where the correction coefficient $K = 1.13$. The equation (1) used in the first method can be rewritten in the form of Eq. (8) and the correction coefficient K can be calculated as a combination of known physical parameters and the value of the coefficient β. This value of the correction coefficient K in the first method was found to be 0.9, which is within 20% of the value obtained using the second method, thus providing an independent test of the validity of the second modeling-based approach. In further experiments in the presence of a phantom of ribs, the second method was used to recalculate the temperature rise distribution in IR camera images to the intensity distribution.

4. Correlation of modelled and IR measured intensity distributions

An example of comparing the intensity distributions, modelled, and estimated from the measurements, is shown in Fig.5. The intensity distributions are presented in the focal plane (z=130 mm) in the presence of the phantom imitating the rib cage for a single focus at (0, 0, 130 mm): (a) - simulated; (b) - measured by the IR-method; (c) – 1D distributions of the corresponding quantities over the vertical coordinate y perpendicular to the direction of ribs for theory (black) and experiment (red). Measurements were carried out for the acoustic power of 11 W and time of heating of 0.2 s. Theory and experiment are in reasonably good agreement. The discrepancy in the level of intensity may be caused by relatively high noise level in the IR measurements and low precision in positioning of the absorber.

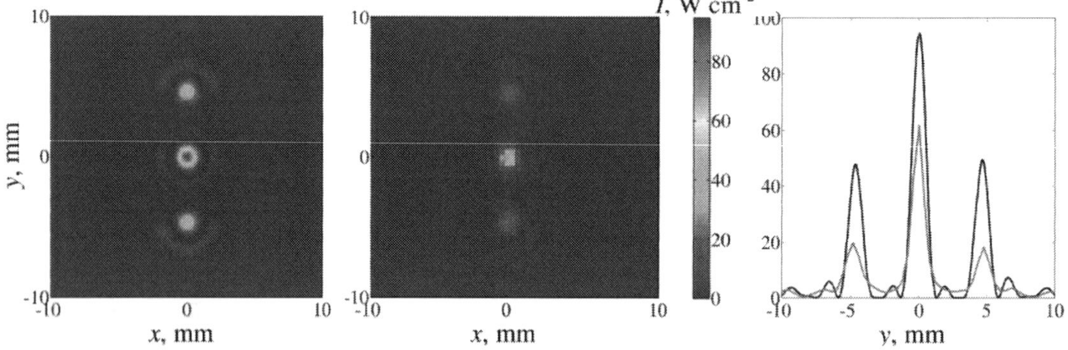

Figure 5. Intensity distributions in the focal plane in the presence of the phantom of ribs for a single focus positioned at (0, 0, 130 mm): (a) - simulated; (b) - measured by IR-method. (c) – 1D distributions.

5. Conclusions

Two independent methods to estimate free field intensity distributions from IR camera measurements were proposed and tested in a field of a 2D 1-MHz randomized phased array. Corresponding correction factors between the intensity and temperature were obtained and introduced in the conversion of temperature images to intensity. The results of the two calibration methods were found to agree to within 20% of each other. Good correlation between the measured and modeled results in both spatial distributions and the absolute values of intensity was demonstrated. This suggests that measurement of temperature changes using an IR-camera is a rapid and effective method to obtain spatial distributions and estimates of absolute values of ultrasound intensity in HIFU experiments.

Acknowledgements
This work was supported by Cancer Research UK-C18509/A78, NIHR BRC, RFBR 09-02-00066, 10-02-91062-NTsNI, UMNIK, and NIH R01EB007643; and by the National Measurement Office of the UK Department for Business, Innovation and Skills.

References

[1] Hand J, Shaw A, Sadhoo N, Rajagopal S, Dickinson R, Gavrilov L 2009 *Phys. Med. Biol.*, **54**, 5675
[2] Bobkova S, Gavrilov L, Khokhlova V, Shaw A and Hand J 2010 *Ultrasound Med Biol* **36**(6) 888
[3] Khokhlova V A, Miller N, Ollos R, Martin R W, Bailey M R, Mohammadian Y and Naghavi M 2001. *Proc. 17th International Congress on Acoustics* (Rome, Italy, Sept 2-7, 2001)
[4] Zderic V, Foley J, Luo W and Vaezy S 2008 *Med. Phys.* **35**(10) 4292
[5] Shaw A and Nunn J 2010 Phys.Med.Biol., **55** (11) N321
[6] Gavrilov LR, Hand JW 2000 IEEE Trans Ultrason Ferroelectr Freq Ctrl **47**(1):125
[7] Filonenko EA, Gavrilov LR, Khokhlova VA, Hand JW 2004 *Acoustical Physics* **50**(2) 222

Journal of Physics: Conference Series **279** (2011) 012020 doi:10.1088/1742-6596/279/1/012020

Estimation of the tissue lesion induced by a transmitter with aluminium lens

Tingbo Fan, Dong Zhang, Xiufen Gong

Institute of Acoustics, Key Laboratory of Modern Acoustics(Nanjing University), Ministry of Education, Nanjing University, Nanjing 210093, China

dzhang@nju.edu.cn

Abstract. A virtual source model is developed to describe the nonlinear acoustic field generated by a transmitter with aluminium lens. This model assumes the geometry focal length of the virtual concave transducer to be equal to the acoustic focal length of the real transducer. Thus the acoustic field in the focal area can be accurately presented by that generated by the virtual source. This model in combination with the bio-heat equation is applied to the size estimation of tissue lesion created by high intensity focused ultrasound. A transmitter with aluminium lens is used to create tissue lesions for bovine livers in vitro under three different focal acoustic intensities of 7, 13.1, 25.4 kW/cm2. The predicted tissue lesion volumes coincide well with those previously reported measured results for 7 kw/cm2, but are smaller those in the measurement for 13.1 and 25.4 kW/cm2 owing to effect of vapor bubbles.

1. Introduction

High Intensity Focused Ultrasound (HIFU) is a noninvasive treatment that delivers high intensity ultrasound energy into the targeted area of tissue within the body without damage to intervening tissue. The principle of this technique is to raise the tissue temperature to relatively high values and cause thermal coagulation and ablation of cells by using an ultrasonic focusing transducer. HIFU devices are currently under investigation as an effective treatment modality for ablating solid tumors of the prostate, liver, breast, kidney, and brain. In order to provide a focus with prescribed depth and shape in HIFU therapy, single element therapy transducers are designed using a flat piezoelectric element with a lens, a spherically curved element with concavity in the propagation direction, or a combination of transducer curvature and a lens. The first type of transmitter has been widely used in the clinical applications owing to its facile manufacture. Previously, Wang et al used a 1.6 MHz flat annular piezoelectric element and an aluminium lens as a HIFU transmitter, and three kinds of acoustic power were applied to create tissue lesions in bovine liver tissues [1]. As we know, it is highly desirable in HIFU therapy to produce lesions with well-controlled spatial characteristics. To achieve this purpose, it is important of well understanding of the acoustic field generated by the transmitter.

The Westervelt equation is usually used to describe the finite amplitude ultrasound propagation in a medium [2]. Some approximations can be employed to reduce the complexity of numerical calculation. The widely accepted Khokhlov-Zabolotaskaya-Kuznetsov (KZK) equation can be derived under paraxial approximation [3]. It can accurately describe the acoustic field generated from a transducer with half aperture angle less than 16° [4]. For strongly focused finite amplitude sound beams, the spheroidal beam equation (SBE) is developed for describing the sound propagation [5]. Unfortunately, since the half aperture angle of this kind of transducer study is larger than 16°, the KZK equation is not suitable for describing the acoustic field. On the other hand, as the phase on the lens surface is not uniform, the SBE model cannot be directly used to describe the acoustic field. For this reason, we put forward a virtual source approach to predict the acoustic field generated from this transmitter. To examine the validity of this theoretical model, we apply the same transmitter used in the previous experimental study [1] to the numerical simulation. Three different kind of focal acoustic intensities are used in to create tissue lesions for 20s, and the predicted tissue lesion volumes are compared with the published in vitro results.

2. Methods

Figure 1 illustrates a 1.6 MHz single element transmitter (diameter a_2=7.5 cm) used in the simulation, whose configuration is the same as that in the previous experiment [1]. It consists of a flat annular piezoelectric element and an aluminium lens. An aperture (a_1=4.0 cm) in the center of the transducer is used to mount the imaging transducer confocally to monitor therapeutic processes. OD=9.3 cm represents geometry focal length of aluminum lens and OF=11.90 cm is acoustic focal length which can be calculated using Rayleigh integral.

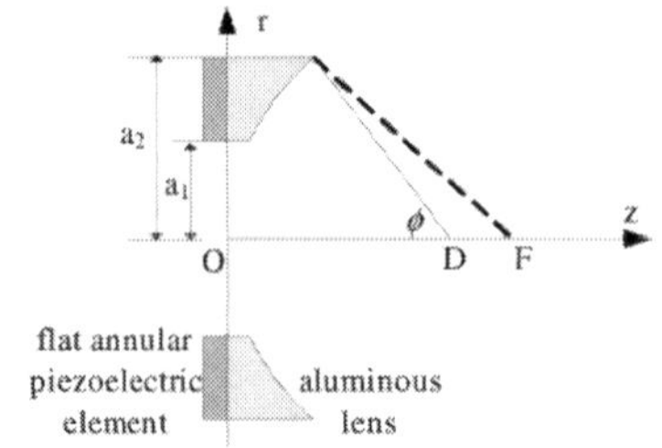

Fig.1 Geometry of the transmitter used in the simulation. The transmitter consists of a flat annular piezoelectric element and an aluminium lens.

Since the half aperture angle of the transducer is larger than 16°, the KZK equation is not suitable for describing the acoustic field. On the other hand, as the phase on the lens surface is not uniform, the SBE model cannot be directly used to describe the acoustic field. For this reason, we put forward a virtual source approach to predict the acoustic field generated from this transmitter. Points D and F represent the geometry focuses of the real source and virtual source respectively. The inner and outer apertures of the virtual source are a_{v1}=4.07 cm and a_{v2}=8.09 cm respectively, which can be calculated easily using geometric relations. The geometry focal length of the virtual concave transducer is set to be equal to the acoustic focal length of the real transducer with lens, and the phase on the virtual source is assumed to be uniform. In this way, the SBE model can be applied to predict the acoustic field generated by the virtual source. Under this hypothesis, only the acoustic field in the focal area can be approximately presented by the virtual acoustic field. The acoustic field generated by the virtual source is described using SBE model.

The temperature rise in soft tissues can be modeled by a diffusion type equation. The Pennes bio-heat transfer equation (BHTE) is used to describe the thermal effects of ultrasound in tissue [6], [7].

$$\frac{\partial T}{\partial t}=\frac{1}{\rho_0 C_t}\nabla(k_t\nabla T)-\frac{W_b C_b}{\rho_0 C_t}(T-T_a)+\frac{Q_v}{\rho_0 C_t}, \tag{1}$$

where ρ_0 is the ambient tissue density and T is the temperature in the tissue; the thermal conductivity and specific heat at constant pressure are k_t and C_t. The second term on the right-hand side accounts for heat loss via tissue perfusion. The perfusion constant for a given tissue is W_b and C_b is specific heat of blood. T_a is the ambient temperature. The last term is heat source due to the acoustic field, which can be calculated by:

$$Q_v = 2\alpha I = \frac{1}{\rho c_0}\sum_{n=1}^{N} 4\alpha_0 n^{\mu} |C_n|^2 \quad (2)$$

where N is the harmonic number used in calculation, C_n is the nth harmonic complex amplitude of pressure.

Equation (1) is solved in cylindrical coordinate using the operator-splitting paradigm [8]. Constant flux boundary conditions are assumed for the edges of the computational domain to simulate the extended thermal region beyond the computational boundaries of simulations. The step sizes are set to be 0.1mm along the wave propagation direction and 0.06 mm perpendicular to the propagation direction. The concept of thermal dose (TD) developed by Saparteo and Dewey provides a quantitative description on the heating of tissue [9]. The expression of the TD can be written as:

$$TD(x,y,z) = \int_{t=t_0}^{t=t_1} R^{43-T(x,y,z)} dt \quad (3)$$

where TD is the thermal dose in equivalent minutes at 43°C, t_0 and t_1 are the starting and ending time of the ultrasound exposure; R is 0.5 if T>43°C, and 0.25 if T<43°C. Results of previous studies on the effects of hyperthermia have shown that the threshold for 100% necrosis in different tissues ranges 50-240 minutes [10], and we consider 240 minutes as the threshold for lesion formation in this work.

3. Results

To examine the validity of this virtual source model, we first calculate the linear acoustic field in pure water using the virtual source model, and make a comparison with that from Rayleigh integral. In the numerical computation, the acoustic parameters for water [11] are: ρ_0=1000 kg/m3, c_0=1486 m/s, α =0.025 Np/m/MHz^2, and the acoustic nonlinearity is assumed to be zero. Figure 2 presents the axial and radial pressure profiles normalized by the peak pressure at the transducer surface, where 3a and 3b are axial and radial distributions respectively, and solid lines are calculated by Rayleigh integral, and dash lines are calculated by the virtual source model. Figures 2c and 2d compares the two-dimensional (2D) acoustic spatial distributions between Rayleigh integral and virtual source model. The good agreement of the linear acoustic field in the focal area indicates that it is possible to simulate the temperature distribution and lesion formation on the basis of the virtual source model.

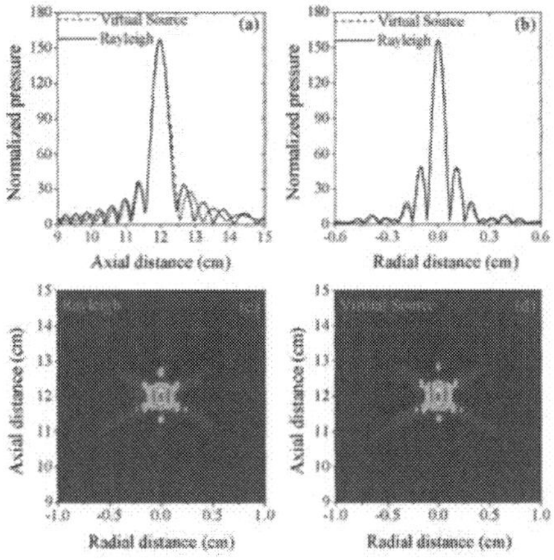

Fig. 2 The normalized pressure profiles: (a) axial profile, (b) radial profile at focal plane; The two-dimensional acoustic spatial distributions: (c) by Rayleigh integral, and (d) by virtual source model.

In the previous study [1], the transmitter (Fig. 1) was used to be the HIFU source, and three different focal acoustic intensities of 7, 13.1, 25.4 kW/cm^2 were applied to create tissue lesions in bovine liver tissues in vitro. For further examination of the validity of the virtual source model, we calculate the thermal dose and induced tissue lesions, and make a comparison with the published experimental results. In the numerical calculation, the values of the acoustic parameters are [11]: ρ_0=1000 kg/m3, c_0=1486 m/s, α =0.025 Np/m/MHz^2, β=3.5, for water and ρ_0=1214 kg/m3, c_0=1614 m/s, α =4.3 Np/m/MHz$^{1.266}$, β=4.775, for liver tissue; the thermal properties are: kt=0.508 W/m/°C, Ct=3810 W.s/m3/°C and Wb=0 for bovine liver tissue in vitro. Changes in the acoustic parameters of tissue due to HIFU-induced heating are not considered in the simulation.

Figure 3 gives the calculated lesion volume created by HIFU at various focal acoustic intensities for 20s. For comparison, the experimental results (from Ref. 1) are also plotted as solid points in this figure. We can find that in Fig. 3a, the calculated result coincides well with the experimental result; however deviations are found from 15 to 20 s in Fig. 3b and from 10 to 20 s in Fig. 3c. Especially in Fig. 6c, the measured lesion volume is almost twice than the calculated volume at 20 s. The deviations probably come from the reason of boiling bubbles which enlarge the lesion volume.

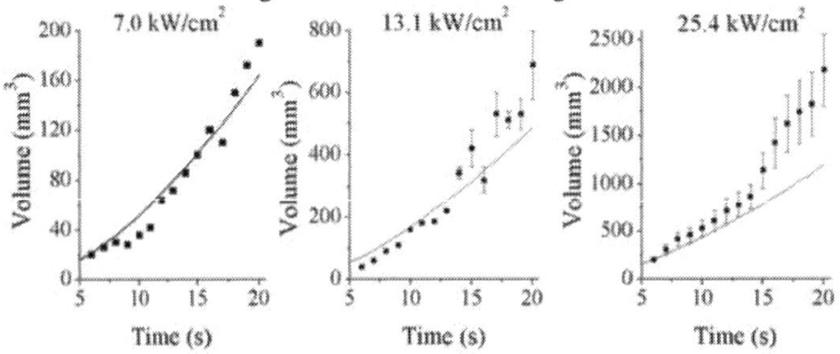

Fig. 3 Comparison of the lesion volume gathered in experiment (dot line) to calculated result (solid line) for intensity is (a) 7.0, (b) 13.1 and (c) 25.4 kW/cm2 during 20 s exposure.

Previous studies indicated that boiling is approached when tissue temperature reaches up to 100°C, and millimeter-sized boiling bubbles are observed [12][13]. HIFU-induced boiling is the responsible for distortion of lesion shape which changes to drop-shape, and results in rapid growth from the focus towards the transducer. Simulated results indicated that tissue temperature in focal region rises up to 100°C in 1.82, 0.41, and 0.07 s for the used focal acoustic intensities of 7.0, 13.1, and 25.4 kW/cm2 in this work. The divergence between measured and simulated volume mainly comes from the boiling reason. For relative large intensity, the focal temperature rises up to boiling point more quickly and thus impact on lesion volume is more serious.

4. Conclusions

In this paper, a virtual source model is proposed to predict the nonlinear acoustic field generated from a transmitter with aluminum lens. The geometry focal length of the virtual concave transducer is set to be equal to the acoustic focal length of the real transducer. As a result, the acoustic field in the focal area can be presented by that generated by the virtual source. Since the acoustic power concentrates mostly in focal region and the induced temperature rise during HIFU treatment depends on the focal heat deposition; the deviations of near and far field acoustic distributions does not impact the temperature rise distribution in HIFU therapy prediction, this theoretical model can be applied to the tissue lesion predictions in HIFU treatment. For examine its validity, the linear acoustic field in focal region calculated from this model is in good agreement with that obtained by Rayleigh integral. Further, we extend this model to predict the nonlinear acoustic field and the induced tissue lesion generated by the transmitter with aluminum lens. The lesion volume acquired from simulation coincides well with that acquired from experiment at relative low focal acoustic intensity; however, with the increment of focal acoustic intensity, the divergence between experiment and calculation becomes more obvious owing to the effect of boiling bubbles. Also it should be noted that the further complete model should consider the behavior of temperature effects on tissue physical parameters.

5. Acknowledgements

This work is supported by the National Basic Research Program 973 (Grant No. 2010CB732600) from Ministry of Science and Technology, China, .and the National Natural Science Foundation of China (10774071 and 10974093).

References

[1] Z. B. Wang, J. Bai, F. Q. Li, Y. H. Du, S. Wen, K. Hu, G. H. Xu, P. Ma, N. G. Yin, W. Z. Chen, F. Wu and R. Feng, "Study of a "biological focal region" of high-intensity focused ultrasound," Ultrasound Med. Biol., vol. 29, pp. 749-754, May, 2003.

[2] P. J. Westervelt, "Parametric acoustic array," J. Acoust. Soc. Am., vol. 35, pp. 535-537, Apr., 1963.

[3] S. I. Aanonsen, T. Barkve, J. N. Tj□tta and S. Tj□tta, "Distortion and harmonic generation 'm the nearfield of a finite amplitude sound beam," J. Acoust. Soc. Am., vol. 75, pp. 749-768, Mar., 1984.

[4] T. S. Hart and M. F. Hamilton, "Nonlinear effects in focused sound beams," J. Acoust. Soc. Am., vol. 84, pp. 1488-1496, Oct., 1988.

[5] T. Kamakura, T. Ishiwata, K. Matsuda, "Model equation for strongly focused finite-amplitude sound beams," J. Acoust. Soc. Am., vol. 107, pp. 3035-3046, Jun., 2000.

[6] H. Pennes, "Analysis of tissue and arterial blood temperature in the resting human forearm," J. Appl. Phys., vol. 1, pp. 93-122, Aug, 1948.

[7] M. C. Kolios, A. E. Worthington, M. D. Sherar and J. W. Hunt, "Experimental evaluation of two simple thermal models using transient temperature analysis", Phys. Med. Biol., vol. 43, pp. 3325-3340, Nov., 1998.

[8] Y. Jing and R. O. Cleveland, "Modeling the propagation of nonlinear three-dimensional acoustic beams in inhomogeneous media," J. Acoust. Soc. Am., vol. 122, pp. 1352-1364, Sep. 2007.

[9] S. A. Sapareto and W. C. Dewey, "Thermal dose determination in cancer therapy," Int. J. Radiat. Oncol. Biol. Phys., vol. 10, pp. 787-800, Jun. 1984.

[10] N. J. McDannold, R. L. King, F. A. Jolesz and K. H. Hynyen, "Usefulness of MR imaging-derived thermometry and dosimetry in determining the threshold for tissue damage induced by thermal surgery in rabbits," Radiology, vol. 216, pp. 517-523, Aug. 2000.

[11] F. A. Duck, Physical properties of tissue. New York: Academic Press, 1990.

[12] T. D. Khokhlova, M. S. Canney, D. Lee, K. I. Marro, L. A. Crum, V. A. Khokhlova and M. R. Bailey, "Magnetic resonance imaging of boiling induced by high intensity focused ultrasound," J. Acoust. Soc. Am., vol. 125, pp.2420-2431, Apr., 2009.

[13] V. A. Khokhlova, M. R. Bailey, J. A. Reed, B. W. Cunitz, P. J. Kaczkowski and L. A. Crum, "Effects of nonlinear propagation, cavitation, and boiling in lesion formation by high intensity focused ultrasound in a gel phantom," J. Acoust. Soc. Am., vol. 119, pp. 1834-1848, Mar. 2006.

A passive acoustic monitor of treatment effectiveness during extracorporeal lithotripsy

F Fedele[1], K Thomas[2], TG Leighton[3], S Ryves[2], D Phillips[2] and AJ Coleman[1]

[1]Medical Physics Department, Guy's and St Thomas NHS Foundation Trust, Westminster Bridge Road, SE1 7EH, London, UK
[2] Urology Department, Guy's and St Thomas NHS Foundation Trust, Great Maze Pond, SE1 9RT, London, UK
[3]Institute of Sound and Vibration Research, University of Southampton, Highfield, S017 1BJ, Southampton, UK

E-mail: fiammetta.fedele@gstt.nhs.uk

Abstract. Although extracorporeal shockwave lithotripsy (ESWL) has now been in the clinic for at least three decades, there has been little advance in efforts (i) to estimate the efficacy of the treatment whilst it is in progress, or (ii) to determine the end-point of a treatment session in terms of the degree of stone fragmentation achieved. Previous *in vitro* experimentation and clinical trials have shown that a passive acoustic monitor has the potential to provide evidence of the effectiveness and end-point of lithotripsy. The system exploits secondary emissions generated during shock-tissue interaction, whose features depend on the quality of tissue at the beam focus. This prototype was developed into the first commercially available clinical ESWL treatment monitor (Precision Acoustic Ltd, Dorchester, UK), and a unit has been acquired and tested in the clinical routine by urologists at Guy's and St Thomas NHS Trust in March 2009. This paper critically assesses the performance of the new system for the first 25 treatments monitored. The ESWL monitor correctly predicted the treatment outcome of 15 of the 18 treatments that were followed-up clinically. In addition, it was noted that the measure of treatment effectiveness provided by the monitor after 500 shocks was predictive of the final treatment outcome ($p< 0.001$). This suggests that the system could be used in pre-assessment; indicating if the stone is susceptible to ESWL or if the patient should be sent for surgery.

1. Introduction

Extracorporeal shockwave lithotripsy (ESWL) has been used since the 1980s for the non-invasive treatment of urinary stones [1, 2]. A patient's stone is fragmented by means of thousands of ultrasound shock waves, administered with at a rate of about 1-2 per second [1, 2]. Even after three generations of commercial lithotripters there has been little advance in providing the operator feedback on the procedure success [1, 2].

In the early nineties Coleman et al. identified a characteristic double-burst acoustic emission structure (Figure 1) arising form the focus of lithotripters, driven by the shock-tissue interaction [3-4]. A research team formed by members of both Guy's and St Thomas NHS Foundation Trust (GSTT) and the University of Southampton has since focused on analyzing the characteristics of these double-burst

emissions by means of both computational fluid dynamics models (CFD) [5, 7-8] and in vitro experiments [4,5, 10-15]. These studies proved that the emissions, which are formed by a mixture of components [13, 16-18], showed features dependent on stone targeting and fragmentation [9-14]. In particular, one feature, the time between the two bursts of emission, known as the collapse time (t_c), was shown to be sensitive to targeting, whilst the ratio of the amplitude of the two bursts, m_2/m_1, could be used to classify different stages of stone fragmentation [13-15]. This knowledge was exploited to design a first passive prototype acoustic system (PAS) for monitoring lithotripsy procedures in collaboration with Precision Acoustics Ltd. (PA, Dorchester, UK) [5, 10-12]. The prototype design was completed and tested during a two stage clinical trial on 79 patients from January 2006 and June 2007 [5, 13-15]. After some re-design of its components to facilitate further use in the clinical environment, a first commercial prototype [19-21], was made available to the Urology Department of GSTT, for clinical evaluation in March 2009. A comprehensive report on the development of the PAS system into a commercial treatment monitor for ESWL can be found at http://www.isvr.soton.ac.uk/fdag/Litho_07/litho_07(main).htm.
This paper reports on the first attempts to use the first commercial clinical prototype routinely in the clinic.

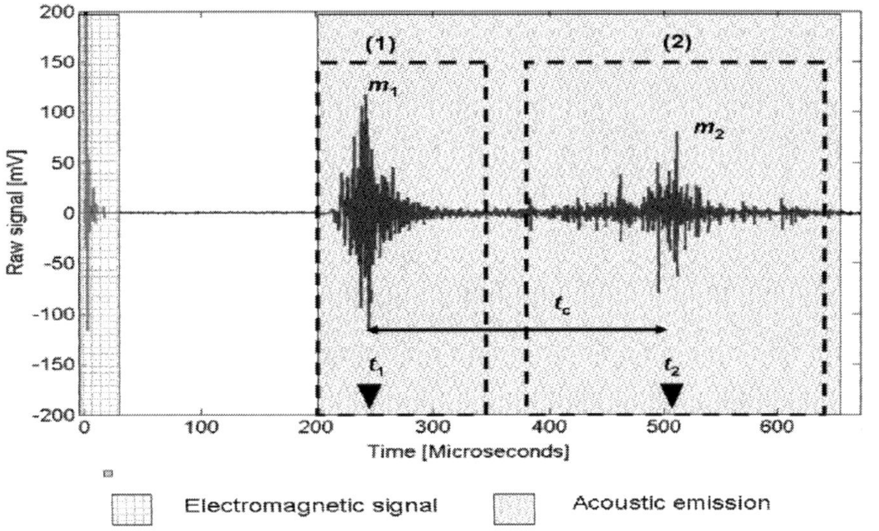

Figure 1. Typical signal acquired for a single ESWL treatment shock. An electromagnetic triggering signal, is followed after about 200 μs by an acoustic emission, which lasts about 500 μs. The main features of this double burst emission are labelled: m_1: maximum amplitude of the first burst, m_2: maximum amplitude of the second burst, t_1: central time of the first burst, t_2: central time of the second burst; t_c: interval between the two bursts, named collapse time. The method of calculating the values of these parameters is detailed in references [13, 14].

2. Material and methods

The commercial clinical prototype monitor, based on the PAS system described in details in references [13-14], was used by the Stone unit at GSTT to monitor 25 treatments between March and June 2009. After obtaining the approval of GSTT Ethical Committee, 25 patients with a BMI of 27±4 were recruited and administered a dose of ~3000 shocks using a Storz Modulith SLX-MX lithotripter (Storz Medical, Tägerwilen, Switzerland) at a power settings of 5±1 and a rate of 120 shocks per minute. The monitored treatments were, for most patients, their first or second session (20/25). Unlike clinical trials of the previous monitor prototypes [5, 13-15], this study included both kidney (17) and ureteric stones (8). The ESWL treatment monitor (Figure 2) consists of four main components, which have previously been described in detail [13-14]:

- (Figure 2a) A passive acoustic probe, which includes a broadband acoustic sensor and a commercial 8dB preamplifier (PAL)
- (Figure 2b) A data conditioning module, which includes high pass noise filtering at 300 kHz, a DC coupler and a rechargeable batteries unit
- (Figure 2c) A commercially available digital scope for data acquisition and digitalization (TiePie Handyscope 3, TiePie engineering, WL Sneek, Netherlands)
- (Figure 2d) A laptop-based MATLAB™ user interface for system control and on-line monitoring (figure 2d shows a screen-shot, with explanatory labels on each area of the screen).

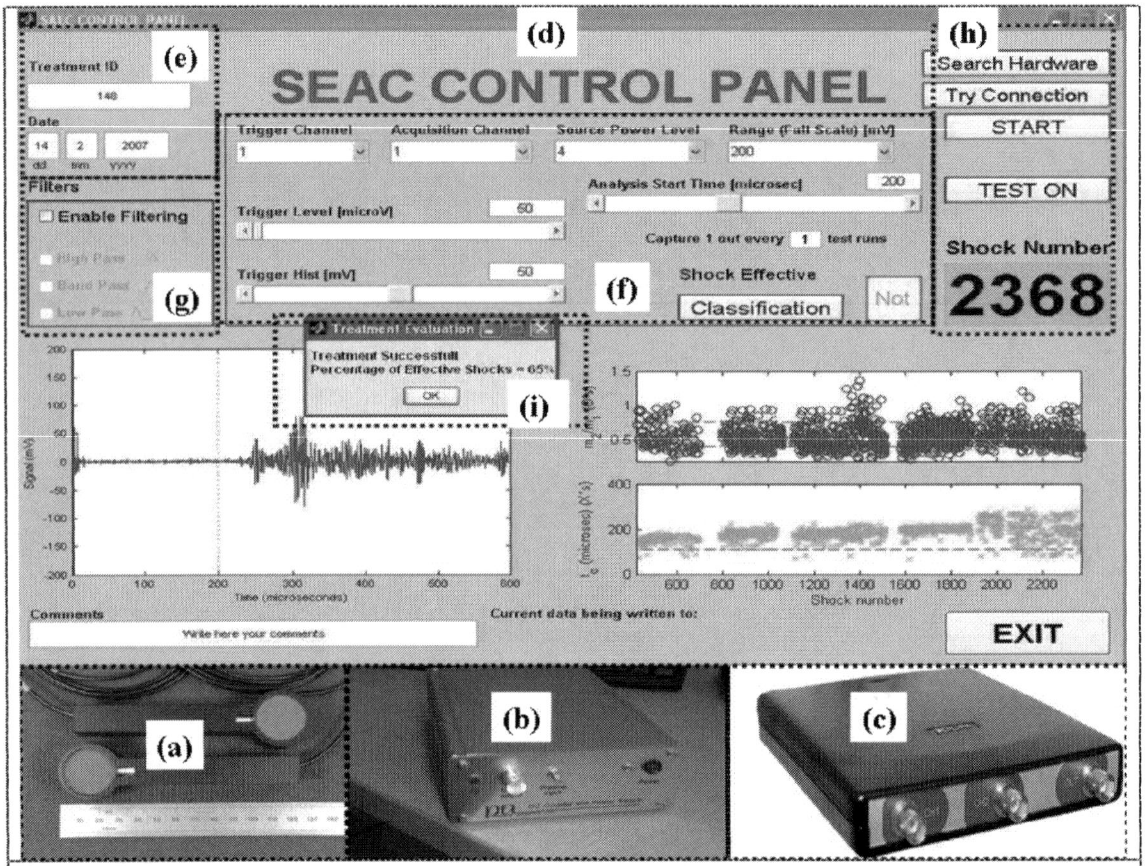

Figure 2. ESWL treatment monitor components: (a) The acoustic probes, (b) The data conditioning module, (c) the TiePie digital scope, (d) the MATLAB™ software interface (screen shot shown here). The interface (d) displays, for every shock, the shock number (panel h), the acoustic emission (left chart), the value of its main features: the ratio of the two peak amplitudes m_2/m_1 (top right chart) and the collapse time t_c (bottom right chart), and the estimated effectiveness of that particular shock (panel f). It also includes some control panels: (e) treatment details, (f) controls of the digital acquisition, (g) bank of optional digital filters and (h) a section to test the software-hardware synchronisation. Clicking with the PC mouse on the classification button in (f) reveals screen (i), which displays an estimate of the cumulative percentage of effective shocks, which is the treatment score (*TS*) used to classify a treatment. A treatment with a final *TS* of 50% or higher is considered successful. This may be done at any time during the treatment, and is always done at the end of treatment.

The passive probe (Fig 2a) is placed on the patient's flank at the beginning of the procedure. It monitors the emission generated after each shock. The data is acquired and processed on-line in real time to provide an estimate of the effectiveness of each shock and a cumulative percentage of effective shocks (Figure 2i) is used as the acoustic treatment score (*TS*) by which the treatment outcome is classified. That is to say, if at the end of the session (~3000 shocks) at least half of the monitored shocks were considered to be effective (TS_0=TS(3000)≥50%) the treatment is acoustically classified as having been '*successful*' [12-14, 18-20]. Treatments with TS_0≤30% are considered '*unsuccessful*' and those with 30<TS_0≤49% are classified as 'borderline' [12].
In this trial *TS* was estimated after each block of 500 shocks, and the correlation between the score after an initial test dose of 500 shocks, *TS*(500), and the final score TS_0 was examined. The outcome of each monitored treatment was established at the patient's follow-up appointment 1-3 weeks after the procedure. A team of 3 clinicians (blinded to the classification results from the acoustic monitor) were asked to agree and assign to each treatment an X-Ray based clinical treatment score (CTS_2, [12-14, 18-20]). CTS_2, based on the comparison of the pre–treatment stone X-Ray against the follow-up X-Ray, could range from 0 (no change in the stone) to 5 (stone disappeared). Treatments with a CTS_2≥3 (the stone was halved in size or smaller) were considered as successful. The quality of the raw acoustic data collected (which are automatically stored by the system) was also examined.

3. Results

Five treatments were excluded on the basis that the raw acoustic data showed one of the following issues: flat noisy signals (1/5) denoting coupling problems, clipped signals (2/5) and signals (2/5) associated with the saturation of the preamplifier. The system was used by a variety of users: urology consultants, research urology fellows, radiologists and clinical nurses. All these issues could be linked to the difficulty nurses had in using the system. Of the 20 cases with adequate signal, 18 could be assessed clinically.
The monitor results (Table 1) provide agreement with the clinical classification in 14/18 cases and, in particular it correctly classified 11/12 unsuccessful treatments (specificity = 92%) and 3/6 successful treatments (sensitivity = 50%). Two 'successful' treatments were misclassified and one 'successful' and one 'unsuccessful' treatment were classified as 'borderline'. The high specificity confirmed the ability of the system to identify stone not suitable for ESWL.
The two subsets of cumulative scores *TS*(500) and TS_0=TS(3000) (Figure 3) are highly correlated (R^2=0.83) and do not differ statistically (p<0.001), showing that the evaluation of the results from the monitor after a test dose of only 500 shocks, is a good predictor of treatment outcome.

4. Discussion and future work

This first clinical experience showed the ability of the commercial prototype ESWL treatment monitor to predict treatment outcome early-on during the procedure, and confirmed its particular suitability to identify unsuccessful treatments, which most likely can be associated with hard stones that are refractory to ESWL [20-21].
On the other hand, some clinical users found the system difficult to use and felt that they would have preferred a different method of visual feedback. These suggestions were exploited to redesign the visual appearance of the system. Figure 4 and Figure 5 illustrate the new design. Figure 4, in particular shows the simplified monitoring screen that will give a simpler feedback on targeting and percentage of effective shocks.

Acknowledgments
This study was generously supported by a Guy's & St Thomas' Charity GIFFT scheme (GIFFT01). The authors also wish to thank Ms Naomi Smith for her invaluable help in the first phase of this study and Mr Chris Page at the Design Department of Brunel University for his help with visual display.

Table 1. Classification of the acoustic treatment cumulative score at the end of the treatment TS_0=TS(3000) versus the treatment clinical follow-up score CTS_2. *TP*, true positive (treatment classified correctly as *successful*) *FP*, false positive (treatment classified incorrectly as *successful*), *FN*, false negative (treatment classified incorrectly as *unsuccessful*), *TN*, true negative (treatment classified correctly as *unsuccessful*).

		Treatments (as assessed at follow-up by the clinical score CTS_2)	
		Successful	*Unsuccessful*
Classification of the monitor based on the acoustic treatment score TS_0=TS(3000)	*Successful*	*TP* = 3	*FP* = 1
	Unsuccessful	*FN* = 3	*TN* = 11
	Sensitivity = 50%	Specificity= 92%	

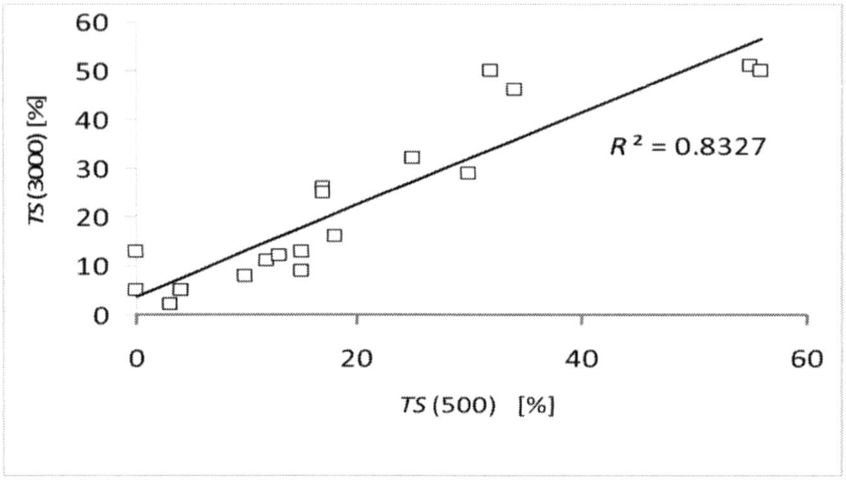

Figure 3. Correlation between the cumulative treatment score after a test dose of 500 shocks TS(500) and that at the end of the treatment TS_0=TS(3000).

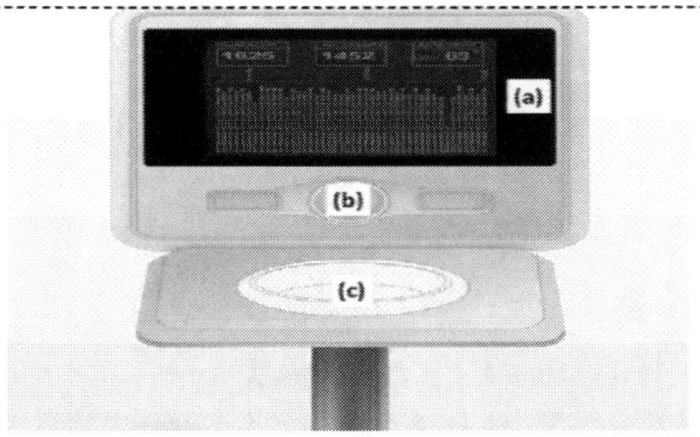

Figure 4. Visual design of the new monitoring system. (a) Monitoring screen. (b) Navigation controls. (c) Probe testing station. Picture courtesy of Chris Page, Brunel University.

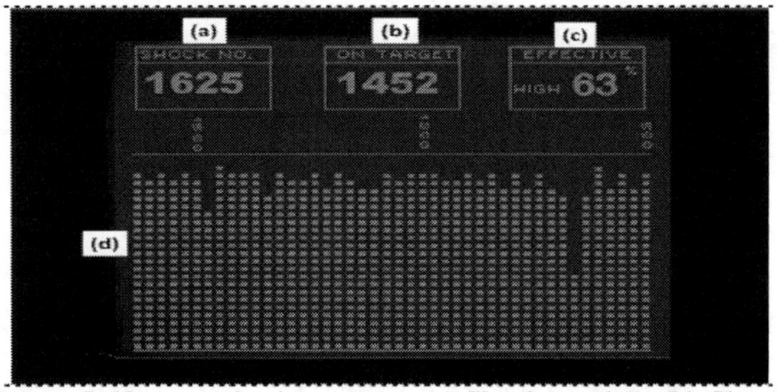

Figure 5. New monitoring screen. (a) Shock number. (b) Shocks on target. (c) Cumulative percentage of effective shocks *TS*. (d) Feedback on targeting. A shock with good targeting (t_c> 100 μs) is shown as a white dot. Shocks are grouped in columns of 25. Picture courtesy of Chris Page, Brunel University.

References

[1] Leighton TG and Cleveland RO 2010 *Proc. IMechE.* **224**(2), 317-42
[2] Skolarikos A, Alivizatos G and de la Rosette J 2006 *Eur. Urol.* **50**(5), 981-90
[3] Coleman AJ, Choi M, Saunders JE and Leighton TG 1992 *Ultrasound Med. Biol.* **18**(3), 267-81
[4] Coleman AJ, Whitlock M, Leighton TG and Saunders JE 1993 *Phys. Med. Biol.* **38**, 1545-60
[5] Leighton TG, Fedele F, Coleman AJ, McCarthy C, Jamaluddin AR, Turangan CK, Ball GJ, Ryves S, Hurrell AM, De Stefano A and White PR 2008 *Hydroacoustics* **11**, 159-180
[6] Cunningham KB, Coleman AJ, Leighton TG and White PR 2001 *Acoust. Bull.* **26**(5), 10-6
[7] Ball GJ, Howell BP, Leighton TG and Schofield MJ 2000 *Shock Waves* **10**, 265-76
[8] Turangan CK, Jamaluddin AR, Ball GJ and Leighton TG 2008 *J. Fluid. Mech.* ***598**, 1-25*
[9] Leighton TG, White PR and Marsden MA 1995 *Acta Acoustica* **3**, 517-29
[10] Fedele F, Coleman AJ, Leighton TG, White PR and Hurrell AM 2004 *Acoust. Bull.* **29**, 34-9
[11] Fedele F, Coleman AJ, Leighton TG, White PR and Hurrell AM 2004 In Proceedings of the First Conference in *Advanced Metrology for Ultrasound in Medicine. J. Phys.: Conf. Ser.* 1,134-9
[12] Fedele F, Coleman AJ, Leighton TG, White PR and Hurrell AM 2004 In IFMBE Proceedings of MEDICON and HEALTH TELEMATICS 2004: *X Mediterranean Conference on Medical and Biological Engineering, IFMBE Proceedings Series* 6 CD paper 052, p. 4
[13] Fedele F 2008 PhD Thesis University of Southampton, UK.
[14] Leighton TG, Fedele F, Coleman AJ, McCarthy C, Ryves S, Hurrell AM, De Stefano A and White PR 2008 *Ultrasound Med. Biol.* **34**(10),1651-65
[15] Leighton TG, Fedele F, Coleman AJ, McCarthy C, Ryves S, Hurrell AM, De Stefano A and PR White 2008 In Proceedings of the Second International Urolithiasis Research Symposium 256-77
[16] Zhu S, Cocks FH, Preminger GM and Zhong P 2002 *Ultrasound Med. Biol.* **28**(5), 661-71
[17] Bailey MR, Pishchalnikov YA, Sapozhnikov OA, Cleveland RO, McAteer JA, Miller NA, Pishchalnikova IV, Connors BA, Crum LA and Evan AP 2005 *Ultrasound Med. Biol.* **31**, 1245-56
[18] Owen NR, Bailey MR, Crum LA, Sapozhnikov OA and Trusov LA 2007 *J. Acoust. Soc. Am.* **121**, EL41-6
[19] Coleman A, Fedele F and Leighton TG 2009 *US Patent 12/415,733* filed 31/03/2009
[20] Smith NC, Fedele F, Thomas K, Hegarthy N, Glass J and Coleman A 2009 *Eur. Urol. Suppl.* **8**(4), 232
[21] Smith NC, Fedele F, Ryves S, Leighton TG, Coleman AJ and Thomas K 2009 *BJU Internat.* ***103**(4), 32*

Biological response *in vitro* of skeletal muscle cells treated with different intensity continuous and pulsed ultrasound fields

Viviane M. Abrunhosa[1,2], Claudia S. Mermelstein[2], Manoel L. Costa[2], Rodrigo P.B. Costa-Felix[1]

[1]*Laboratory of Ultrasound, Directory of Scientific and Industrial Metrology (DIMCI), National Institute of Metrology, Standardization, and Industrial Quality (Inmetro), Av. Nossa Sra das Graças, 50 (Prédio 1), Duque de Caxias, RJ, Brazil, ZIP 25250-020*

[2]*Laboratory of Muscle Differentiation and Cytoskeleton, Biomedical Sciences Institute, Federal University of Rio de Janeiro (UFRJ), Cidade Universitária, Rio de Janeiro, RJ, Brazil, ZIP 21949-590*

[1]rpfelix@inmetro.gov.br

Abstract. Therapeutic ultrasound has been used in physiotherapy to accelerate tissue healing. Although the ultrasonic wave is widely used in clinical practice, not much is known about the biological effects of ultrasound on cells and tissues. This study aims to evaluate the biological response of ultrasound in primary cultures of chick myogenic cells. To ensure the metrological reliability of whole measurement process, the ultrasound equipment was calibrated in accordance with IEC 61689:2007. The skeletal muscle cells were divided in four samples. One sample was used as a control group and the others were submitted to different time and intensity and operation mode of ultrasound: 1) 0.5 W/cm^2 continuous for 5 minutes, 2) 0.5 W/cm^2 pulsed for 5 minutes, 3) 1.0 W/cm^2 pulsed for 10 minutes. The samples were analyzed with phase contrast optical microscopy before and after the treatment. The results showed alignment of myogenic cells in the sample treated with 0.5 W/cm^2 continuous during 5 minutes when compared with the control group and the other samples. This study is a first step towards a metrological and scientific based protocol to cells and tissues treatment under different ultrasound field exposures.

1. Introduction

Skeletal muscle's function are locomotor activity, postural behavior and assist breathing. After direct trauma or resulting from indirect causes such as neurological dysfunction or innate genetic defects, the skeletal muscle is affected by injury (Charge & Rudnicki, 2004).

In physiotherapy, the ultrasound wave has been used over then 50 years, such as therapeutic appeal, for the treatment of injuries. Commonly, this has been used in rehabilitation for care skeletal muscle disordered. Depending on the treatment and degree of heating in tissue, this equipment can operate in two forms: athermal or thermal. The athermal ultrasound wave is characterized by increase the temperature in tissue less than 1° C, for that, the pulsed mode of application is used. The thermal effects produced in tissue are usually applied in continuous mode application (Locke & Nussbaum, 2001). However, the degree of heating depends not only in the intensity of the ultrasound wave but also in the duration of application, the correct setting of these parameters is important to avoid irreversible changes in the biological tissue (Zeqiri, 2007).

The ultrasound wave produces an acoustic vibration and this effect induces changes in the cell membrane, such as an increase in cell permeability, and changes in ionic concentrations within the cell (Locke & Nussbaum, 2001; Ter Haar, 1998). Moreover, it has been shown that the ultrasound wave induces an increase in protein synthesis and in the levels of intracellular calcium, promotes fibroblasts activity and angiogenesis, accelerates the healing of bone fractures, among other functions. Despite these effects, there is no scientific evidence to explain how the ultrasound wave interacts with tissue (Nishikori *et al.*, 2001). To avoid damage in the biological tissue, the absolute maximum effective intensity should be less than or equal to 3.0 W/cm^2 (IEC 61689, 2007). Laboratories with recognized expertise in the area have procedures and systems for reliable measurement evaluate this and other metrological parameters of physiotherapy equipment (Alvarenga & Costa-Felix, 2009). There is no agreement in the literature on the best way to treat each lesion. The present study was performed to examine the interference of different ultrasound treatment conditions on a model of *in vitro* growth of primary skeletal muscle cells.

2. Experimental Procedure

2.1. Primary skeletal muscle cell culture

Primary cultures of mononucleated cells were prepared from pectoral muscles of 11-day-old chick embryos (Mermelstein et al., 2005) from Tolomei's farm (Rio de Janeiro, Brazil). Fragments of breast muscle were incubated in a plate for 15 min under a humidified 5% CO2 atmosphere at 37°C with calcium-magnesium-free solution (CMF) containing 0.25% trypsin (Sigma, USA), to assist the dissociation of cells. The contents of the plate was placed in a tube with the culture medium 8-1-0.5 (80% minimum essential medium – MEM, 10% horse serum and 0.5% chick embryo extract), and the cells were centrifuged. Trypsin activity was stopped by addition of this culture medium. Cells collected in the pellet were dispersed by repeated pipetting with culture medium. The resulting suspension was filtered, and the cells were plated at an initial density of 5×10^5 cells/35-mm culture dishes (Corning, EUA) previously coated with rat-tail collagen. Cells were grown under a humidified 5% CO2 atmosphere at 37°C for 24 hours, following which ultrasound wave treatment was performed.

2.2. Treatment with ultrasound waves

Before treatment, ultrasonic transducer was evaluated according to IEC 61689:2007. The transducer diameter of this equipment was 35 mm and operating at frequency of approximately 1 MHz. The coupling between the transducer and culture dish was achieved by the use of vaseline. The cell cultures were divided in four different treatment groups, in which the duration and the intensity of the ultrasound waves were different. One group was used as a control (with no treatment) and the others received treatments according to table below. Cells that were grown for 24 hours were subjected to ultrasound wave treatment.

Table 1 – Protocols for ultrasound wave treatments

Mode of Operation	Intensity (W/cm^2)	Time of Treatment (Minutes)
Pulsed	0.5	5
Pulsed	1.0	10
Continuous	0.5	5

Figure 1 – Experimental setup showing how the ultrasound wave transducer was applied to cultures cells.

2.3 Phase contrast optical microscopy
To investigate the possible changes in cell morphology and differentiation, skeletal muscle cells (treated and untreated) were analyzed in an inverted optical microscope (Axiovert 100, Carl Zeiss, Germany). Cells were fixed with 4% paraformaldehyde in phosphate buffered saline (PBS) for 10 min at room temperature, followed by membrane permeabilization with 0.5% Triton X-100 in PBS (3 times for 10 min each) at room temperature (Mermelstein et al., 2005). Phase contrast images were acquired immediately before, immediately after, and 24 and 48 hours after the ultrasound wave treatments.

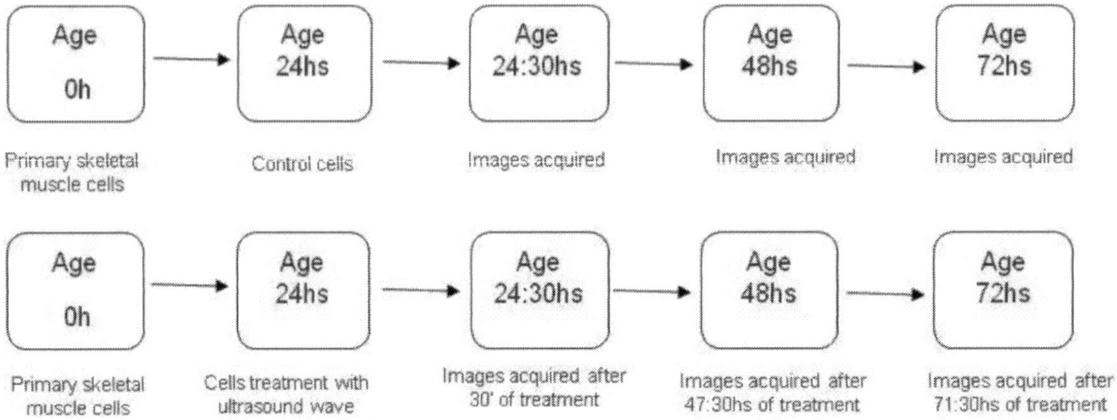

Figure 2 – Fluxogram showing when the images were acquired with and without ultrasound treatment.

3. Preliminary results and discussion
Chick embryonic skeletal muscle cells grown in culture were analyzed after the different ultrasound wave treatments. It was observed that low intensity-pulsed ultrasound wave induces cell proliferation, whereas low intensity-continuous ultrasound wave induces muscle differentiation. Furthermore, high intensity-pulsed ultrasound wave induced cellular death. Control and treated cells are currently being analyzed by indirect immunofluorescence microscopy, SDS-PAGE electrophoresis and Western blotting, in order to study the molecular and cellular basis associated with the observed changes in the cell cultures.

Journal of Physics: Conference Series **279** (2011) 012022 doi:10.1088/1742-6596/279/1/012022

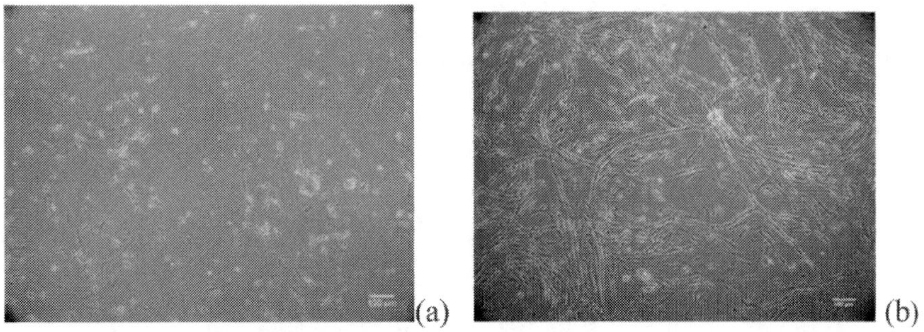

Figure 3 – (a) Control cell culture grown for 24 hours. (b) The same culture as shown in (a) was observed after 72 hours.

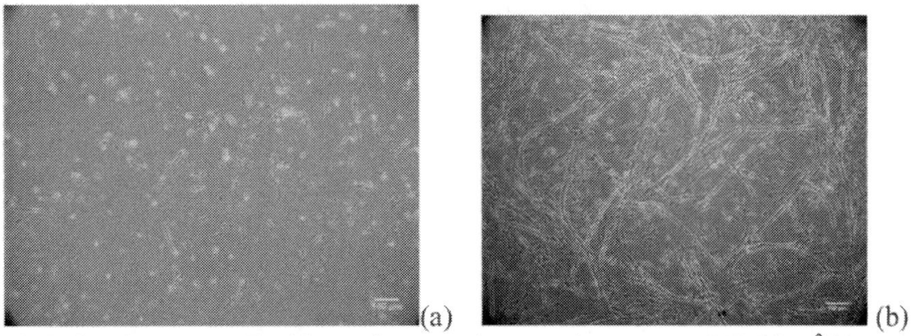

Figure 4 – (a) Cell culture treated with continuous ultrasound treatment (0.5W/cm^2 intensity) during 5 minutes. (b) The same culture as shown in (a) was observed after 48 hours of ultrasound treatment.

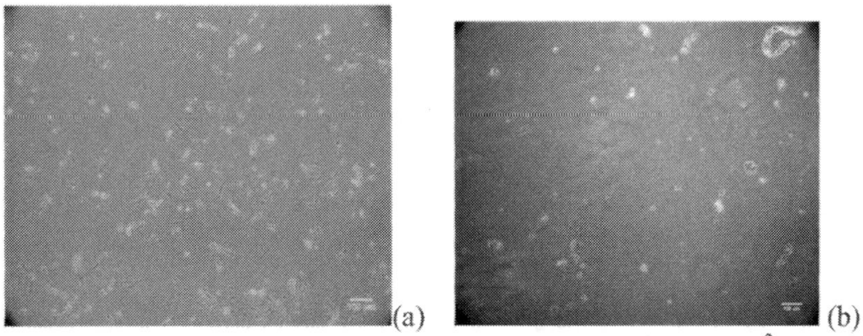

Figure 5 – (a) Cell culture treated with pulsed ultrasound treatment (1.0 W/cm^2 intensity) during 10 minutes. (b) The same culture as shown in (a) was observed after 48 hours of ultrasound treatment.

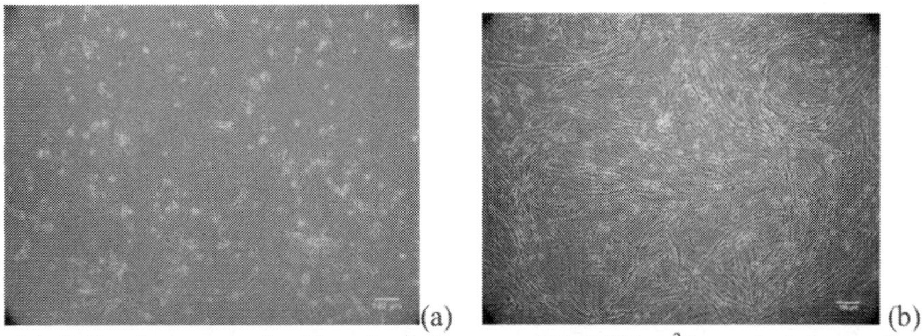

Figure 6 – (a) Cell culture treated with pulsed ultrasound (0.5 W/cm^2 intensity) during 5 minutes. (b) The same culture as shown in (a) was observed after 48 hours of ultrasound treatment.

4. Conclusion

This work concludes that depends on operated mode of equipment, the intensity adjusted and the duration of treatment the biological response of ultrasound in primary cultures of chick myogenic cells will be different.

Acknowledgements

This work was suported by National Institute of Metrology, Standardization, and Industrial Quality (Inmetro) and Fundação Carlos Chagas Filho de Apoio à Pesquisa do Estado do Rio de Janeiro (FAPERJ)

References

[1] Charge S B P, Rudnicki M A 2004 Cellular and molecular regulation of muscle regeneration *Physiol. Rev.*, **84** 209-238

[2] Locke M, Nussbaum E 2001 Continuous and pulsed ultrasound do not increase heat shock protein 72 content *Ultrasound in Med. & Biol.* **27 10** 1413-1419

[3] Zeqiri B 2007 Metrology for ultrasound applications. *Biophysics & Molecular Biology* **93** 138-152

[4] Ter Haar G 1998 *Electrophysical Principles. In: Kitchen S, Bazin S Electrotherapy Clayton* (São Paulo: Manole) p 3-30

[5] Nishikori T, Ochi M, Uchio Y, Maniwa S, Kataoka H, Kawasaki K, Katsube K, Kuriwaka M 2001 Effects of low-intensity pulsed ultrasound on proliferation and chondroitin sulfate synthesis of cultured chondrocytes embedded in Atelocollagen gel *Journal of Biomedical Materials Research Part A* **56 2** 201-205

[6] IEC 61689:2007, Ultrasonics – Physiotherapy systems – Field specifications and methods of measurements in the frequency range 0,5 MHz to 5 MHz

[7] Alvarenga A V, Costa-Félix R P B 2009 Uncertainty assessment of effective radiating area and beam non-uniformity ratio of ultrasound transducers determined according to IEC 61689:2007, *Metrologia* **46** 367-374.

[8] Mermelstein C S, Portilho D M, Medeiros R B, Matos A R, Einicker-Lamas M, Tortelote G G, Vieyra A, Costa M L 2005 Cholesterol depletion by methyl-β-cyclodextrin enhances myoblast fusion and induces the formation of myotubes with disorganized nuclei. *Cell Tissue Res* **319** 289–297

Speed of sound and density of ethanol-water mixture across the temperature range 10 to 50 degrees Celsius

J Tong [1], M J W Povey [2], X Zou [2], B Ward [1] and C P Oates [1]

[1] Regional Medical Physics Department, Freeman Hospital, Newcastle upon Tyne, NE7 7DN, UK
[2] Proctor Department of Food Science, University of Leeds, LS2 9JT, UK

E-mail: jie.tong@nuth.nhs.uk

Abstract. Properties such as the speed of sound and density of ethanol-water mixtures are of concern in applications such as the calibration of ultrasound imaging machines, and the study of molecular interactions in binary solutions. The speed of sound and density in ethanol-water mixtures up to a concentration of 40% (w/w) were measured across the temperature range 10 to 50 °C using an ultrasound resonator and a densitometer, with accuracies of ±0.25 m/s and ±0.05 kg/m^3, respectively. Measured data were interpolated and presented in forms of multi-polynomial functions of the concentration and the temperature. The functions may serve as a database for reference and also a source to derive parameters such as the temperature coefficient of the speed, the thermal expansion coefficient, and the isentropic compressibility of the mixture, which are commonly used in physiochemical and molecular acoustics studies.

1. Introduction

Ethanol-water mixtures can be used in ultrasound test objects as sound propagation media to assess medical ultrasound scanners. A mixture with a speed of sound of 1540 m/s is often used for testing the measurement tools provided by the imaging system [1]. Mixtures of other speed of sound, obtained by altering the ethanol concentration in the mixture, can be used in phantoms for assessing bladder volume. The speed of sound and the density in the mixture are also of interest in the study of molecular interactions [2]. It is therefore desirable to have a data source which gives the speed of sound and the density in the mixtures by simply specifying their concentration and temperature. Since the speed of sound has often been measured in-house, there are few numerically accessible data available in the literature [3-5]. Giacomini [3] measured the speed of sound in some ethanol-water mixtures with an uncertainty of about ±2 m/s; Martin and Spinks [4] measured the speed of sound in some mixtures with an uncertainty of about ±1.5 m/s. However, both the measurements were made at a few scattered ethanol concentrations and temperatures, which made the prediction of the speed within the concentration range and the temperature range less reliable. In addition, the data were presented in graphs instead of numerical values. Graph reading errors, which could be up to ±5–10 m/s and ±2–4 m/s in the two papers respectively, significantly reduce the usability of the data. Furthermore, there were no corresponding density data available from either set of authors.

In this work, we measured the speed of sound and density in ethanol-water mixtures at higher accuracies. The speed of sound and density were obtained in mixtures of concentrations of 0–40% (w/w) across the temperature range 10–50 °C. The results were presented as numerical expressions in the form of multi-variant polynomial functions to provide a reference for not only the ultrasound quality assurance community but also those in physiochemical or molecular acoustics study.

2. Methods and materials

2.1. Density measurement

The densities of the mixtures were measured using an `Anton Paar DMA 4500M` densitometer (Anton Paar GmbH, Austria), based on the oscillating U-tube method [6]. A U-shaped borosilicate glass tube is filled with the measured sample and is then exited so that it vibrates at its characteristic frequency. This characteristic frequency is mass-sensitive and thus related to the density of the material in the tube. By determining the two characteristic frequencies when the tube is filled with the sample and with a standard material, the density of the sample ρ is obtained using the relation $\rho = (K_A \times f_{sample}^{2} \times f_{ref}^{-2} \times g_1) - (K_B \times g_2)$. Where f_{sample} and f_{ref} are the characteristic frequencies of the tube with the sample and with the standard, respectively; and K_A, K_B, g_1 and g_2 are constants determined from the physical properties of the tube. During the measurement, the sample temperature was controlled to an accuracy of $\pm$ 0.03 °C, guaranteeing a density measurement accuracy of $\pm$ 0.05 kg/m^3 [6].

2.2. Speed of sound measurement

The speed of sound was measured using an ultrasound resonator, the `ResoScan Research System` (IF Instruments GmbH, Germany). This instrument is a fixed-path ultrasound interferometer based on the method developed by Egges et al [7]. Two piston transducers are held parallel to one another to form an interferometer. Ultrasound standing waves are generated between the two transducers and the resonant frequency is identified. The speed of sound c is obtained by determining the resonance frequency f_n at which the sound path length l is equal to a whole number n of half-wavelengths $\lambda/2$, with the relation $c = 2lf_n/n$.

The system had a sample cell with a volume of 170 μL and the sound path of 7 mm. The small sample volume made accurate temperature control possible, with a stability of 10^{-3} °C and an accuracy of 0.05 °C when running at temperatures of 5–85 °C. Working in the frequency range 7.5–9.5 MHz and with techniques to minimise amplitude and phase ambiguities, the system gave the speed of sound with an accuracy of about ±0.25 m/s and a resolution of 0.001 m/s [8].

2.3. Materials and measurement procedures

Ethanol (99.1% v/v) was obtained from Rathburn Chemicals (Mfg) Ltd, Walkerburn, UK. Deionised distilled water was made in-house. The ethanol-water mixtures were made by mixing a certain amount of the ethanol pre-determined by weight with the water without further purification. The mixtures were left to stand still for about 1–2 hours before samples were taken for the measurements.

To determine the ethanol concentration of a mixture, first the density was measured and its value at 20 °C was taken. Then the value was used with the ethanol tables of the International Union of Pure and Applied Chemistry (IUPAC) provided with the densitometer, to give the concentration as both a weight percentage (% w/w) and a volume percentage (% v/v). With the density measured within $\pm$ 0.05 kg/m^3, the concentration was obtained with an accuracy of about ±0.025% (w/w) or ±0.031% (v/v).

Seventeen mixtures with ethanol concentrations ranging from 3–40 % (w/w) were prepared, each of them with a volume of more than 500 ml. The density was measured in the mixture samples across the temperature range 10–50 °C with a step size of 2 °C. The speed of sound was measured in the mixture samples across the same temperature range but with a temperature changing rate of 0.3 K/minute, with a measurement rate of 60 measurements per minute. Each sample was measured twice in a temperature rising and falling cycle. The averaged value at the same temperature in the cycle was taken for the speed value. The total measurement time for each sample was about 4–5 hrs.

3. Results and discussions

3.1. Density results

The density of the in-house water was measured to verify its purity. At 20 °C, the measured value was 998.21 kg/m^3, which is in agreement with the value 998.2063 kg/m^3 from the literature [5]. This ensured the usability of the water.

Densities of 17 mixtures, together with those of water and ethanol, were obtained. The concentrations, obtained using the IUPAC ethanol table with a density value at 20 °C, were obtained as both a weight percentage ξ (in % w/w) and a volume percentage φ (in % v/v) at 20°C. The two concentrations were related via the expression

$$\varphi = 0.016 + 1.254\xi - 9.963\times10^{-4}\xi^2 - 1.836\times10^{-5}\xi^3\,, \qquad 0<\xi<40 \tag{1}$$

This expression was obtained using the Mathcad® software package (Parametric Technology Corporation, MA, USA) with a least-squares polynomial fit algorithm [9]. An error estimate ε, defined as $\varepsilon = \{\sum_i[(y_i - Y_i)/y_i]^2/m\}^{0.5}$ for assessing the goodness of fit (eliminating any singular points), was found to be of the order of 10^{-4}. In the definition, y_i and Y_i are the ith measured and fitted data, respectively, and m is the total number of the measured data.

In order to present the density values numerically, a least-squares multi-variant polynomial fit algorithm in Mathcad® [9] was used to obtain the density as a function of the concentration ξ (in % w/w) and the temperature T (in °C). With a polynomial of order 7, the density was given in kg/m^3 in the form

$$\rho(\xi,T) = \sum_{i=1}^{36} A_i \xi^{M_i} T^{N_i}\,, \qquad 0\le\xi<40,\quad 0\le T\le 50 \tag{2}$$

where A, M and N were matrices of 36 components, whose values are shown in Table 1.

The polynomial of order 7 was used for two reasons: 1) the order should not be too large so that few polynomial terms were involved; 2) the order should be large enough so that reasonable accuracy of the interpolation prediction could be maintained. With an order of 7, the error estimate for Eq. (2) was about 10^{-6}.

With the same polynomial order, density values $\rho_p(T)$ (in kg/m^3) for the `pure` ethanol (98.49% w/w or 99.1% v/v) were given in the form:

$$\rho_p(T) = 810.410 - 0.690979T - 1.94981\times10^{-2}T^2 + 1.28718\times10^{-3}T^3 - 4.95359\times10^{-5}T^4 + 1.08911\times10^{-6}T^5 - 1.27659\times10^{-8}T^6 + 6.17318\times10^{-11}T^7,\quad 10\le T\le 50 \tag{3}$$

The error estimate for Eq. (3) was about 10^{-6}.

3.2. Speed of sound results

The speed of sound was obtained at frequencies between 7.5 MHz and 8.5 MHz. Within this frequency range, no speed dispersion greater than 0.05 m/s was found for mixture samples.

The same polynomial algorithm with a polynomial of order 10 was applied to the measured data to get the speed of sound in the form:

$$c(\xi,T)=\sum_{i=1}^{66}D_i\xi^{K_i}T^{L_i}\,,\qquad 0\le\xi\le 25,\quad 0\le T\le 50 \tag{4}$$

where D, K and L were matrices of 66 components, whose values are shown in Table 1.

Table 1. Matrix components in the polynomial expressions of the density and the speed of sound.

i	Ai	Mi	Ni	Di	Ki	Li	i	Di	Ki	Li
1	-1.31979E-10	1	6	2.293523E-13	1	9	37	2.581214E+00	0	2
2	9.26559E-11	0	7	-3.334069E-14	0	10	38	-9.778700E-02	1	2
3	-1.84296E-08	0	6	4.347779E-12	0	9	39	-3.409837E-03	2	2
4	1.48789E-06	0	5	2.748470E-10	0	8	40	9.554803E-05	3	2
5	2.68566E-08	1	5	-6.740290E-11	1	8	41	-2.029789E-06	4	2
6	-1.17238E-10	2	5	4.872988E-14	2	8	42	7.303260E-08	5	2
7	-6.27703E-05	0	4	-9.379410E-08	0	7	43	-2.678939E-09	6	2
8	-2.24365E-06	1	4	8.441089E-09	1	7	44	5.681490E-11	7	2
9	2.47408E-08	2	4	-7.191991E-12	2	7	45	-4.742113E-13	8	2
10	-1.34332E-10	3	4	-3.459921E-14	3	7	46	-1.154564E+01	0	1
11	1.51236E-03	0	3	8.520253E-06	0	6	47	2.115519E-01	1	1
12	9.82635E-05	1	3	-5.876577E-07	1	6	48	5.354679E-02	2	1
13	-2.06834E-06	2	3	-2.717664E-11	2	6	49	-8.378834E-03	3	1
14	2.53304E-08	3	3	1.566856E-11	3	6	50	1.154267E-03	4	1
15	-1.19569E-10	4	3	-7.961328E-14	4	6	51	-9.631472E-05	5	1
16	-2.65277E-02	0	2	-4.175723E-04	0	5	52	4.720320E-06	6	1
17	-2.38970E-03	1	2	2.487626E-05	1	5	53	-1.331721E-07	7	1
18	8.12064E-05	2	2	6.317368E-08	2	5	54	2.000458E-09	8	1
19	-1.09917E-06	3	2	-2.116633E-09	3	5	55	-1.238626E-11	9	1
20	-3.79088E-10	4	2	1.542227E-11	4	5	56	1.446424E+03	0	0
21	7.42769E-11	5	2	-2.889305E-14	5	5	57	1.142117E+01	1	0
22	1.72377E-01	0	1	1.246830E-02	0	4	58	-8.523948E-01	2	0
23	3.03939E-02	1	1	-6.572525E-04	1	4	59	3.394837E-01	3	0
24	-1.91310E-03	2	1	-4.957552E-06	2	4	60	-6.665556E-02	4	0
25	-2.95143E-06	3	1	1.370765E-07	3	4	61	7.478822E-03	5	0
26	1.02853E-06	4	1	-1.363176E-09	4	4	62	-5.137953E-04	6	0
27	-6.76603E-09	5	1	8.968693E-12	5	4	63	2.187272E-05	7	0
28	-7.56401E-11	6	1	-4.644391E-14	6	4	64	-5.617519E-07	8	0
29	9.99613E+02	0	0	-2.323159E-01	0	3	65	7.949837E-09	9	0
30	-2.06524E+00	1	0	1.065812E-02	1	3	66	-4.745228E-11	10	0
31	4.69556E-02	2	0	1.790475E-04	2	3				
32	6.00674E-04	3	0	-4.763008E-06	3	3				
33	-6.78157E-05	4	0	5.512563E-08	4	3				
34	1.18896E-06	5	0	-2.392248E-10	5	3				
35	-1.13524E-08	6	0	-6.812776E-12	6	3				
36	9.52109E-11	7	0	1.027370E-13	7	3				

In Eq. (4) the valid concentration range was narrowed down to 0–25% (w/w) to avoid larger errors of the interpolation prediction, which was induced by the fact that there were only three measured mixtures in the range of $25 < \xi < 40$ and also by the fact that the polynomial prediction, by its nature, fluctuates when raw data spacing becomes large. In the Eq. (4), the error estimate for the speed data fitting was of the order of 10^{-3}. The maximum residual between the measured speed and the predicted value from Eq. (4), defined as $\chi = \max\left|c - c(\xi,T)\right|$, was 1.51 m/s.

Values from Eq. (4) at $\xi = 0$ were compared with those given by Grosso (cited as Table 1 in the literature [4]) who measured the speed of sound in water at various temperatures with an accuracy of 10^{-3}%. The standard deviation of the difference between Grosso's values and those from Eq. (4) was 0.097 m/s. This shows that the expression could give reasonably accurate predictions, especially at or near to the measured data points.

Similar to the density, the speed of sound for the `pure` ethanol was given as

$$c_p(T) = 1219.90 - 0.02312T - 0.41830T^2 + 0.02660T^3 - 9.91594\times10^{-4}T^4 + 2.05985\times10^{-5}T^5 - 2.34839\times10^{-7}T^6 + 1.10971\times10^{-9}T^7,\quad 10 \le T \le 50 \tag{5}$$

The error estimate for the expression was of the order of 10^{-5}.

3.3. Discussions

The advantage of Eqs. (1)-(5) is the convenience for accessing numerical values of the speed of sound and the density of ethanol-water mixtures. For example, given a speed of sound of 1540 m/s and a tolerance of ±2 m/s, one can easily determine that, working in the temperature range of 19–21 °C, the mixture which has the specified speed of sound must lie in a diamond-shaped area in the speed against concentration and iso-temperature curve surrounded by four points (weight concentration %, temperature °C): (6.8, 21), (7.33, 21), (7.77, 19) and (7.28, 19). Expressed in the volume concentration (% v/v) at 20 °C, the four points are (8.49, 21), (9.15, 21), (9.69, 19) and (9.09, 19).

Martin and Spinks [4] pointed out that, at 20 °C a mixture of 9.5 ± 0.25 % (v/v) has a speed of sound of 1540 m/s. Using Eq. (4), the mixture of the concentration 9.5 % (v/v) at 20 °C is of the speed of sound of 1542.56 m/s. Since Martin and Spinks did not specify the temperature at which the mixture concentration was obtained, the difference between their results and our results here could be partially due to the uncertainty of the volume concentration, which varies as the temperature changes.

Another advantage is that numerical values for various derivatives of the speed of sound, and the density, can be easily derived from the equations. Those derivatives, such as the temperature coefficient of the speed of sound, the thermal expansion of the mixture, the isentropic compressibility, the excess volume of the mixtures, etc, can be important parameters in the physiochemical study or the molecular acoustics study of the mixtures [2, 10].

4. Conclusions

The speed of sound and density of ethanol-water mixtures with concentrations of up to 40% (w/w) were measured across the temperature range 10–50 °C. The measurement accuracies were of ±0.25 m/s and ±0.05 kg/m^3, respectively. The values were expressed as numerical equations, with prediction uncertainties better than about 10^{-3} and 10^{-6}, respectively.

5. References

[1] International Electrotechnical Commission (IEC) 2006 IEC 61391-1 Ed.1.0: Ultrasonics – pulse-echo scanners – part 1: Techniques for calibrating spatial measurement systems and measurement of system point-spread function response. (IEC, Geneva, Switzerland)

[2] Rowlinson J S and Swinton F L 1982 *Liquids and liquid mixtures (Butterworths Monographs in Chemistry 3rd Ed)* (London: Butterworth Scientific).

[3] Giacomini A 1947 Ultrasound velocity in ethanol-water mixtures *J. Acoust. Soc. Am.* **19(4)** 701-2.

[4] Martin K and Spinks D 2001 Measurement of the speed of sound in ethanol-water mixtures *Ultrasound in Med. & Bio.* **27(2)** 289-91.

[5] Lide D R 2000 *CRC Handbook of chemistry and physics 82nd Ed* (Florida: CRC Press Inc.)

[6] Anton Paar GmbH 2010 *Anton Paar DMA4500M Instruction Manual* (Graz, Austria: Anton Paar GmbH).

[7] Eggers F and Funck T 1973 Ultrasonic measurements with millilittlre liquid samples in the 0.5-100 MHz range *Rev. Sci. Instr.* **44** 969-977.

[8] TF Instruments GmbH 2007 *ResoScanTM research system operations manual* (Heidelberg, Gemerny: TF Instruments GmbH)

[9] Parametric Technology Corporation 2007 *Mathcad® data analysis extension pack* (Massachusetts: Parametric Technology Corporation)

[10] Sarvazyan A P 1994 *New frontiers of molecular acoustics in liquids J. Acoust. Soc. Am,* **96(5)** 3225-30.

Journal of Physics: Conference Series **279** (2011) 012024 doi:10.1088/1742-6596/279/1/012024

The effects of focused transducer geometry and sample size on the measurement of ultrasonic transmission properties

T J Atkins[1,2], V F Humphrey[2], F A Duck[1] and M A Tooley[1]

[1]Department of Medical Physics and Bioengineering, Royal United Hospital, Combe Park, Bath BA1 3NG, UK

[2]Institute of Sound and Vibration Research, University of Southampton, Southampton SO17 1BJ, UK

Email: timothy.atkins@nhs.net

Abstract. The response of two coaxially aligned weakly focused ultrasonic transducers, typical of those employed for measuring the attenuation of small samples using the immersion method, has been investigated. The effects of the sample size on transmission measurements have been analyzed by integrating the sound pressure distribution functions of the radiator and receiver over different limits to determine the size of the region that contributes to the system response. The results enable the errors introduced into measurements of attenuation to be estimated as a function of sample size. A theoretical expression has been used to examine how the transducer separation affects the receiver output. The calculations are compared with an experimental study of the axial response of three unpaired transducers in water. The separation of each transducer pair giving the maximum response was determined, and compared with the field characteristics of the individual transducers. The optimum transducer separation, for accurate estimation of sample properties, was found to fall between the sum of the focal distances and the sum of the geometric focal lengths as this reduced diffraction errors.

1. Introduction

Spherically focused ultrasonic transducers are used in many applications including materials characterization and medical imaging. In a biomedical context, coaxially aligned weakly focused transducers are often used to measure the acoustic properties of tissue *in-vitro* and of tissue-like media [1]. An advantage of using focused transducers in this context is that it helps to reduce the required sample size and increase sensitivity; this paper quantifies the effect of sample size on diffraction error.

A number of workers have investigated the output of a pair of coaxial planar circular transducers as a function of separation distance [2, 3]. The resulting diffraction loss curves are useful for selecting separation distances in immersion measurements in order to minimise diffraction loss errors [1].

The present work explores the equivalent geometry and sample size issues related to measurement of sample properties, using coaxially aligned spherical transducers with weak focusing, by modelling and measuring the combined diffraction effects for combinations of three specific transducers.

2. Theoretical expression for receiver response

Consider an unapodized spherically concave source with radius a_1 and radius of curvature D_1 producing sound at angular frequency ω in cylindrical coordinate system (z, r, φ). The radiation is

Journal of Physics: Conference Series **279** (2011) 012024 doi:10.1088/1742-6596/279/1/012024

generated by uniform motion normal to the transducer surface. A second transducer is located coaxially with the source at a separation $Z = z_1 + z_2$ (see figure 1).

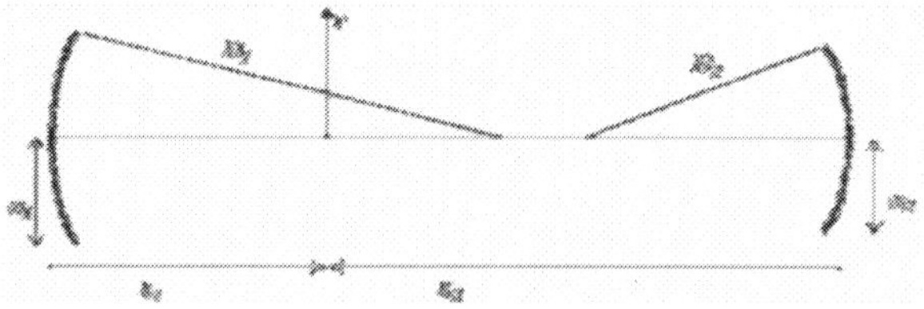

Figure 1. Cross section view of two coaxially aligned cylindrically symmetric concave transducers, showing coordinates and symbols used in the analysis.

Following Gavrilov *et al* [4], the sound pressure averaged over the active area of the receiver, $p_{rec}(Z)$, can be calculated from the convolution of the sound pressure distribution functions of the two transducers over an infinite plane between the transducers. This receiver response can be estimated by

$$p_{rec}(Z,R) = \frac{2}{P_2 a_2^{\ 2}} \int_{r=0}^{R} p_1(r,z_1) p_2(r,z_2) r dr \tag{1}$$

where $p_1(r, z_1)$ is the complex sound pressure produced by the radiator at radial coordinate r from the z axis in a plane at a distance z_1 from the source. Likewise $p_2(r, z_2)$ is the complex sound pressure produced by the second transducer operating in transmit mode at a distance z_2 from the source when P_2 is the average pressure amplitude across the surface of the second transducer.

For such a system the measured free field receiver response will be proportional to the calculated value $p_{rec}(Z)$ in the limit $R \to \infty$. This solution is invariant over the choice of integration plane. For a given transducer separation Z, z_1, and hence z_2, may be varied without affecting the result.

Restricting the integration to a finite radial distance R introduces an error into the calculation of $p_{rec}(Z)$, and hence the measured receiver response. This enables the effect of using finite size samples on the received sound pressure to be investigated by analysing the radius of the restricted integral region required to accurately estimate the receiver response.

With P_0 as the average pressure amplitude at the surface of the transducer, Gavrilov *et al*[4] give a solution for the sound pressure of a focusing source of radius a and focal length D in the focal plane as

$$p(r) = -iP_0 \frac{a}{r} \exp(ikD) J_1\left(\frac{kar}{D}\right). \tag{2}$$

Lucas and Muir [5] provide a more general solution at a distance z from a source as

$$p(r,z) = -ik\frac{P_0}{z}\exp(ikz)\exp\left(\frac{ikr^2}{2z}\right)\int_0^a \exp\left[\frac{ikx^2}{2}\left(\frac{1}{z}-\frac{1}{D}\right)\right] J_0\left(\frac{krx}{z}\right) x dx. \tag{3}$$

A significant feature of equation (3) is the presence of the phase term $\exp(ikr^2/2z)$. This indicates that the phase is not constant across the field. The phase term is not important when considering point measurements in a field; however, in this case the subsequent convolution of two of these functions means that the phase needs to be carefully considered. It is omitted from equation (2) as this is an approximation based on the geometrical optics work of Debye [6].

Introducing a cylindrical sample into the measurement system during attenuation measurements can be considered to split the integration for the receiver response, $p_{rec}(Z)$, into two parts. The first part consists of an integration from the axis up to the sample radius s, for which there will be an attenuation introduced, and an integration beyond the sample radius s. The introduction of the sample may change the phase of the ultrasonic wave through the sample area.

Let the average receiver pressure without the sample be $p_{rec}(Z)$ and the sample have a transmission coefficient T (allowing for attenuation and transmission at the sample interfaces). The error in estimating T by the measured transmission coefficient T' can be shown to be

$$T' - T = (1 - T)\,[1 - p_{rec}(Z,s)/p_{rec}(Z)]. \tag{4}$$

The magnitude of the error in estimating T by T' is therefore minimised by ensuring $p_{rec}(Z, s)/p_{rec}(Z)$ is close to unity. In this paper the ratio $|p_{rec}(Z, R)|/|p_{rec}(Z)|$ will be used to illustrate how close $p_{rec}(Z, R)$ is to the infinite value $p_{rec}(Z)$. To assist in the analysis R_{lim} is defined as the minimum integration limit required to ensure that the ratio of $|p_{rec}(Z, R)|/|p_{rec}(Z)|$ is greater than a specified value for all $R > R_{lim}$. Similar analysis and integration limits are used for the phase variation of $p_{rec}(Z, R)/p_{rec}(Z)$.

The maximum error introduced by using a finite sized sample can be calculated from the ratio $|p_{rec}(Z, R)|/|p_{rec}(Z)|$ and the sample transmission coefficient; alternatively the minimum sample size required to ensure that these errors are below a specified level can be calculated. This analysis does not include the effects of sample impedance on the diffracted wave or sample inhomogeneity.

Three weakly-focused transducers of approximate centre frequency 3.5 MHz were selected for this investigation; Table 1 shows the properties of the transducers estimated from acoustic measurements of the transducers using an NPL Beam Calibrator [7]. Each transducer was modelled using equation (2) or by numerically integrating equation (3). The models for $p(r, z)$ were then used in equation (1) to calculate the receiver response $p_{rec}(Z)$ and the estimate for a restricted integration plane $p_{rec}(Z, R)$.

Table 1. Measured transducer characteristics used in theoretical model.

Transducer	Radius of curvature (D) / cm	Transducer radius (a) / cm	Acoustic focal length / cm	-6dB Beamwidth at acoustic focus / cm
A	7.6	0.61	5.0	0.27
B	11.6	0.62	6.0	0.39
C	13.5	0.91	9.5	0.35

3. Results

3.1. Debye approximation at sum of geometric foci

Consider a pair of transducers separated by the sum of their radii of curvature. With a substitution of $X = ka_1r/D_1$ and $\beta X = ka_2r/D_2$ at the plane of their coincident geometric foci, equation (2) becomes

$$p_{rec}(D_1 + D_2, R) = -2P_0\frac{a_1}{a_2}\exp[ik(D_1 + D_2)]\int_{X=0}^{X=R}\frac{J_1(X)J_1(\beta X)}{X}dX. \tag{5}$$

The integral $I = \int_{X=0}^{X=\infty}\frac{J_1(X)J_1(\beta X)}{X}dX$ has a value of $\beta/2$ for $0 < \beta \le 1$ and $1/2\beta$ for $\beta \ge 1$.

The integrand and the integral I are plotted in figures 2 and 3 for $\beta = 1$ and $\beta = 0.669$ respectively. For $\beta = 1$ (matched transducers) the integral is a monotonically increasing function which reaches a value of 0.5 when integrated over an infinite range. For $\beta \neq 1$ (non-identical transducers) the received signal is less than that for matched transducers; however the received signal approaches the value for infinite R faster than for identical transducers.

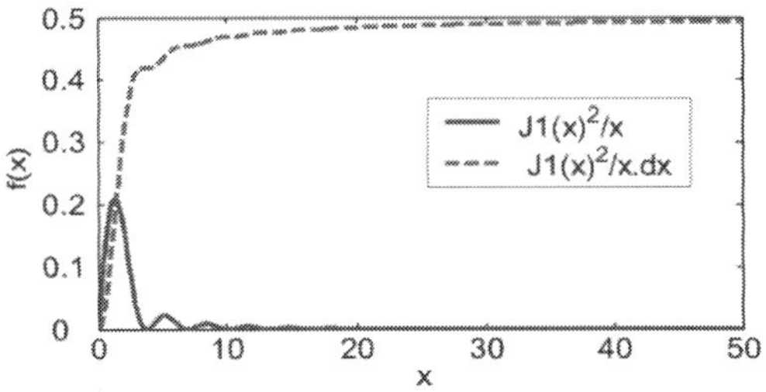

Figure 2. Plot of the functions $\frac{J_1^2(X)}{X}$ and $\int\frac{J_1^2(X)}{X}dX$ used in the calculation of the receiver response for coaxially aligned transducers (X is dimensionless).

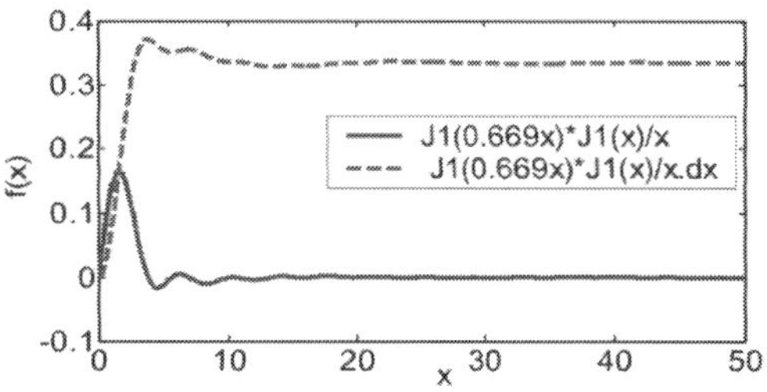

Figure 3. Plot of the functions $\frac{J_1(X)J_1(\beta X)}{X}$ and $\int\frac{J_1(X)J_1(\beta X)}{X}dX$ used in the calculation of the receiver response for coaxially aligned transducers with $\beta = 0.669$ (X is dimensionless).

3.2. Location of diffraction peak

To reduce the effect of diffraction error on attenuation measurements the transducers should be located at a position of diffraction maxima or minima [1]. This ensures that the diffraction changes caused by the introduction of a medium of different sound speed are minimized when a sample is placed between the transducers. For coaxially aligned strongly focused transducers this has conventionally been accepted as requiring a transducer separation of the sum of the geometric foci [1, 8]. For weakly focused transducers, however, the diffraction peak does not occur when the transducers are separated by the sum of their geometric foci since the geometric and acoustic foci do not coincide. Figure 4 shows the modelled variation in the integrated response varies with separation for transducers A and C. The sum of the geometric foci is shown as a vertical line on the figure. The transducer separation has to be reduced from the sum of the geometric foci to reach the position of the diffraction maximum. This agrees with measurements of the receiver output obtained using the coaxial transducers A and C, shown in figure 5. The sum of geometric foci is indicated at 21 cm and the sum of acoustic foci at 14.5 cm. This shows that the position of the diffraction peak occurs at a position between the sum of acoustic and the sum of the geometric focal lengths; this was seen for all three transducer pairings.

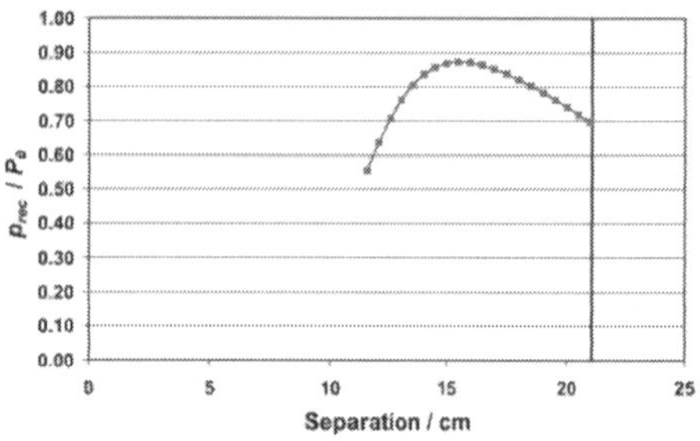

Figure 4. Example of the modelled variation in the received response p_{rec}, normalized to the average pressure amplitude across the surface of the transmitting transducer P_0, as a function of the separation between the transducers A and C. The sum of geometric foci is shown at 21 cm.

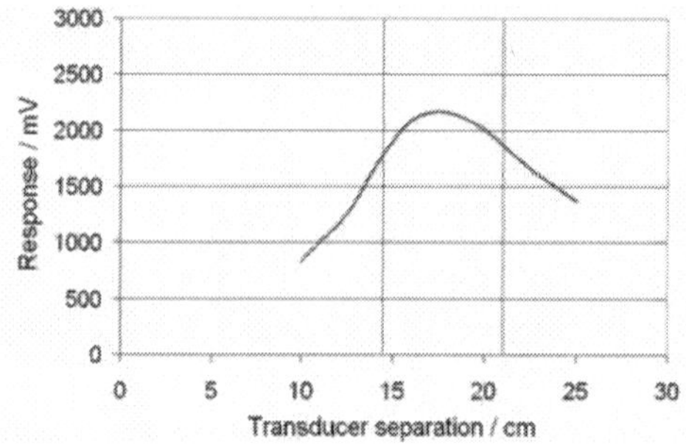

Figure 5. Example of the measured variation in the received response p_{rec}, as a function of separation between the transducers A and C. The sum of geometric foci is shown at 21 cm and the sum of acoustic foci shown at 14.5 cm.

3.3. Full expression for identical and non-identical transducers

One of the properties of equation (1) is that for integration to infinity the average pressure $p_{rec}(Z)$ (and hence receiver output) is not affected by the integration plane that is chosen. Figure 6 shows how the magnitude of $p_{rec}(Z, R)$ varies with R for transducers A and C for different distances of the integration plane from the receiver. When the integration limit is greater than approximately 2.0 cm, the magnitude of $p_{rec}(Z, R)$ is the same for all integration planes.

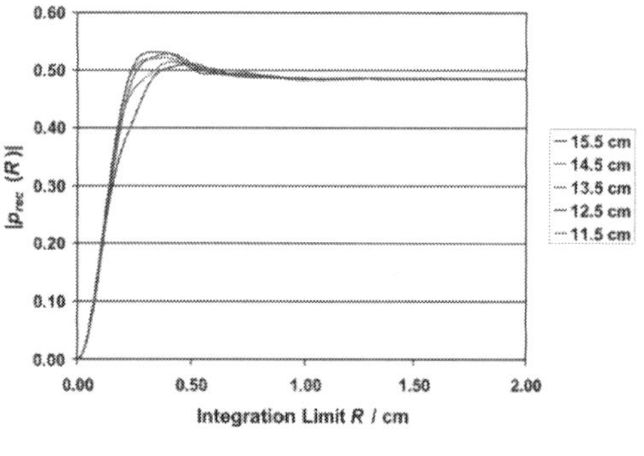

Figure 6. Plot showing the variation in received pressure $p_{rec}(R)$ with integration limit R for different distances of integration plane from the receiving transducer for transducers A and C.

Figure 7 demonstrates the effect of the integration radius on the solution at the diffraction peak when equation (3) is used in the model. For different values of the $|p_{rec}(Z, R)|/|p_{rec}(Z)|$ ratio there are different transducer combinations which give the smallest value of R_{lim}. For each transducer separation, the choice of integration plane affects the value of R_{lim}, and the displayed standard deviations quantify the spread obtained if the integration plane is varied over a 4 cm range. Figure 8 shows similar data for the phase difference between the restricted and infinite integration.

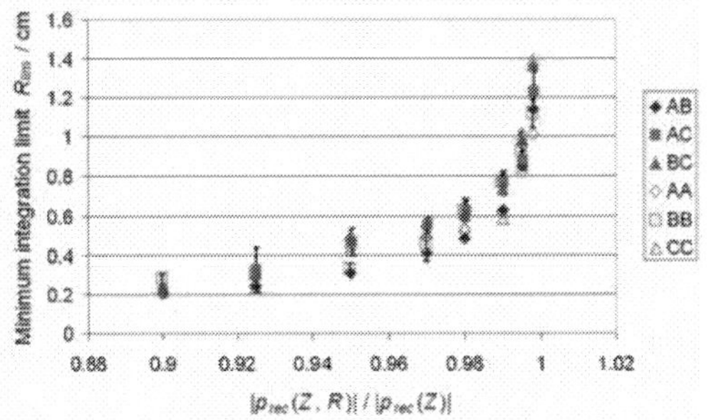

Figure 7. Plot showing the minimum integration limit R_{lim} required for selected ratios of $|p_{rec}(Z, R)|/|p_{rec}(Z)|$ for matched transducer pairs (AA, BB, CC) and transducers of different geometry (AB, AC, BC) at the position of diffraction loss peak, calculated using equation (3). Example error bars (1 SD) are shown for transducer pairings AC and BB.

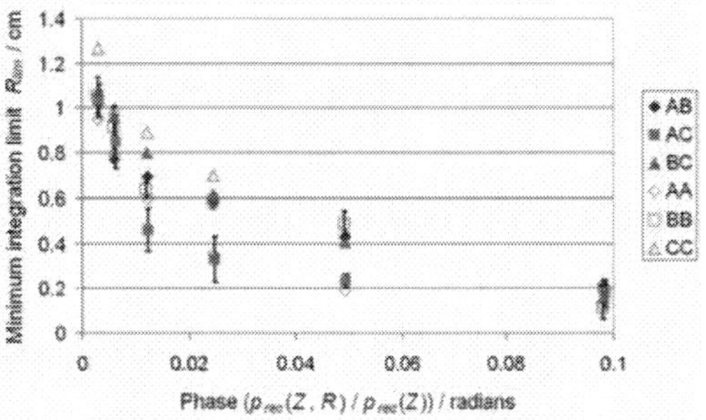

Figure 8. Plot showing the minimum integration limit R_{lim} required for selected phase differences between $p_{rec}(Z, R)$ and $p_{rec}(Z)$ for matched transducer pairs (AA, BB, CC) and transducers of different geometry (AB, AC, BC) at the position of diffraction loss peak, calculated using equation (3). Example error bars (1 SD) are shown for transducer pairings AC and BB.

4. Discussion

This work highlights the fact that the Debye solution quoted by Gavrilov *et al* [4] is an approximation obtained in the geometrics optics limit. It agrees with fuller formulations in terms of the focal plane amplitudes but not phases. The results presented here show it is important to include the transverse phase term $\exp(ikr^2/2z)$ when modelling opposed pairs of weakly focused transducers, as the calculation of the receiver response involves an integration of the product of two complex sound pressure distribution functions over the radial coordinate r. Analysis of the two solutions used in this paper indicates that the phase term reduces the size of the integration limit R_{lim} required to ensure an accurate estimation of p_{rec} by a restricted integration. The phase term does introduce a phase difference between the restricted integral $p_{rec}(Z, R)$ and the infinite integral $p_{rec}(Z)$. However, the phase differences are small and should not significantly affect results (figure 8).

The difference between the integration results for a limited radius and infinite radius gives an indication of the error that is likely to be obtained by using a coaxial circular sample of a given radius in immersion measurements of acoustic attenuation. The results obtained here can be used to estimate the minimum size of sample required to measure a given attenuation to a given precision.

The criterion quoted in the literature [1, 8] to minimize errors in measurement due to diffraction losses is to place opposed transducers at the diffraction peak. For the transducers used in this paper both the theoretical model and measurements indicate that the separation between the transducers should be reduced from the position of the sum of the geometric focal lengths (figures 4 and 5). This is due to the transducers being weakly focused as opposed to strongly focused transducers where the geometric and acoustic foci would coincide. Placing weakly focused transducers at the sum of their geometric foci would lead to increased diffraction errors.

When using coaxially aligned spherically focused transducers to measure ultrasonic attenuation properties the measurement system needs to be designed to minimize uncertainties in the measurement of the attenuation of the sample. The analysis presented here for the average receiver pressure

$p_{rec}(Z, R)$ shows that if R is small then $p_{rec}(Z, R)$ is not an accurate estimate of $p_{rec}(Z)$. To provide an accurate estimate of $p_{rec}(Z)$ (small error) the integration must be performed over a large radial distance ($R > R_{lim}$).

When measuring ultrasonic attenuation, it has been shown [9] that the optimum attenuation for measurement is approximately 1 neper (8.68 dB). The errors introduced by sample size effects calculated here may have large effects on such measurements of attenuation. Therefore relatively large samples are required to ensure sample size effects do not contribute a significant error. For the transducers used in this paper sample sizes of approximately 0.8cm radius are required to ensure the ratio of $|p_{rec}(Z, R)|/|p_{rec}(Z)|$ is greater than 0.98 (figure 7). This is equivalent to a sample diameter of 4 to 6 times the focal beam diameter. The actual effect this will have is dependent on the transmission properties of the sample itself (see equation (4)). The phase difference between $p_{rec}(Z, R)$ and $p_{rec}(Z)$ will be negligible with samples of this size.

When the difference between the estimated and actual value of the average receiver pressure is small (ratio $|p_{rec}(Z, R)|/|p_{rec}(Z)|$ is close to one) there does not seem to be an obvious choice of integration plane to minimize the required sample size (figures 6 and 7). The choice of integration plane is analogous to the physical positioning of the sample between the two transducers, indicating that for relatively large samples, the axial positioning of the sample is not critical.

5. Conclusions

Two models have been used to calculate the average pressure on the receiver of a pair of coaxial focused transducers. Results for three specific transducers operating at 3.5 MHz have been utilized to analyze the effects of sample size and geometry on the measurement of ultrasonic properties.

A potential source of error in attenuation measurements due to finite sample size has been discussed. This source may introduce significant errors into the measurement of attenuation and should be considered when designing attenuation measurement experiments. For the transducers used in this paper sample sizes of approximately 0.8cm radius, equivalent to a sample diameter of 4 to 6 times the focal beam diameter, are required to ensure an ratio $|p_{rec}(Z, R)|/|p_{rec}(Z)|$ is greater than 0.98. This will limit the errors due to sample size effects introduced into attenuation measurements.

Both the theoretical model and measurements of specific transducers indicate that for weakly focused transducers the optimum separation between the transducers should be reduced from the sum of the geometric focal lengths in order to minimise diffraction loss corrections.

Acknowledgements

This work was funded under the UK Department of Health NEAT programme project J008.

References

[1] Bamber J C 2004 Attenuation and absorption *Physical Principles of Medical Ultrasonics* ed C R Hill, J C Bamber and G R ter Harr (Chichester: J Wiley) pp93-166

[2] Rhyne T L 1977 *J. Acoust. Soc. Am.* **61** 218-324

[3] Harrison J A, Cook-Martin G N and Challis R E 1984 *J. Acoust. Soc. Am.* **76** 1008-1022

[4] Gavrilov L R, Dmitriew V N and Solontsova L V 1988 *J. Acoust. Soc. Am.* **83** 1167-1179

[5] Lucas B G and Muir T G 1982 *J. Acoust. Soc. Am.* **72** 1289-1296

[6] Debye P 1909 *Ann. Phys.* **335** 775-776

[7] Preston R C 1988 *IEEE Trans. Ultrason. Ferroelect. and Freq. Contr.* **35** 122-139

[8] Penttinen A and Luukkala M 1977 *J Phys. D: Appl. Phys.* **10** 665-669

[9] Kalashnikov A N and Challis R E 2005 *IEEE Trans. Ultrason. Ferroelect. And Freq. Contr.* **52** 1754-1768

Ultrasonic immersion probes characterization for use in non-destructive testing according to EN 12668-2:2001

C E R Silva[1], A V Alvarenga[1], R P B Costa-Felix[1]

[1]Laboratory of Ultrasound, Directory of Scientific and Industrial Metrology (DIMCI), National Institute of Metrology, Standardization, and Industrial Quality (INMETRO), Av. Nossa Sra das Graças, 50 (Prédio 1), Duque de Caxias, RJ, Brazil, ZIP 25250-020

[1]cesilva@inmetro.gov.br

Abstract. Ultrasound is often used as a Non-Destructive Testing (NDT) technique to analyze components and structures to detect internal and surface flaws. To guarantee reliable measurements, it is necessary to calibrate instruments and properly assess related uncertainties. An important device of an ultrasonic instrument system is its probe, which characterization should be performed according to EN 12668-2. Concerning immersion probes beam profile, the parameters to be assessed are beam divergence, focal distance, width, and zone length. Such parameters are determined by scanning a reflector or a hydrophone throughout the transducer beam. Within the present work, a methodology developed at Inmetro's Laboratory of Ultrasound to evaluate relevant beam parameters is presented, based on hydrophone scan. Water bath and positioning system to move the hydrophone were used to perform the scan. Studied probes were excited by a signal generator, and the waterborne signals were detected by the hydrophone and acquired using an oscilloscope. A user-friendly virtual instrument was developed in LabVIEW to automate the system. The initial tests were performed using 1 and 2.25 MHz-ultrasonic unfocused probes (Ø = 1.27 cm), and results were consistent with the manufacturer's specifications. Moreover, expanded uncertainties were lower than 6% for all parameters under consideration.

1. Introduction

The normal practice in most ultrasound non-destructive testing is the calibration of probes and ultrasonic measurement systems to ensure the accuracy and repeatability of the test. Ultrasonic transducers play a key role in any ultrasonic measurement system since they generate and receive the ultrasonic waves. To quantitatively describe the effect of transducers on the measured signals in an ultrasonic test, it is necessary to characterize both the transducer's transmitting and receiving properties. It is shown that these properties can be completely determined by four types of quantities: (1) the transducer's electrical impedance; (2) its sensitivity; (3) its radiation impedance; and (4) the beam parameters, such as: focal length, distance, and width; and beam divergence. It is also known that all of these properties can be obtained through a series of calibration methods depends on kind of probe [[1]]. The aim of this work was implements a method to assess the ultrasound immersion probes beam parameter in accordance with EN12668-2. This one can potentially influence the outcome of an inspection if it were not in accordance with the standard cited above. The initial tests using 1 and

Journal of Physics: Conference Series **279** (2011) 012025 doi:10.1088/1742-6596/279/1/012025

2.25 MHz-ultrasonic unfocused immersion probes (Ø = 1.27 cm) showed that results are consistent with the manufacturer's specifications.

2. Theorical Approach

Testing beam parameters, for immersion probes applied in non-destructive testing, is defined in EN 12668-2:2001 standard – Non-Destructive Testing – Characterization and Verification of Ultrasonic Examination Equipment – Part 2: Probes [1], specifically in the 7.7 subtitle - Beam parameters for immersion probes. The measurement technique consists of studying the probe acoustic beam in water, using a reflector target or hydrophone receiver. Labus chose to use a hydrophone receiver whereas the signals captured by hydrophone have better relation signal-noise than the one obtained by reflector target.

Parameters should be determined by mapping the immersion probes as follows: axial profile - focal distance and length of the focal zone; transverse profile - focal width, X and Y directions as well as beam divergence.

The focal distance is given as [2]:

$$F_D = |Z_P - Z_0|, \tag{1}$$

where Z_P is the last maximum position of ultrasound beam (Vp), and Z_0 is the coordinate of the probe or its acoustic lens (non-focused probe).

The focal length is given by [2]:

$$F_L = |Z_{L2} - Z_{L1}|, \tag{2}$$

where Z_{L1} and Z_{L2} are the beam axis coordinates where Vp is reduced by 3 dB. Focal distance and focal length shall be within ± 15% of the manufacturer's specifications.

The focal widths on X-axis and Y-axis at focal point are given by the differences [2]:

$$W_{x1} = |X_2 - X_1| \text{ and} \tag{3}$$

$$W_{y1} = |Y_2 - Y_1|, \tag{4}$$

where X_1 and X_2 (Y_1 and Y_2) are the X (Y) transverse axis coordinates where Vp is reduced by 3 dB. The focal widths shall be within ± 15% of the manufacturer's specifications.

The beam divergence is only required for probes that have unfocusing means, such as acoustic lens or curved piezoelectric elements. These are estimated by the measurement of focal width on F_D and Z_{L2} given as (5) and (6):

$$\Omega_x = \frac{360}{2\pi} \cdot \arctan\left[\frac{(W_{x2} - W_{x1})}{2(Z_{L2} - F_D)}\right], \tag{5}$$

$$\Omega_y = \frac{360}{2\pi} \cdot \arctan\left[\frac{(W_{y2} - W_{y1})}{2(Z_{L2} - F_D)}\right], \tag{6}$$

where W_{x2} and W_{y2} are the focal width determined on X-axis and Y-axis on Z_{L2} position. The angles of divergence shall not differ from the manufacture's specified values by either ± 10% or 1°, whichever is the largest.

3. Experimental Procedure

Labus is equipped with a water bath measuring 1700 mm × 1000 mm × 800 mm, which is large enough for most usual measurements and calibrations in the megahertz frequency range. The specified positioning system, used to move the transducer (or hydrophone) in the water bath, allows movement of 300 mm along the X and Y axes, and of 600 mm along the Z axis (Newport Corporation, Irvine,

CA, USA). The X and Y axis present resolution and repeatability better than 1.25 μm, whilst Z achieves a maximum of 5.0 μm. Additionally, there is a 360° rotation system, with a 0.01° resolution.

The typical system configuration used during the mapping acquisition comprises a personal computer connected to an oscilloscope, a signal generator, and movement controllers [4]. To integrate all system components, and also to provide a user-friendly interface, a virtual instrument (VI) was developed in LabVIEW (National Instruments Corporation, Austin, TX, USA) [4]. The VI allows movement control along all the axes, acquisition of waterborne signals, and the calculation of essential parameters to assess and calibrate US transducers. In addition, the software automatically performs the raster scans necessary to calculate the parameters related to the beam parameters for immersion probes applied in non-destructive testing, based on EN12668-2.

The F_D, F_L, W_{x1}, W_{x2}, Ω_x and Ω_y parameters were determined on two probes of 1.27 cm diameter, and frequencies of 1.0 and 2.25 MHz. The probes were excited by using a 20-cycle burst of a sine wave generated by a function generator AFG 3252 (Tektronix, Beaverton, Oregon, USA), and waterborne signals were acquired using an oscilloscope TDS 3032B (Tektronix, Beaverton, Oregon, USA). Needle hydrophones were used in the mapping procedure and, for this, active elements of 0.5 mm were applied (Precision Acoustics Ltd., Dorchester, Dorset, UK).

After aligning the probe and the hydrophone, the scan along the beam axis using steps of 1.0 mm was carried out to determine the position of the last maximum pressure of the transducer beam (F_D). Then, the probe was mapped over two transverse axes to beam axis (X and Y), at F_D position, with steps of 0,1 mm, in a dimension enough to obtain signals lower than 10 dB of the amplitude value found at F_D. The same procedure was carried out at Z_{L2} position. All mappings were recorded, and employed to calculate the previously cited parameters. Details about the measurement procedure are described below.

3.1 Virtual Instrument

Figures 1 and 2 show the VI windows. The controls of the function generator and oscilloscope are presented in [4]. The "Mapeamento" (Mapping) tab (figure 1) is used to define a beam mapping in one or two dimensions, covering the region defined by the user. On the left lower corner, there are radio buttons to select what kind of mapping (line or plane), and controllers to define the step and length, in millimetres, of the displacement in each axis. Note that unused axis is set disable. The files obtained from mappings are analyzed in the "END" tab (figure 2), where files are uploaded and showed, and the values of focal distance (F_D), focal length (F_L), focal width on F_D (W_{x1} and W_{y1}), focal width on Z_{L2} (W_{x2} and W_{y2}) and divergence angles (Ω_x and Ω_y) are calculated and presented.

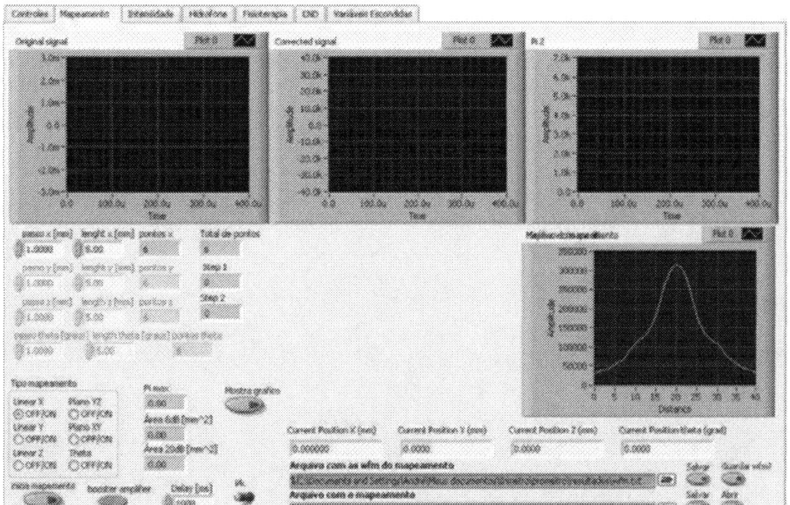

Figure 1. "Mapeamento" tab, where all mappings are carried out.

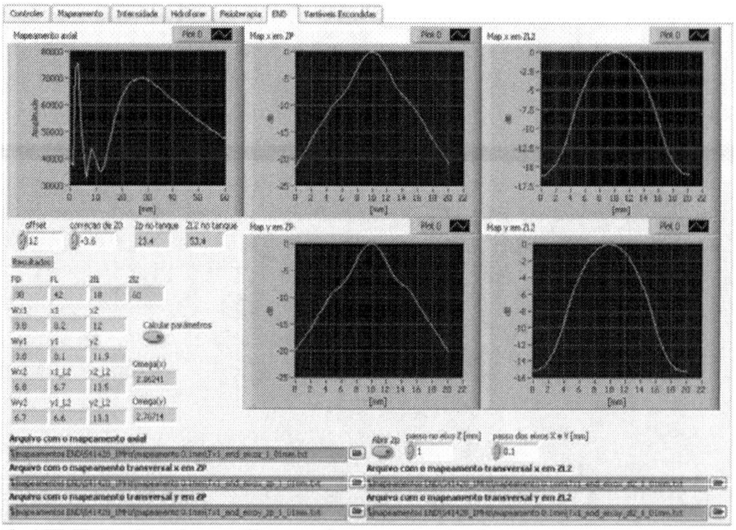

Figure 2. "END" tab, where all mappings and calculated parameters are presented.

4. Determination of standard uncertainty of Type A and Type B

Herein, the finite resolution of the positioning system used to perform the raster scans is assumed to present a rectangular distribution. Hence, the Type B uncertainty of s ($u_{s(Type\,B)}$) is estimated by dividing the equipment resolution (1.25x10^{-4} cm) by $2\sqrt{3}$. The Type A uncertainty ($u_{s(Type\,A)}$) of the position system is estimated, for each of the three axes, as the standard deviation of the mean of four measurements of their linear translation. This consideration, takes in account that the F_D conventional value can have a uniform probability between two mapping consecutive points. Type A uncertainty for F_D ($u_{F_D(Type\,A)}$) measurements was estimated as the standard deviation of the mean of 4 measurements, divided by $\sqrt{4}$

The Type B uncertainty of Z_{L2} ($u_{Z_{L2}(Type\,B)}$) and Z_{L1} ($u_{Z_{L1}(Type\,B)}$) were estimated considering the used step (s) divided by $2\sqrt{3}$, taking the same consideration made in F_D. The Type A uncertainty of Z_{L2} ($u_{Z_{L2}(Type\,A)}$) and Z_{L1} ($u_{Z_{L1}(Type\,A)}$) were estimated as the standard deviation of the mean of 4 measurements, divided by $\sqrt{4}$.

In W_{x1}, W_{x2}, W_{y1}, W_{y2} uncertainty determination, the Type B for X_2 ($u_{X_2(Type\,B)}$), X_1 ($u_{X_1(Type\,B)}$), Y_2 ($u_{Y_2(Type\,B)}$), Y_1 ($u_{Y_1(Type\,B)}$), X_2_2 ($u_{X_2\_2(Type\,B)}$), X_1_2 ($u_{X_1\_2(Type\,B)}$), Y_2_2 ($u_{Y_2\_2(Type\,B)}$) and Y_1_2 ($u_{Y_1\_2(Type\,B)}$) were estimated considering the used step (s) divided by $2\sqrt{3}$, taking the same consideration made in F_D estimative. The respective Type A uncertainty was estimated as the standard deviation of the mean of 4 measurements, divided by $\sqrt{4}$.

Lastly, Ω_x and Ω_y uncertainty determination considered the uncertainty obtained of the used parameters to its determination.

5. Results and discussion

The performance of the implemented system was evaluated by determination of FD, FL, Wx1, Wx2, Wy1, Wy2 e Ωx and Ω_y values, as well as respective measurement uncertainty, to 1.0 and 2.25 MHz NDT probes. Tables 1 and 2 show the evaluated parameters and the respective uncertainty to 1 and 2.25 MHz probes. The obtained values to F_D and F_L are different from the theoretical values available in the manufacturer's technical notes [5]. It is important to note that if the obtained result F_D (29.50 mm) was applied to calculate F_L by equation $F_L = F_D.[2/(1+0.5)]$ found at manufacturer's technical notes [5], the F_L would be 39.33 mm, which would be coherent with the obtained value of 41.25 mm, considering the estimated uncertainty. Therefore, the measured values of F_D and F_L to 1 MHz probe should be reassessed. The tested values and uncertainty calculated to 2.25 MHz probe are in accordance with the theoretical values presented manufacturer's technical notes [5] and in EN 12668-2, respectively. The W_{x1}, W_{x2}, W_{y1}, W_{y2} e Ω_x and Ω_y parameter required in actual standard were not found manufacturer's technical notes [5], however, the estimated uncertainties were in accordance with the ones stated in EN 12668-2.

Table 1. Parameter values, and respective values of uncertainty, for the 1 MHz NDT probe (Ø = 12.7 mm), panametrics A303S, serial number 541428.

Parameter	Unit	Theoretical values [5]	Results	Expanded Uncertainty	Expanded Uncertainty[%]
F_D	[mm]	26.49	29.50	0.87	2.9
F_L	[mm]	35.32	41.25	0.97	2.4
W_{x1}	[mm]		3.78	0.10	2.6
W_{y1}	[mm]		3.80	0.09	2.4
W_{x2}	[mm]		6.73	0.14	2.1
W_{y2}	[mm]		6.73	0.10	1.5
Ω_x	[°]		2.84	0.13	4.6
Ω_y	[°]		2.81	0.08	2.8

Table 2. Parameter values, and respective values of uncertainty, for the 2.25 MHz NDT probe (Ø = 12.7 mm), panametrics V306, serial number 536444.

Parameter	Unit	Theoretical values [5]	Results	Expanded Uncertainty	Expanded Uncertainty [%]
F_D	[mm]	60.30	60.50	0.87	1.4
F_L	[mm]	80.67	83.00	4.25	5.1
W_{x1}	[mm]		3.40	0.13	3.8
W_{y1}	[mm]		3.45	0.15	4.3
W_{x2}	[mm]		6.55	0.35	5.3
W_{y2}	[mm]		6.60	0.17	2.6
Ω_x	[°]		1.44	0.05	3.5
Ω_y	[°]		1.44	0.05	3.5

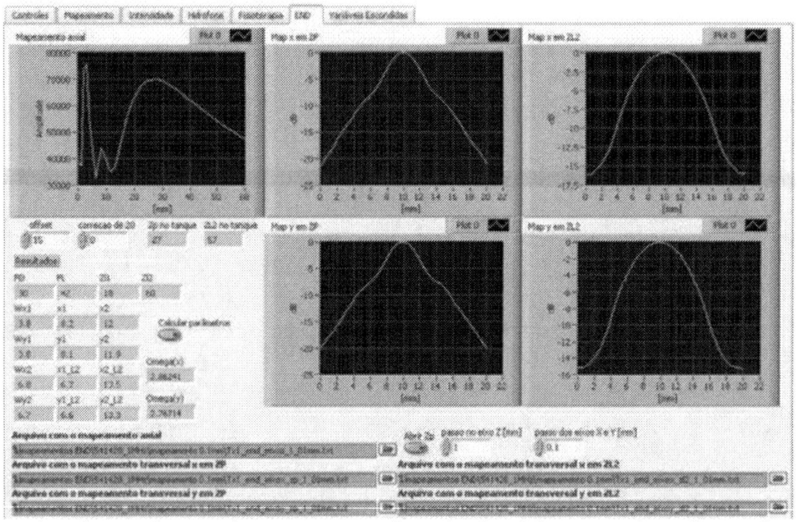

Figure 3. Example of mappings and parameters value obtained from test 1 performed on 1 MHz NDT probe (Ø = 12.7 mm), panametrics A303S, serial number 541428.

6. Conclusion

The mapping system of ultrasound immersion NDT probe was assessed in determining the focal distance, length and width and divergence beam for 1 and 2.25 MHz probes. The values obtained to 1 MHz probe were different to the theoretical values found in the manufacturer's technical notes and should be reassessed. Nevertheless, results achieved to 2.25 MHz probe were in accordance to the values presented manufacturer's technical notes. Finally, the presented estimated uncertainties were in accordance with EN 12668-2:2001 standard.

References

[1] Schmerr L W, Lopez-Sanchez A, Huang R, 2006, Complete ultrasonic transducer characterization and its use for models and measurements, Ultrasonics, **44**, e753–e757.

[2] EN 12668-2, Non Destructive Testing – Characterization and Verification of Ultrasonic Examination Equipment – Part 2: Probes.

[3] Alvarenga AV, Cardona M A R, Costa-Félix, R.P.B., 2006, Acoustic pressure field mapping system at laboratory of ultrasound of Inmetro, In: Proceedings of XVIII IMEKO World Congress Metrology for a Sustainable Development, Rio de Janeiro, p. 1-6, 17–22 Sept.

[4] Alvarenga A V, Costa-Félix, R P B, Oliveira, E G, 2009, Uncertainty assessment of effective radiating area and beam non-uniformity ratio of ultrasound transducers determined according to IEC 61689:2007", Metrologia, **46,** 367-374.

[5] Ultrasonic Transducers Technical Notes, Panametrics-NDT - Olympus. p. 1-11, 2003. Available in www.olympusNDT.com / [acessed on 16/Apr/2010].

Measurement system for experimental determination of acoustic properties of gels at INRIM

C Musacchio, G Durando, A Bernardi and A Troia
INRIM - Istituto Nazionale di Ricerca Metrologica-, Strada delle Cacce 91, 10135 – Torino, Italy

E-mail: c.musacchio@inrim.it

Abstract. Due to the large diffusion of ultrasonic technique in medicine, use of tissue mimicking materials to understand physical interaction of ultrasound has grown in importance. From this view measuring acoustic properties of such materials is crucial. This work shows the settlement at INRIM laboratory of a simple system for acoustic properties measurements with particular interest in attenuation of ultrasound. The system is based on substitution method and Fourier analysis and allow the evaluation of speed of sound, impedance and attenuation in the frequency range between 3 MHz and 12 MHz.

1. Introduction

The characterization of ultrasonic fields generated in water by transducers as provided by national and international standards is surely a fundamental tool for the control of device performance and for patient safety protection, but the information they give is not complete. As a matter of fact, they describe the morphology and intensity of the field, but not the effect induced by this field in real tissue.

A new approach is required, based on simulation of physical effects and the quantitative measure of them through in vitro measurements. So, the issue is to realize new materials for simulation of acoustic properties of real tissue, new parameters which really refer to biophysical effects intensity, and a robust methodology for measuring them. At the moment a really great interest is present in international scientific community and in standardization board towards research in this field [1-8].

The first problem for this approach would be the realization of suitable tissue mimicking phantoms. They are object made with materials whose physical properties are as close as possible to those of real tissues. [9]

Some object of this kind already exists and have been realized for the control of performance of medical ultrasonic devices. As an example they exist for sale phantoms for the control of image quality for scanners. They are phantoms, usually made of agar gel or other kind of gels. Materials similar to tissue mimicking object However, for the realization to be studied in detail: the realization of new measurement method.

This work describe a study of tissue mimicking materials and the realization of a quite simple set-up and a easy method for testing these materials and in particular for the determination of their main acoustic properties, i.e. speed of sound, acoustic impedance and absorption coefficient [10].

The measurement system designed and the activity conducted can be considered a good and promising starting point for further studies.

2. Materials and methods

Acoustic impedance and attenuation coefficient of a tissue mimicking gel have been measured using a substitution method and a trough transmission technique. In the developed system the unknown acoustic properties of a gel sample are obtained from the comparison with the properties of water. In extreme synthesis the method requires two ultrasonic transducers, one for the generation of a wave pulse and the other for the detection of the wave after its propagation trough the sample. Substituting the unknown sample with water allows to estimate its acoustic properties by comparing the received signal when the sample is inserted in the propagation path and when the wave travels only trough water.

The samples used for the measurement system testing are polyacrylamide gels. They have been realized in our laboratory starting from information found in scientific literature [10]. The recipes and production system has been improved in order to realize gels with different acoustic characteristics.

2.1. Measurement set up

The measurement set up is quite simple and it is composed of two wide band ultrasonic transducers, one used as a transmitter and the other as a receiver, a signal generator, to drive the transmitter, and a digital oscilloscope for acquire both the electrical signal sent at the transmitter input and the electrical signal produced by the receiver. All the instruments are remote controlled with a personal computer. The software for the instrument control, the signal generation and acquisition has been developed.

The signals acquired are then processed by an other home made software to calculate acoustic impedance and absorption coefficient of the sample. Figure 1 shows a picture of the measurement set up.

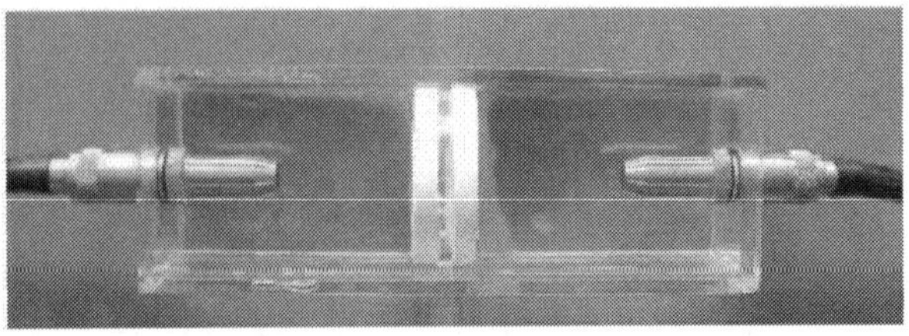

Figure 1. Picture of the measurement box with the sample between the two transducers

The measurement process involves a two steps procedure: first a measure with water, and then a second measure with the gel sample.

The two transducers are placed at both end of a little water tank, one in front of the other. A sinusoidal burst with a limited number of cycles at the working frequency of the transducer is generated. Driven by the generator, the transducer emits pulsed ultrasonic waves at the set pulse frequency. Emitted and recived signals, are acquired by the digital oscilloscope and stored on pc. In order to reduce the noise of the signals each acquisition is the result of a large number of average and the acquisition is repeated about ten times so that the subsequent analysis is conducted over a larger number of signals.

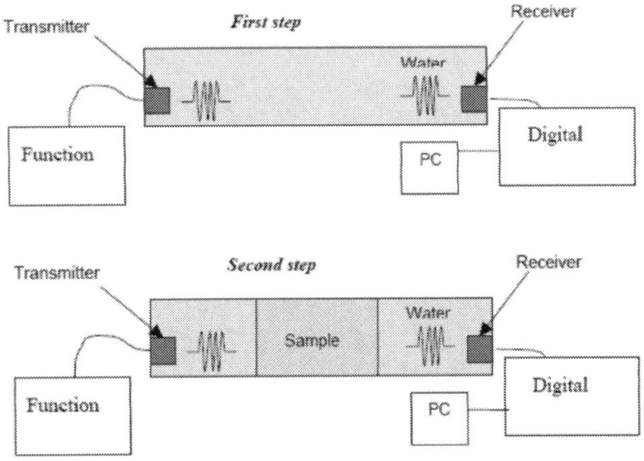

Figure 2. A schematic representation of the procedure of measurements. The acoustic properties of the sample are measured by a comparison with the known properties of water

Because of the dependence of the acoustic properties of materials from their temperature the process has been designed so that its duration is quite short (about few minutes) so that environmental influence variables has not time enough to change significantly. However during the whole measurement the temperature of the water tank is monitored continuously with a thermocouple.
When the measure is completed the gel sample is inserted in the water tank between the transmitter and the receiver and the described process is repeated (Figure 2).

2.2. Signal processing and data analysis

The signal processing and the data analysis is carried out off line by a software, developed in LabView 7.1. First of all software calculates the time of flight of the wave in water and in presence of the sample. The cross correlation of the two time functions (transmitted and received signals as acquired by the oscilloscope)is calculated and the time delay that maximizes such a function is taken as the flight time. So the speed of sound in the sample, c_S can be calculated from the measured times of flight as:

$$c_S = \frac{c_{H2O} \cdot d}{d + (t_S - t_{H2O}) \cdot c_{H2O}} \quad (1)$$

where c_{H2O} is the speed of sound in water, d is the sample length, t_{H2O} is the signal time of flight measured in water and t_S is the time of flight in presence of the sample. c_{H2O} is calculated from the water temperature according to existing theoretical models.
Once the sound speed in sample is known, its impedance Z_S can be immediately obtained from the formula:

$$Z_S = \rho_S \cdot c_S \quad (2)$$

with ρ_S being the measured sample density.
The reckoning of the absorption coefficient requires the comparison between the amplitude of the signal received when the whole path is water with the amplitude of the signal when the sample is present. The signal analysis is conducted in the frequency domain so that it is possible to estimate the absorption behaviour at different frequencies with a single calculation.
Without considering in detail all the phenomena occurring, it is possible to consider all the system as a black box which transform an electrical signal into an other. According to this simplified model, in the frequency domain, it is possible to express the component of the amplitude of the received signal at the frequency f, $A_{rw}(f)$, as:

$$A_{rw}(f) = A_{tw}(f) \cdot F_{tw}(f) \quad (3)$$

with $A_{rw}(f)$ being the Fourier transform of the received signal, $A_{tw}(f)$ is the Fourier transform of the generated signal, and $F_{tw}(f)$ is the transfer function of the system. The last term of the equation is certainly an abstraction, but it is possible to imagine that it contains the effect of many physical phenomena, such as the transduction efficiency of the transmitter and the receiver, the geometrical spreading of the ultrasonic wave, the diffraction effect, the absorption in water, usually considered negligible for so short distances, and so on.
In a similar way when the sample is inserted, the amplitude of the received signal can be expressed as:

$$A_{rS}(f) = A_{tS}(f) \cdot F_{tS}(f) \quad (4)$$

where $A_{rS}(f)$ is the Fourier transform of the received signal with the sample between the transmitter and the receiver, $A_{tS}(f)$ is the transform of the signal generated by the function generator and $F_{tS}(f)$ is the transfer function of the system.
In $F_{tS}(f)$ two very important effects are present that missed in the former term: a loss in amplitude due to the impedance de-coupling between the two different propagation media and the attenuation in gel.
The amplitude of the transmitted wave, A_T, between two media is the result of the product between the amplitude of the incident wave, A_I , and the transmission coefficient,$T_{1\text{-}2}$, according to the equation:

$$A_T = T_{1-2} A_I = \left(\frac{2Z_2}{Z_2 - Z_1} \right) \qquad (5)$$

where Z_1 and Z_2 are the acoustic impedance of the two media

Therefore, if we neglect all the effects but the loss due to de-coupling the relationship between the received signal with water only and with gel sample would be:

$$A_{rS} = A_{rw}(f) \cdot T_{S-w} T_{w-S} = A_{rw}(f) \cdot \left(\frac{4Z_w Z_S}{(Z_w - Z_S)^2} \right). \qquad (6)$$

The value of attenuation coefficient in water can be neglected and the only attenuation in gel can be taken in consideration. The attenuation coefficient of gel, α, is expressed in dB/cm.

If we now summarize all the consideration made about the main effects experienced by the wavelets travelling along the gel it is possible to suppose a relationship between the transfer functions which accounts for the effect of the impedance mismatch between water and gel sample as well as of the attenuation into the gel.

$$F_{tS} = F_{tw} \cdot T_{S-w} \cdot T_{w-S} \cdot 10^{\frac{\alpha d}{20}} \qquad (7)$$

Since the received signal can be influenced by an accidental variation of the electrical signal sent to the transmitted, in order to correct for this effect, it has been preferred to compare not directly the Fourier transform of the received signals, but the frequency response of the system ,it is to say the frequency components of the received signals normalized for the frequency components of the corresponding input signal. Then attenuation coefficient is calculated as:

$$\alpha(f) = \frac{20}{d} \log_{10} \left[\frac{\dfrac{A_{rs}(f)}{A_{ts}(f)}}{\dfrac{A_{rs}(f)}{A_{ts}(f)} \cdot \dfrac{4Z_w Z_S}{(Z_w - Z_S)^2}} \right] \qquad (8)$$

3. Measurement results

Gel samples used for the measurements have been realized according to the procedure assessed and their main acoustic parameters have been calculated. First test have been done with simple polyacrylamide gel. In figure 3 the results of a set of five measurements of the attenuation coefficient, conducted at INRIM laboratories over the same sample, are reported. From the data it is possible to recognise the dependence of the coefficient from the frequency. It can be seen as the data dispersion is quite limited in the frequency range 3-15 MHz, but they dramatically spread when the frequency increases, due to the low signal to noise ratio at the higher frequencies. So the attenuation coefficient. has been evaluated for the frequency range between 3 MHz and 12 MHz.

Measurements have been done on gel samples obtained with different recipes. In particular polyacrylamide gels have been realize with kieselguhr mixed in the material during preparation process in order to obtain different attenuation coefficient.

Measurement results of attenuation coefficient are reported in figures 4 and 5. Results are compared to those made at NPL on similar samples produced at INRIM laboratory and sent to NPL for characterization. NPL results are marked as "reference values".

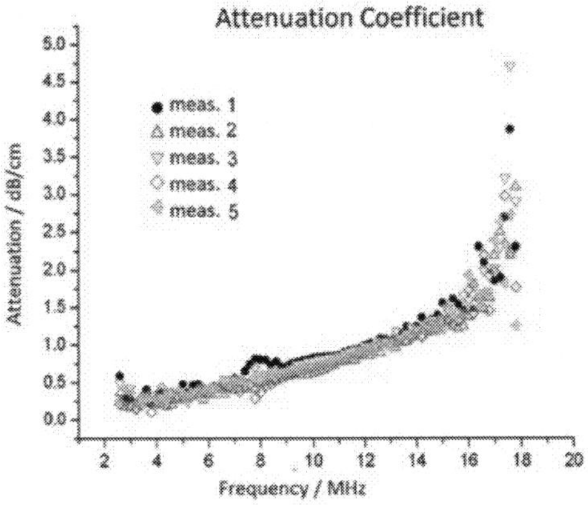

Figure 3. Results of some preliminary measurements of attenuation coefficients in the frequency range 2.5–18 MHz. Measurements executed at the INRIM laboratories

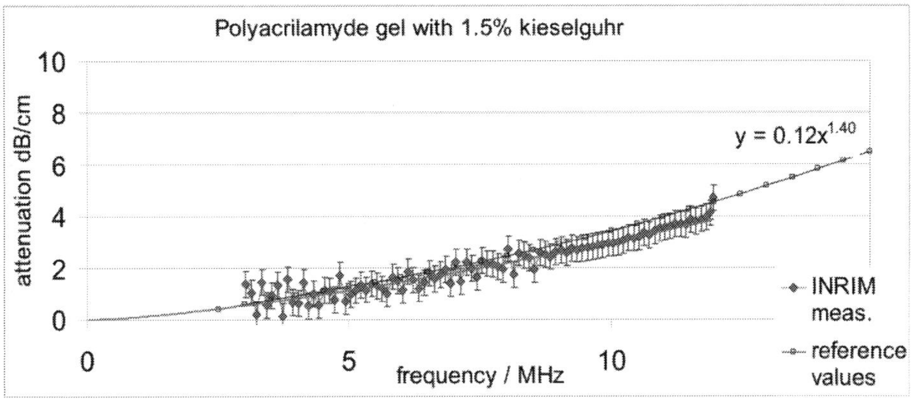

Figure 4. Results measurements of attenuation coefficients in the frequency range 3 – 12 MHz. of sample with 1.5% weight fraction mixed .

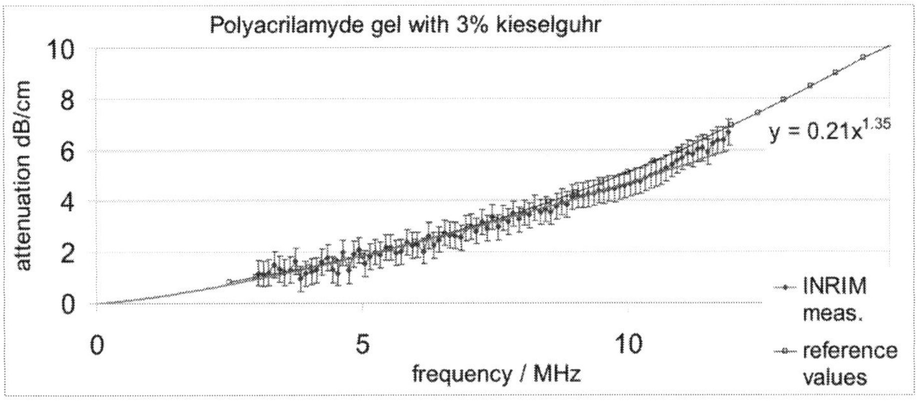

Figure 5. Results measurements of attnuation coefficients in the frequency range 3 – 12 MHz. of sample with 3% weight fraction mixed

Power law curve has been use to fit the data and to obtain the value of the attenuation coefficient as a frequency function. The two sample under investigation differ only for the weight fraction of kieselghur that is 1.5% in the firs case and 3% in the second. The attenuation coefficient found are respectively 0.12 dB7cm7MHz$^{1.4}$ and 0.21 dB7cm7MHz$^{1.4}$.

4. Conclusion and future development

The present set up allows measurement of attenuation coefficient in a frequency range from 3 MHz to 12 MHz. INRIM data and reference data are quite in a good agreement, especially for lower frequencies (below 6 MHz), while at higher frequencies the two data sets tend to diverge with systematically higher values for the NPL data. The difference is significant in particular for the less attenuating sample. This discrepancy could be explained considering that the two samples analysed in the two different laboratories can present some differences due to their maintenance, but a complete explanation of possible causes of discrepancy should be needed. Anyway the measurements can be considered reliable considering the uncertainty of the INRIM values that is about 10 %.

The measurement system can be used for simple, easy and quick measurement of characterization of gel samples.

The method described has shown to be promising, but surely need some further development. One problem is the too low signal to noise ratio for measurements relative to absorption coefficient in the spectral band far from the acoustic working frequency of the transducer. The problem could be solved by using other kind of transducers, at the moment not present in the laboratory, with a broader band or using a set of transducers with different acoustic working frequency able to cover a wider range.

The frequency range under investigation is of interest in medical application, One of the future targets will be to succeed in characterizing the acoustic properties of the gels in used for simulating physical effects induced by High Intensity Focuses Ultrasound fields.

The system realized for acoustic characterization of material will be useful for the aim of studying heating effects on tissue mimicking phantom induced by High Intensity Focuses Ultrasound fields.

5. References

[1] Shanna Lochhead et al "A gel phantom for the calibration therapy" 2004 IEEE Ultrasonics Symposium1481-1483

[2] Burlew MM, Madsen EL, Zagzebski JA, Banjavic RA, Sum SW. A new ultrasound tissue-equivalent material. Radiology 1980;134: 517–520.

[3] Lafon C, Vaezy S, Noble M, et al. New synthetic tissue-mimicking phantom for high intensity focused ultrasound. Proceedings of the International Congress of Acoustic, Rome, Italy, 2001.

[4] Wu J. Tofu as a tissue-mimicking material. Ultrasound Med Biol 2001;27:1297–1300. 1422 Ultrasound in Medicine and Biology Volume 30, Number 10, 2004

[5] E.L. Madsen et al. "Tissue-Mimicking Oil-in-Gelatin Dispersions for Use in Heterogeneous Elastography Phantoms" Ultrasonic Imaging 25, 17-38 (2003)

[6] A. F. Prokop et al. "Polyacrylamide gel as an acoustic coupling medium for focused ultrasound therapy" Ultrasound in medicine and biology, volume 29 number 9, 1351–1358 (2003)

[7] Kondo, T.; Kitatuji, M.; Kanda, H. "New tissue mimicking materials for ultrasound phantoms" Ultrasonics Symposium, (2005) IEEE; 3 (18-21), 1664 – 1667

[8] J.J. Ammann, G. Donoso , et al. "Tissue-mimicking materials assessment through ultrasound velocity measurement" Medical Imaging 2003: Ultrasonic Imaging and Signal Processing. Edited by Walker, William F.; Insana, Michael F. Proceedings of the SPIE, Volume 5035, pp. 440-451 (2003).

[9] F.A. Duck, "Physical properties of tissue. A comprehensive reference book", New York: Academic Press, 1990.

[10] El. Madsen et al " Interlaboratory comparison of ultrasonic backscatter attenuation and sound speed measurements" J. Ultrasound Medicine, vol. 18 615-631 (1999).

Acknowledgements

This work was supported by the European Community's Seventh Framework Programme, ERA-NET Plus, under Grant Agreement No. 217257 (EURAMET joint research project).

Journal of Physics: Conference Series **279** (2011) 012027 doi:10.1088/1742-6596/279/1/012027

The importance of expression of uncertainty of acoustical parameters of ultrasonic phantoms

L E Maggi[1,2,3], A B B Souza[1], R M Ichinose[1], W C A Pereira[1], M A von Kruger[1], R P B Costa-Felix[3]

1 Programa de Engenharia Biomédica / COPPE – UFRJ, Rio de Janeiro, Brasil.

2 Departamento de Fisioterapia / Universidade Estadual de Goiás, Goiânia, Brasil.

3 Ultrasound Laboratory, Diavi/Dimci/Inmetro, Duque de Caxias, RJ, Brasil.

[1] luis.maggi@gmail.com

Abstract. The measurement of uncertainties in scientific experiments improves greatly quality and reliability of the results. However, in many cases, experimental results are only expressed by its average value and standard deviation. The longitudinal velocity and attenuation coefficient are acoustic parameters commonly used to characterize biological tissues and materials. In this work it is studied the uncertainty in experiments designed to evaluate these parameters on two different materials (silicone rubber and PVCP). The uncertainties were studied following the Guide to the Expression of Uncertainty in Measurement and calculated by a program in Labview8.6. One setup was developed to measure the acoustic parameters by a transmission/reception technique. Five signals of each medium (water and materials) were collected. The attenuation coefficient was calculated using the relation between the amplitude spectrum peak of the water signal and its respective point on the spectrum of the material signal. The longitudinal velocity was calculated using the time delay between signal peaks (from water and from the material). The individual uncertainties of each part of setup were estimated and these values permitted to identify which were the sources of uncertainty that most contributed to increase the value of associated uncertainty. It permitted to improve experiment's quality and reliability.

1. Introduction

Ultrasonic Phantoms are materials intended to mimic acoustic properties of biological tissue (propagation velocity, attenuation coefficient and acoustic impedance [1,2,3] and are used to calibrate and evaluate the quality of ultrasound equipment among other applications. To develop these phantoms it is necessary to measure the acoustic properties of their constituents. The Laboratory of Ultrasound (LUS) of the Biomedical Engineering Program (PEB) from the Federal University of Rio de Janeiro (UFRJ) and the Ultrasound Laboratory from Diavi/Dimci/Inmetro are implementing a methodology for measuring the attenuation coefficient and longitudinal velocity of the testing objects along with their uncertainties expression.

Several published studies such as Prokop et al. (2003) [1], Takegami et al. (2004) [2] and Zell et al.

(2007) [3] present the attenuation coefficient and acoustic velocity just as mean values and standard deviations. The measurement of uncertainties in scientific experiments greatly improves quality and reliability of the results. Publications such as the Guide to the Expression of Uncertainty in Measurement (GUM) by ISO (Adopted by INMETRO [4]) provide a methodology to calculate the uncertainty.

The aim of this work was to apply the method described in GUM to express the uncertainties of measurement of attenuation coefficient, obtained by the method implemented in LUS.

2. Materials and Methods

2.1. Materials

Two different materials were used as testing objects: RTV615 silicone rubber (GE) and PVCP (M-F Manufacturing Co). The RTV615 is a silicone product, consisting of a solution of two viscous components (RTV 615 A: RTV 615 B) when mixed in a 10:1 proportion, they initiate a healing process, even at room temperature. To prepare a PVCP phantom, the liquid is placed in a beaker and then put into a vacuum chamber to remove air bubbles. The healing of both phantoms were made in containers with 2.20 x 3.05 x 4.68 cm.

2.2. Measurements

The experiment to measure ultrasound attenuation coefficient and velocity in phantoms is divided in two stages. In the first stage, the phantom is positioned between two similar transducers, such that one of these emits a pulse signal (15 cycles and voltage of ± 10 V_{pp}) and the other receives the signal transmitted through the phantom (Figure 1). In the second stage, the phantom is removed and the signal measurements are made in a water layer with same phantom thickness. The measurements performed in water are used as reference.

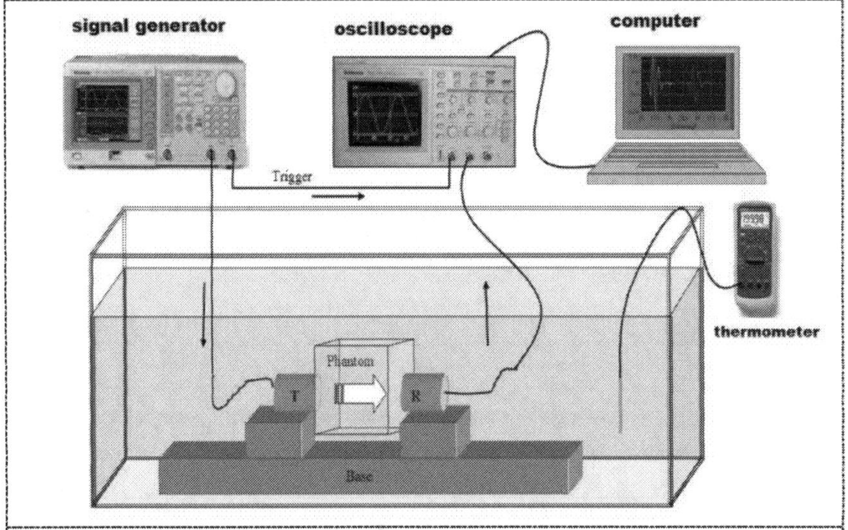

Figure 1. Design of experimental setup.

The functions representing the measurement of attenuation can be seen in equation (1), α is the phantom attenuation coefficient, V_{pha} and V_{ref} represent the amplitudes of ultrasound signal in phantom and in water respectively. Two different amplitudes were used to compare (in spectral - FFT and time

- RMS domains) using a program developed in Labview 8.6 (AcousticTrans6 in Figure 2). D is the phantom thickness. Nonlinear propagation effects (not investigated in this study) were assumed to be negligible for the α measurement because pressure amplitudes were small and propagation paths were short.

$$\alpha = \frac{20\log\frac{V_{pha}}{V_{ref}}}{D} \tag{1}$$

In equation (2), v_{ref} represents the ultrasound velocity in water according to equation (3), where T is water temperature [5]. t_{ref} and t_{pha} are respectively the time of flight (TOF) of the ultrasound signal into the water layer and into the phantom.

$$v_{pha} = v_{ref}\frac{t_{ref}}{t_{pha}} \tag{2}$$

$$v_{ref} = 1.40238744 \times 10^{3} + 5.03836171\,T - 5.81172916 \times 10^{-2}\,T^{2} + 3.34638117 \times 10^{-4}\,T^{3} - 1.48259672 \times 10^{-6}\,T^{4} + 3.16585020 \times 10^{-9}\,T^{5} \tag{3}$$

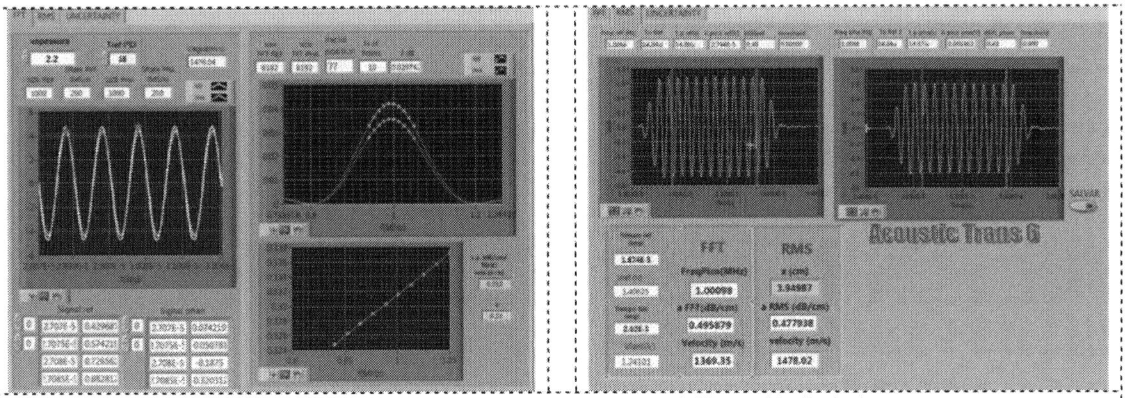

Figure 2. Windows of the software AcousticTrans6 in Labview.

Uncertainty types A and B of each source that contributes to the expression of attenuation uncertainty were calculated and applied in equation (4), where x could be either thickness (D) or amplitudes of ultrasound signal in phantom (V_{pha}) or in water (V_{ref}), according to ISO GUM 2003[4].

$$u^{2}(x) = u_{A}^{2}(x) + u_{B}^{2}(x) \tag{4}$$

2.3. Combined standard uncertainty

It is the quadratic sum of the product of each uncertainty component with their respective sensitivity coefficient. This coefficient is defined as the partial derivative of the respective measurand (attenuation) with respect to the uncertainty component. The terms used in the calculation of sensitivity coefficients can be seen in equation (5):

$$\frac{\partial \alpha}{\partial V_{pha}} = \frac{20\log e}{D \cdot V_{pha}} \quad ; \quad \frac{\partial \alpha}{\partial V_{ref}} = -\frac{20\log e}{D \cdot V_{ref}} \quad ; \quad \frac{\partial \alpha}{\partial D} = -\frac{20\log \frac{V_{pha}}{V_{ref}}}{D^2} \tag{5}$$

2.4. Expanded uncertainties

The expression of the expanded standard uncertainty ($U = k \cdot u_c$) considered the confidence level of 95%. The coverage factor k depends on the effective degrees of freedom that were calculated by equation (6):

$$v_{eff} = \frac{u_c^{\,4}}{\sum \frac{u_i^{\,4}}{v_i}} \tag{6}$$

Where u_c is the combined standard uncertainty of the measurand whose effective degrees of freedom are calculated; u_i, is components of combined standard uncertainty; and degrees of freedom v_i is component of combined standard uncertainty in question.

3. Results and Discussions

The results were based on data from experiments with silicone phantoms and manufactured in PVCP at LUS / PEB / COPPE / UFRJ. Measurements were made with transducers of 1 MHz. It was considered irrelevant uncertainties due to variation of temperature, relative humidity and no homogeneity of the phantoms in the calculation of the uncertainty. The results of measurements performed using silicone phantoms are on Table 1, while the ones with PVCP phantoms are on Table 2 and can be seen in Figure 3.

Table 1. Silicone phantom uncertainty measurements including signal amplitudes of V_{ref} and V_{pha}, and thickness.

Greatness	Estimate	Standard uncertainty	Probability distribution	Sensitivity Coefficient	uncertainty contributes	Combined standard uncertainty	Vi	Veff	k	Unc. Exp. 95%
thickness [cm] – type A	2.302	1.80E-03	Normal	-5.30E-001	-9.53E-04	1.10E-01	inf	3.53E+02	2.25	0.247
thickness [cm] – type B		2.89E-04	Rectangular		-1.53E-04		5			
Vref [V] – type A	3.661	1.32E-02	Normal	1.03E+000	1.36E-02		inf			
Vref [V] – type B		5.80E-02	Rectangular		5.98E-02		5			
Vpha [V] – type A	2.650	2.64E-02	Normal	1.42E+000	3.76E-02		inf			
Vpha [V] – type B		5.80E-02	Rectangular		8.26E-02		5			

Table 2. PVCP phantom uncertainty measurements including signal amplitudes of Vref and Vpha, and thickness.

Greatness	Estimate	Standard uncertainty	Probability distribution	Sensitivity Coefficient	Uncertainty contributes	Combined standard uncertainty	Vi	Veff	k	Unc. Exp. 95%
thickness [cm] – type A	2.202	1.80E-03	Normal	-1.80E-001	-3.23E-04	1.22E-01	inf	3.73E+02	2.25	0.274
thickness [cm] – type B		2.89E-04	Rectangular		-5.19E-05		5			
Vref [V] – type A	3.713	3.10E-02	Normal	1.06E+000	3.30E-02		inf			
Vref [V] – type B		7.03E-02	Rectangular		7.47E-02		5			
Vpha [V] – type A	3.359	3.10E-02	Normal	1.17E+000	3.64E-02		inf			
Vpha [V] – type B		7.03E-02	Rectangular		8.26E-02		5			

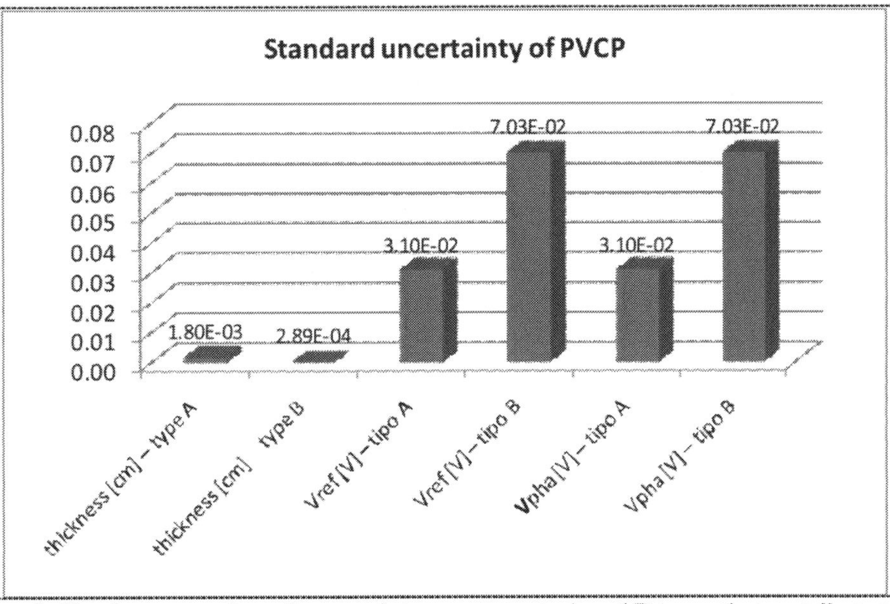

Figure 3. Graph presentation of uncertainty components A and B types that contribute to the uncertainty expression of attenuation coefficient.

Comparing Tables 1 and 2 and the graph in Figure 3 it can be seen that the type B standard uncertainty of V_{ref} and V_{pha} contributes to the combined standard uncertainty of the attenuation coefficient. The value of this quantity was obtained by performing five measurements in the same phantom. Changes in the measurement procedure, for example, increasing the measuring number or even measuring instrument (an instrument with higher accuracy) may allow a reduction in uncertainty.

4. Conclusion

A study of uncertainties in attenuation values of ultrasonic measurement method commonly used in laboratory experiments was conducted. It was found that for the attenuation, the largest uncertainty is due to V_{ref} and V_{pha} measurements. The uncertainties obtained are less than 1% of the value of measurands for this reason it is possible to consider them acceptable for this experiment

5. Reference

[1] Takegami K, Kaneko Y, Watanabe T, Maruyama T, Matsumoto Y, Nagawa H. 2004 Polyacrylamide Gel Containing Egg White as New Model for Irradiation Experiments Using Focused Ultrasound. *Ultrasound In Med. & Biol.* Vol. **30**, No. 10, pp. 1419-22.

[2] Prokop A F, Vaezy S, Noble M L, Kaczkowski P J, Martin R W, Crum L A. 2003. Polyacrylamide Gel as an Acoustic Coupling Medium for Focused Ultrasound Therapy. *Ultrasound In Med. & Biol.* Vol. **29**, No. 9, pp. 1351–58

[3] Zell K, Sperl J I, Vogel M W, Niessner R, Haisch C. 2007. Acoustical Properties of Selected Tissue Phantom. *Physics in Medicine and Biology.* **52** N475–N484.

[4] BIPM 2008 Evaluation of measurement data – Guide to the expression of uncertainty in measurement, JCGM 100: (GUM 1995 with minor corrections), in: www.bipm.org/en/publications/guides/gum.html

[5] National Physical Laboratory Npl. 2000. Underwater Acoustics, Technical Guides - Speed of Sound in Pure Water. Teddington, Middlesex, Uk: Npl;.

Journal of Physics: Conference Series **279** (2011) 012028 doi:10.1088/1742-6596/279/1/012028

Sensibility of hydrous ethanol adulteration detection using ultrasonic parameters validated in a metrological base

Monique K-K Figueiredo[1,2], Rodrigo P B Costa-Felix[1] , Luis E Maggi[1,3], André V Alvarenga[1] and Gilberto A Romeiro[2].

[1]*Laboratory of Ultrasound (LABUS), Directory of Scientific and Industrial Metrology (DIMCI), National Institute of Metrology, Standardization, and Industrial Quality (Inmetro), Av. Nossa Sra das Graças, 50 (Prédio 1), Duque de Caxias, RJ, Brazil, ZIP 25250-020*
[2]*Depto of Organic Chemistry, Institute of Chemistry, Fluminense Federal University (UFF), Campus of Valonguinho, Outeiro de São João Batista, s/n, Niteroi, RJ, Brazil, ZIP 24020150*
[3]*Biomedical Engineering Program, COPPE/Federal University of Rio de Janeiro (UFRJ), Horácio Macedo Ave., n.° 2030, Center of Technology, Block H – room/327 Cidade Universitária, Rio de Janeiro, RJ, Brazil, ZIP 21941-914*

mkfigueiredo@inmetro.gov.br, rpfelix@inmetro.gov.br, lemaggi@inmetro.gov.br, avalvarenga@inmetro.gov.br, gilbertoromeiro@ig.com.br

The aim of this study is to identify possible changes in fuels, in this case hydrous ethanol, through ultrasonic parameters such as attenuation and propagation speed. The system setup and method were implemented at Inmetro's Laboratory of Ultrasound. Experiments and method uncertainties were assessed accordingly to ISO/IEC Guide 98 1:2009 (Uncertainty of measurement – Part 1: Introduction to the expression of uncertainty in measurement). Mixtures of ethanol and water varying from 90% to 94% of alcohol in mass were used as testing samples. Attenuation and propagation speed were accurately measured and uncertainties evaluated. The accordingly to Brazilian biofuel regulations, the concentration of water in hydrous ethanol can be accepted at a maximum concentration of 93.8 and minimum of 92.6 of alcohol in mass. To achieve that figure, a functional combination of tested parameters should be implemented. Those results could be used as a tool to identify adulteration of biofuel, even in analysis performed on site.

1- Introduction

Fuels produced from human grown biomass have high potential and one of the main advantages is reduction of emissions of greenhouse gases. Thus, this may be a safe fuel source, gradually reducing dependence on ex-submersed biomass, such as petroleum, developing a strategic reserve [1].

However, the activity of marketing of fuels, in all stages, the profit margins worked are relatively low. Some owners of gas stations or distributors choose as a solution to rapidly gain: the adulteration of fuels. In the case of ethanol, the most common adulteration of hydrated alcohol is the addition of more water to the formula, resulting in a product off-specification, therefore, unsuitable for use. The immediate victim of adulteration is the consumer who supplies his car with fuel adulterated. However, it is worth nothing as everyone loses with tampering, since the fraud reduces tax revenue, which creates damage to the whole society.

Due to several problems found in ensuring the quality of biofuel used in Brazil, it is necessary to use robust, accurate and nondestructive methods that can be applied in the process line, such as the use of ultrasound.

Ultrasound has been used recursively in several stages of a chemical process, whether to accelerate the reaction [2], as the separation of compounds [3] or even in their identification and analysis [4]. Other related activities, such as flow measurement, they are also suitable for using the ultrasound as a physical principle of process execution. However, by the metrological point of view, there still is some homework to complete. It is essential to support and demonstrate scientifically the advantages and best applications of ultrasound in sonochemical and control of chemical processes wide broadly.

The physical properties of a medium can be obtained through the measurement of acoustic parameters of the medium as well as from the measurement of parameters such as propagation velocity, impedance, attenuation and scattering. It can calculate the density, viscosity, degree homogenization of a mixture, the concentration of solid particles in a liquid, etc. [5] Currently, in the chemical industries, petrochemical, pharmaceutical, just referring to some well-known ones, require a considerable demand for measuring instruments that perform the characterization and discrimination of liquids with high sensitivity and accuracy. Moreover, automation of the process often needs to be measurement of process "in line". To this purpose, the use of ultrasound techniques is interesting, because is a method robust, accurate, and non destructive and can be applied to the process line. [6, 7]

2- Objectives

The aim this paper is identify possible adulteration in fuels such as ethanol, through acoustic measurements such as attenuation and propagation speed using ultrasound, using the method implemented in the Laboratory of Ultrasound (Labus) Inmetro. And get the uncertainties of these measurements as described in the book GUM (Guide to the Expression of Uncertainty in measurement).

3- Experimental

The samples used for measuring the attenuation were inserted into a cylinder 80 mm in height and 35 mm in diameter, its lower end sealed by a PVC film. The reference sample contained only distilled water and sample to be analyzed contained a mixture of ethanol / water in the following proportions: 94.0%, 93.8%, 93.4%, 93.0%, 92.6%, 92 2%, 92.0%, 91.6%, 91.2%, 90.8%, 90.4% and 90%. These percentages were chosen because of the ANP Resolution nº36, de 6.12.2005- DOU 7.12.2005 [8], which specifies that the Hydrated Alcohol Fuel (HAF) must contain an alcoholic strength by °INPM of between 93.8 to 92.6 and so we used that track specific and spread a bit more percentages to better analyze the results.

Each sample was analyzed in five repetitions for that be obtained for a statistical average. Chose to perform dilutions in ethanol, because this is a renewable fuel which developed national technology, while reducing imports petroleum. May thus provide a quality standard for Brazilian fuel.

In a transmission/reception scheme, an arbitrary waveform generator model 33250A (Agilent Technologies, CA, EUA) was used to excite the transmission transducers with 20 V peak-to-peak 20 cycles ultrasonic sine bursts, for each tested frequency. The signal from the reception transducer was digitized with an oscilloscope model DSO6032A (Agilent Technologies, CA, USA). One pair of identical transducers of resonance frequency of 15 MHz (Panametrics-NDT Olympus Corporation, Japan). Figure 1 illustrates the measurement setup. Software was developed in LabVIEWTM 8.5 (National Instruments Corporation, Austin, TX, USA) to automate the measurement, as well as to calculate attenuation and related uncertainties. Temperature was monitored throughout measurements with a calibrated digital thermometer model 34970A (Agilent Technologies, CA, USA).

To calculate attenuation, distilled water was used as reference liquid. For a given frequency, a fixed excitation signal was used firstly with water and then with mixture of ethanol / water, adjusting the same distance between transmitting and reception transducer for both mountings with a manual positioning system (Newport Corporation, Irvine, CA, USA).

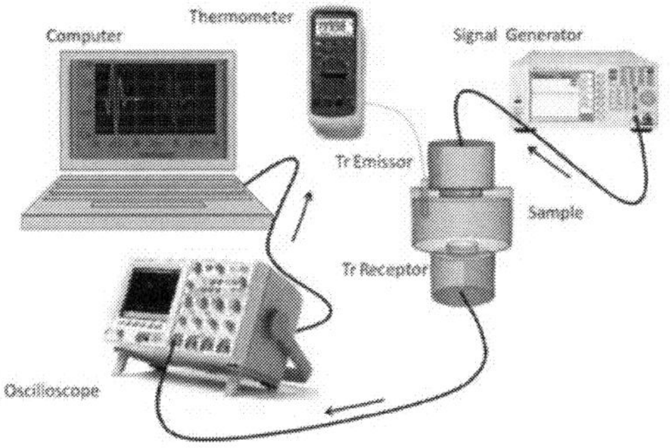

Figure 1. Illustrative figure with the experimental setup, where "Tr Emissor" is the emitting transducer, "Tr Receptor" is the reception transducer, "Sample" is the container for samples of attenuations mediums, "Signal Generator", "Computer", "Thermometer" and "Oscilloscope" are accessories measuring instruments.

All parameters used in the method implemented in Labus were evaluated for their standard uncertainty Type A and Type B as well as the combined standard uncertainty. The aim of classification Type A and Type B is to indicate the two different ways to evaluate the components of uncertainty. In the assessment of Type A, the method of evaluation of uncertainty based on statistical analysis of a number of observations. In the assessment of Type B, the method of evaluation of uncertainty is based on means other than statistical analysis of a series of observations.

The combined standard uncertainty of a measurement result is the standardized uncertainty when this is obtained through the values of several other magnitudes, being equal to the square root of a sum of positive terms, being these the variance or covariance of these other magnitudes, weighted according with as the measurement result varies with variance in these magnitudes. [9]

4- Results:

The chart 1 shows the relationship of attenuation with different ethanol concentrations and their uncertainties. This uncertainty is the expanded uncertainty, where k = 2. [9]

Chart 1: Relationship of attenuation with different concentrations of ethanol / water and their uncertainties.

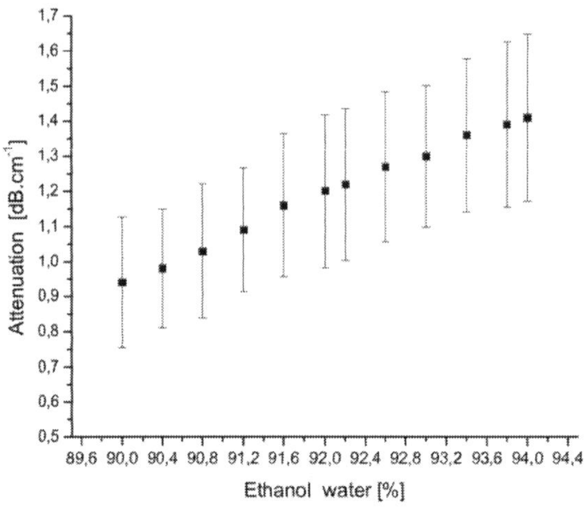

The chart 2 shows the relationship of speed with different ethanol concentrations and their uncertainties. This uncertainty is the expanded uncertainty, where k = 2. [9]

Chart 2: Relationship of the speed of propagation with different concentrations of ethanol / water and their uncertainties.

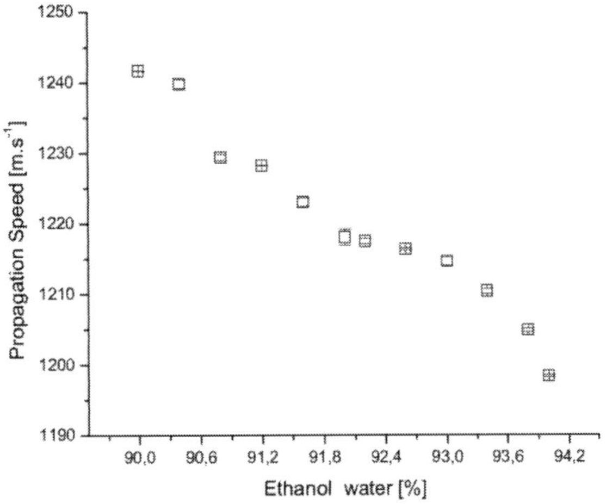

5- Discussion

The results of experimental attenuation in the different fractions of dilution of ethanol, given in [dB.cm^{-1}], where it can observe linearity in the results, since it was observed that the attenuation increases with increasing concentration of ethanol in the sample. And this occurs, due the difference of viscosity between the materials used in the mixture, then the higher the concentration of ethanol higher is the attenuation.

Because the viscosity is one of the main parameters that contribute to attenuation, it is noteworthy also talk that the viscosity is totally dependent on temperature, but the measurements that were carried in Labus, were with control the temperature at which the variation of measurements was between 1.2°C. Thus it can consider that the uncertainty due to temperature is inserted in uncertainty of the sample signal.

It can observe that the largest contribution to the uncertainty of measurement of attenuation is due to measurement of signal amplitude of the sample (ethanol / water), because this causes a certain instability in the sign of amplitude due to interaction between its molecules.

The results of the propagation speed in different fractions of dilution of ethanol using a pair of transducers at a frequency of 15 MHz where it can observe that the propagation speed decreases with increasing concentration of ethanol, because as higher resistance material the greater the time that the sound wave will take to traverse a given space.

It can observe that the largest contribution to the uncertainty of measuring the propagation speed is due to measurement of the thickness of the sample in case (ethanol / water). Yet the uncertainty remains small, allowing a measurement with accuracy. In Chart 2 it was observed a linear trend in the results of measurements of propagation speed.

It was noted that both the parameters used to identify adulteration in fuel ethanol were efficient, managed to distinguish a concentration of the other, but the parameter of attenuation has a value greater of uncertainty due to the instability of the samples.

5- Conclusion

It can be concluded that the methodology used to identify adulteration in fuel ethanol was efficient. And that through the study of uncertainties of measurements of attenuation and propagation speed can be observed that the combined standard uncertainty of measurements of propagation speed was much lower than the standard combined uncertainty in measurements of attenuation. But an analysis complements the other. And such analysis together is good parameters for identifying adulteration in fuels, because the results were consistent and of reliability and other important factor is that these assessments can be performed "in line".

6 - Acknowledgements

The authors thank the "Fundação Carlos Chagas Filho de Amparo à Pesquisa do Estado do Rio de Janeiro (FAPERJ), Convênio Inmetro/CNPq (no. 56129/2007)" for financial support of the project E-26/ 102.554//2008.

8 – References

[1] Almeida, S. C. A.; Clua, E. W. G. "Prospects of the use of biomass for decentralized generation of the energy in Mercosur." In: Mercosur Award of Science & Technology. Category Integration Rio de Janeiro: COPPE/UFRJ, p.25, 2005. (In Portuguese)

[2] Stavarache, C.; Vinatoru, M., and Maeda, Y. "Aspects of ultrasonically assisted transesterification of various vegetableoils with methanol", *Ultrasonics Sonochemistry*, 2007, vol.**14**, pp. 380-386.

[3] Bulent, A., Koc. "Ultrasonic monitoring of glycerol settling during transesterification of soybean oil", *Bioresource Technology*, 2009, vol. **100**, pp. 19-24.

[4] Greenwood, M. S., and Bamberger, J. A. "Measurement of viscosity and shear wave velocity of a liquid/slurry for on-line process control", *Ultrasonics*, 2002, vol. **39**, pp. 623–630.

[5] Higuti, R. T. ; Galindo, B.S. ; Kitano, C. ; Buiochi, F. ; Adamowski, J. C. . "Thermal Characterization of an Ultrasonic Density-Measurement Cell". *IEEE Transactions on Instrumentation and Measurement*, 2007, vol. **56**, p. 924-930.

[6] Higuti, R.T., Buiochi, F., Adamowski, J. C., Montero, E., F., "Ultrasonic density measurement cell design and simulation of non-ideal effects", *Ultrasonics*, 2006, vol. **44**, p.302-309.

[7] Saggin, R., Coupland, J.N., "Rheology of xanthan/sucrose mixtures at ultrasonic frequencies", *Journal of Food Engineering*, 2004, vol. **65**, p.49-53.

[8] NAP (National Agency of Petroleum, Natural Gas and Biofuels) Ordinance No. 36, 6.12.2005-DOU 7.12.2005. (In Portuguese)

[9] Evaluation of measurement data – Guide to the expression of uncertainty in measurement, JCGM 100:2008 (GUM 1995 with minor corrections), www.bipm.org/en/publications/guides/gum.html.

Ultrasonic calibration and certification of V1 and V2 type reference standard blocks for use in Non-Destructive Testing

L E Maggi [1,2], C E R Silva[1], A V Alvarenga[1], R P B Costa-Felix[1]

[1]Laboratory of Ultrasound, Directory of Scientific and Industrial Metrology (DIMCI), National Institute of Metrology, Standardization, and Industrial Quality (INMETRO), Av. Nossa Sra das Graças, 50 (Prédio 1), Duque de Caxias, RJ, Brazil, ZIP 25250-020

[2]Biomedical Engineering Program, COPPE/Federal University of Rio de Janeiro (UFRJ), Horácio Macedo Ave., n.º 2030, Center of Technology, Block H – room/327 Cidade Universitária, Rio de Janeiro, RJ, Brazil, ZIP 21941-914

[1]cesilva@inmetro.gov.br

Abstract. Ultrasonic Non-Destructive Testing (US-NDT) has many industrial applications, such as flaws detection, welding integrity evaluation and solid material thickness measurement. To evaluate the instrumentation conditions, before a measurement begins, a "calibration" should be carried out with the aid of a reference standard block (RSB). Types V1 (ISO 2400) and V2 (ISO 7963) are the most widely used in US-NDT. Due their characteristics, V1 and V2 RSB can be classified as Reference Material, so that they can be certified, resulting in a Certified Reference Material (CRM). This paper describes technical aspects of a V1 and V2 RSB certification, based on ISO 2400, ISO 7963, EN 12223 and EN 27963. According the certification protocol established at Inmetro's Laboratory of Ultrasound, RSB relevant dimensions were determined through calibration. Ultrasound velocity was assessed by using a scope and a signal generator at 5 MHz. Uncertainties were determined including measurement accuracy under repeatability conditions. A complete uncertainty budget was determined for both parameters. Within the specimens used, velocities were in accordance with specifications 5920 ± 30 m/s).

1. Introduction

We live in a world in which technology improves quickly and becomes more present in our daily life. However, this technology can offer risks if is not developed and used properly. Even small defects can be troublesome with potential to generate accidents and injuries, besides causing the customers unsatisfaction. To detect such failures special inspection methods are necessary. Therefore, the Non-Destructive Testing (NDT) arises, defined as a method to evaluate materials, pieces, equipment or detects the failure without causing damage to the inspected element. There are different forms of NDT, such as: Eddy Current, Magnetic Particle, X-Ray, Gamma- Ray and Ultrasound [1].

NDT by ultrasound, especially in thickness measurement and structure failure detection, uses the Pulse-Echo (A Scan-Mode) equipment. In the calibration of this equipment, standard blocks called V1 and V2 are employed. To assure the metrology reliability of all the measurement process, it is necessary that standard blocks are used, preferentially if they were certified such as a CRM (Certified Reference Material) [2]. These follow the principle of CRM and the technical requisites of ISO 2400 [3], ISO 7963 [4], EN 12223 [5] e EN 27963 [6] standards. The longitudinal velocity is the main characteristic to be evaluated.

The aim this work was to calibrate the reference blocks, evaluating its main dimensions and acoustic properties. Therefore, four reference blocks, being two V1 blocks and two V2 blocks were used. The main scales and dimensions and the longitudinal velocity were assessed.

2. Experimental Procedure

The dimensional calibration of two V1 blocks (N° 121 e N°176) (Figure 1A) and two V2 blocks (N° 255 and N°182) (Figure 1B) was carried out in the Dimensional Metrology Laboratory (Lamed)/INMETRO. A system composed by laser interferometer (reference value), a table with a system of horizontal micrometric displacement (maximum range: 300 mm), moved by c.c. motor and a microscope with camera CCD coupled to a computer were used. Specific software to determine the measurements was used. The scale is aligned to the laser and to the microscope in this system.

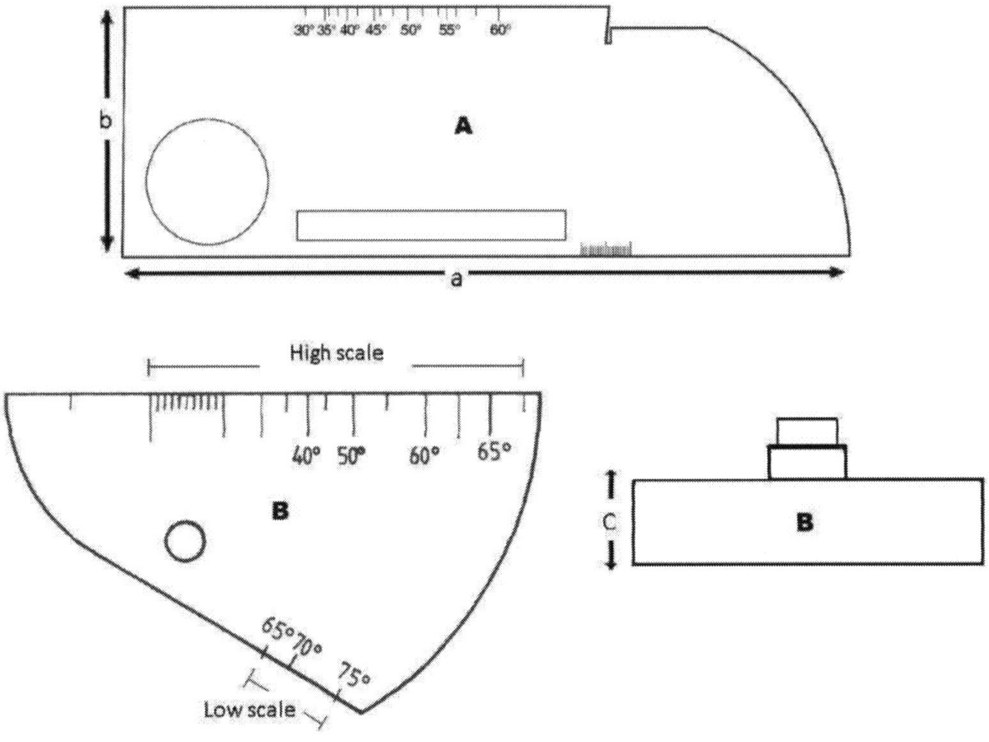

Figure 1. V1 standard block [5] (A) and V2 standard block [6] (B)[5].

In V1 blocks, the specified tolerance is ± 0.10 mm, in accordance with ISO 2400 and EN 12223 standards. V2 blocks, tolerance is ± 0.10 mm, except for the scales length ± 0.50 mm, in accordance EN 27963 standard.

The US GE equipment (figure 2A) in conjunction with a straight beam probe (5MHz Ø=10 mm) (figure 2 B) and a broadband pulse were used to measure the time-of-flight. An attenuator (figure 2C) and a Tektronix oscilloscope (figure 2D) were also used. To acquire the pulse-echo signal and calculate the longitudinal velocity, a Labview program (BlockCalibration6) was developed and it is showed in figure 3. The time difference between the first and second back-wall echoes was measured.

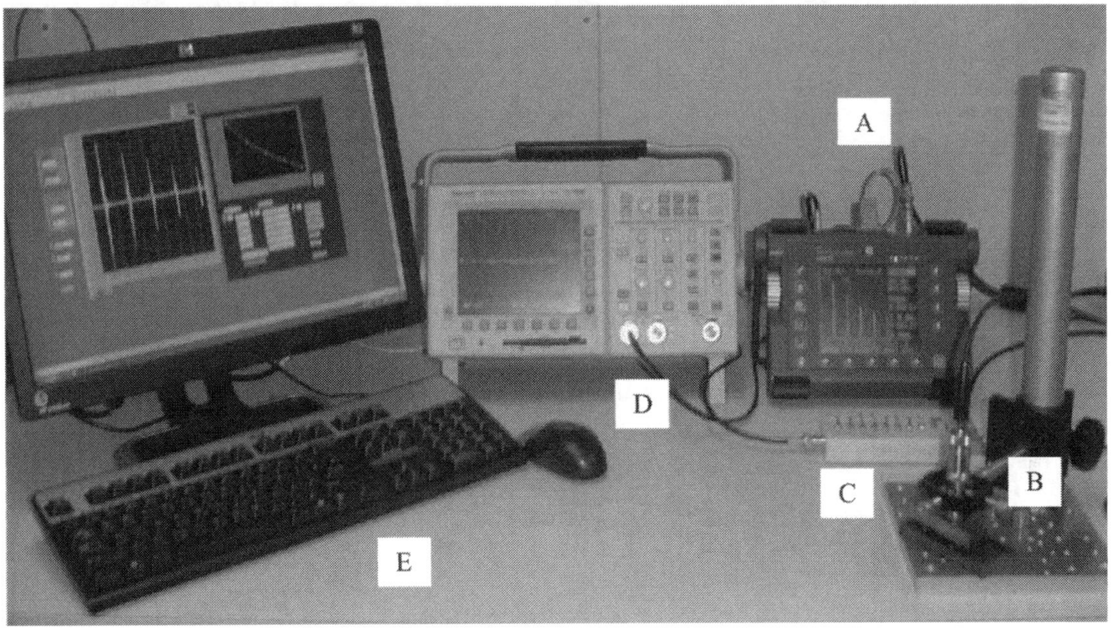

Figure 2. Experimental setup: GE equipment (A), transducer (B), attenuator (C), oscilloscope (D) and computer (E).

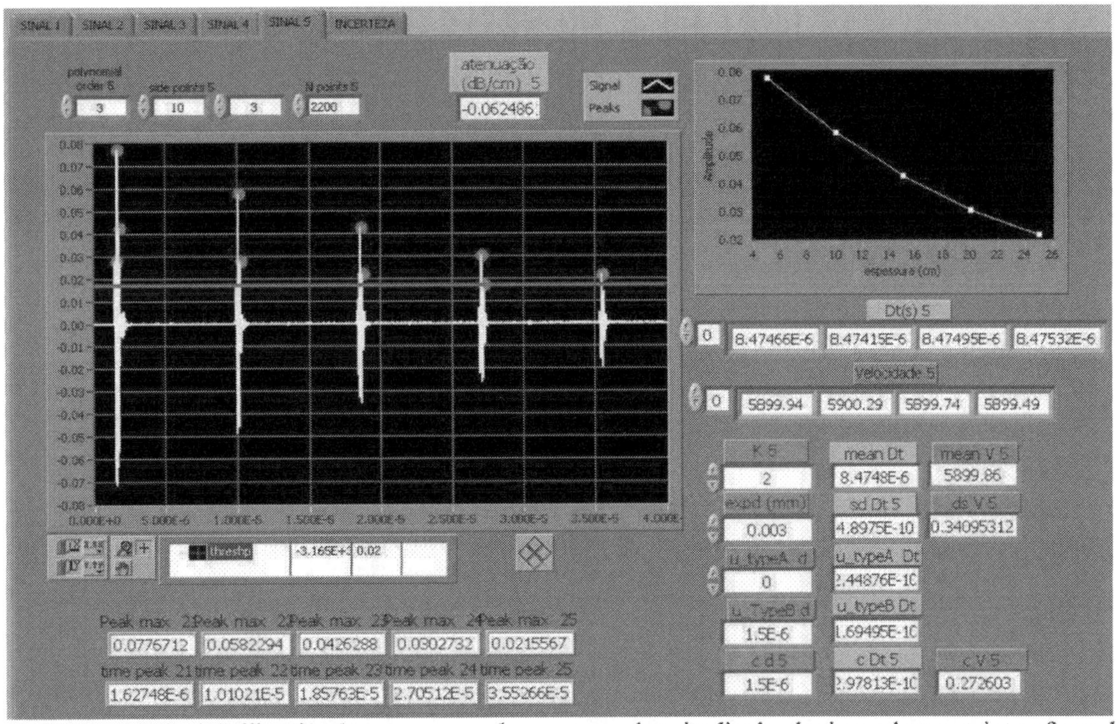

Figure 3. BlockCalibration6, program used to measure longitudinal velocity and uncertainty of standard block.

3. Results and Discussion

The dimensional calibration of both sides (with and without serial number) of V2 standard block scale data (figure 1B) can be seen in table 1. The nominal value scales (present in the EN 12223 standard) were compared with the average calculated values of 3 measurement cycles. The difference between these values was compared with the tolerances found in ISO 7963 standard. (Systematic error means that it would result from an infinite number of measurements of the same measurand carried out under repeatability conditions, minus a true value of the measurand [7]). The main dimensions to calculate V1 and V2 blocks longitudinal velocity are given in table 2.

Table 1. Nominal value and systematic measurement error of V2 standard blocks dimensional scales.

Scales	Nominal value	V2 block - N 255				V2 Block - N 182			
		Side A (w/ serie number)		Side B (w/ serie number)		Side A (without/ serie number)		Side B (without/ serie number)	
		Measured quantity value (mm)	Systematic measurement error + uncertainty*	Measured quantity value	Systematic measurement error+ uncertainty*	Measured quantity value	Systematic measurement error+ uncertainty*	Measured quantity value	Systematic measurement error+ uncertainty*
High	5.00	-5.01	0.02	-5.02	0.04	-5.00	0.01	-4.99	0.02
	5.00	5.01	0.02	5.00	0.01	5.00	0.01	5.00	0.01
	14.00	14.01	0.02	14.00	0.01	14.00	0.01	14.00	0.01
	16.80	16.81	0.02	16.81	0.02	16.79	0.02	16.81	0.02
	20.00	20.01	0.02	20.01	0.02	20.00	0.01	20.00	0.02
	23.80	23.81	0.02	23.81	0.02	23.79	0.02	23.81	0.03
	28.60	28.60	0.01	28.60	0.01	28.59	0.02	28.60	0.02
	34.60	34.62	0.03	34.61	0.02	34.59	0.02	34.59	0.03
	42.90	42.92	0.03	42.92	0.03	42.90	0.01	42.90	0.02
Low	4.60	4.60	0.01	4.61	0.02	4.60	0.01	4.60	0.02
	12.20	12.21	0.02	12.20	0.01	12.20	0.01	12.20	0.02

***measurement uncertainty* = 0,01mm**

Table 2. Nominal value and systematic measurement error + measurement uncertainty of V1 and V2 standard blocks main dimensions.

Block	Serial number	quota	Nominal value	Measured quantity value (mm)	Measurement uncertainty (mm)	*k*	Systematic error+ measurement uncertainty
V1	121/06	a	100	100.002	0.003	2.0	0.005
		b	25	25.036	0.003	2.0	0.039
	0176/09	a	100	99.995	0.002	2.0	0.007
		b	25	25.000	0.002	2.0	0.002
V2	0182/09	c	12.5	12.519	0.003	2.0	0.022
	225	c	12.5	12.503	0.002	2.0	0.005

Thus, scales and dimensions of V1 and V2 blocks were dimensionally verified. In V2 and V1 blocks it could be observed that the systematic measurement error + the measurement uncertainty of the main dimensions were lower than the tolerances established by ISO 7963 and EN12223 standard, which is ± 0.01mm. In V2 blocks it could be observed that the scales systematic measurement error + the measurement uncertainty were lower than the tolerances (± 0.5mm) established by ISO 7963. Nevertheless, in V1 blocks the scale values were not in accordance with EN 12223. Therefore, it is important to notice, that these scale dimensions are not specified in ISO 2400 standard. Difference in the marks positions as well as in their quantity variation could be observed. This prevented V1 block analysis.

Table 3 presents the longitudinal velocity and measurement uncertainty of V1 and V2 standard blocks. As seen, all the measured blocks found longitudinal velocity value lower than the tolerances (0.2 %) i.e. with measurement uncertainty of ±12 m/s for compressional waves as described in ISO 7963 and EN12223.

Table 3. Ultrasound properties and measurement uncertainty of V1 and V2 blocks.

Standard Block	Nominal value (m/s)	Measurement velocity-average (m/s)	Measurement Uncertainty (m/s)	*k*	%	Systematic error+ measurement uncertainty
V1 SN. 121	5920 ± 30	5910.29	6.25	2.36	0.11	14.85
V1 SN. 0176_09	5920 ± 30	5917.16	8.47	2.78	0.14	11.31
V2 SN. 0182/09	5920 ± 30	5910.77	9.52	2.08	0.16	9.52
V2 SN. 225	5920 ± 30	5925.59	5.14	2.78	0.09	10.73

4. Conclusion

This work concludes that V2 blocks in calibrated dimensions are in agreement with the technical specifications of EN 27963 standard. However, V1 blocks standard are not in accordance with EN 12223 standard. Concerning the longitudinal velocity calibration, all blocks are in accordance with EN 27963 and EN 12223 standard.

References

[1] Andreucci R, 2008, Industrial Aplication – Ultrasound Testing, in Portuguese: "Aplicação Industrial – Ensaio por Ultra-som," ABENDE, (São Paulo) 96.

[2] National Institute of Metrology, Standardization and Industrial Quality – (Inmetro). URL: http://www.inmetro.gov.br/ [acessed on 16/Apr/2009].

[3] ISO 2400, Welds in Steel – References Block for the calibration of Equipment for Ultrasonic Examination. First Ed, 1972, 06-15.

[4] ISO 7963, Welds in Steel – References Block N°2 for ultrasonic Examination of Welds. First Ed, 1985, 09-15.

[5] EN 12223, Non destructive testing- Ultrasonic Examination Specification for calibration Block N°1, 2000, 1-10.

[6] EN 27963, Welds in Steel – Calibration Block N°2 for ultrasonic examination of welds, 2000.

[7] Internacional vocabulary of metrology - Basic and general concepts and associated terms (VIM) – JCGM 200:2008, Dec-2008.

Journal of Physics: Conference Series **279** (2011) 012030 doi:10.1088/1742-6596/279/1/012030

Ultrasound Metrology in Mexico: a round robin test for medical diagnostics

R Amezola Luna, A L López Sánchez, A A Elías Juárez
Centro Nacional de Metrología (CENAM, México),
Carretera a Los Cués km.4.5, El Marqués, Querétaro, México. 76246.

E-mail: ramezola@cenam.mx

Abstract. This paper presents preliminary statistical results from an on-going imaging medical ultrasound study, of particular relevance for gynecology and obstetrics areas. Its scope is twofold, firstly to compile the medical ultrasound infrastructure available in cities of Queretaro-Mexico, and second to promote the use of traceable measurement standards as a key aspect to assure quality of ultrasound examinations performed by medical specialists. The experimental methodology is based on a round robin test using an ultrasound phantom for medical imaging. The physician, using its own ultrasound machine, couplant and facilities, measures the size and depth of a set of pre-defined reflecting and absorbing targets of the reference phantom, which simulate human illnesses. Measurements performed give the medical specialist an objective feedback regarding some performance characteristics of their ultrasound examination systems, such as measurement system accuracy, dead zone, axial resolution, depth of penetration and anechoic targets detection. By the end of March 2010, 66 entities with medical ultrasound facilities, from both public and private institutions, have performed measurements. A network of medical ultrasound calibration laboratories in Mexico, with traceability to The International System of Units via national measurement standards, may indeed contribute to reduce measurement deviations and thus attain better diagnostics.

1. Introduction

Ultrasound medical imaging examinations are a useful and reliable tool to perform medical diagnostics. In Mexico, close to 1 M of gyn-obstetric diagnostics by ultrasound are performed in the private sector [1], and it is estimated a similar amount in the public sector. Medical specialists can measure size and depth of human tissues and organs in order to diagnose and recommend a medical treatment. Therefore, the significance of detection and measurement accuracy is related to treatment effectiveness and thus patient's health. Nevertheless, accurateness could be affected by factors such as capabilities of the equipment (ultrasound machine), medical expertise, patient biological characteristics and environmental conditions. Each factor may affect measurement reliability and decrease the quality to diagnose a patient. In order to establish if quality assurance programs, with traceability to national standards, may indeed improve the quality of medical ultrasound imaging diagnostics, the *Centro Nacional de Metrología* (CENAM) started a field study focusing on gynaecology and obstetrics areas. The main goals were to compile the medical ultrasound infrastructure available in a relatively small Mexican province (State of Queretaro) with a population close to 1.6 M [2], in both public and private medical units. Information regarding location and amount of medical units and its specific ultrasound infrastructure for medical diagnostics, as well as how it is being managed and maintained were the starting issues for this study. In addition, considering that reference standards and calibration are metrological concepts not frequently found in the medical sector, it was set to gather information about the use of traceable measurement standards and how quality of ultrasound examinations are assured by medical specialists. In Mexico, usually, calibration of the ultrasound machine is not included during its periodic maintenance program; therefore users do not have enough information to assure its correct operation. This study, as far as the authors know, has not been done in Mexico before; it grew out as a concern of the government of the State of Queretaro about the actual operation of the ultrasound infrastructure available in the State, seeking to improve the quality of ultrasound medical services. To achieve these goals, a round robin test was setup among different medical units in cities belonging to the State of Queretaro. Partial results corresponding to on-site measurements from June 2009 to March 2010 are presented in the following sections.

2. Description of the round robin test

The experimental methodology is based on a round robin test using an ultrasound phantom for medical imaging. The phantom simulates acoustic properties of human tissue and is constructed of rubber-based tissue-mimicking material with nominal sound speed of 1450 m/s at 0.5 dB/cm/MHz at room temperature (23 °C) [3]. The phantom contains line targets and anechoic target structures simulating some biological tissues (see figure 1). During the test, staff from CENAM visited each medical unit and gave details about the measurement procedure. Each physician, using its own ultrasound machine, couplant and facilities, measured the depth, lateral distance or size of a set of pre-defined reflecting and anechoic targets of the reference phantom. Due to time restrictions, measurements taken were mainly limited to the use of one linear transducer and one convex transducer (average time for measuring: 20 min per each transducer).

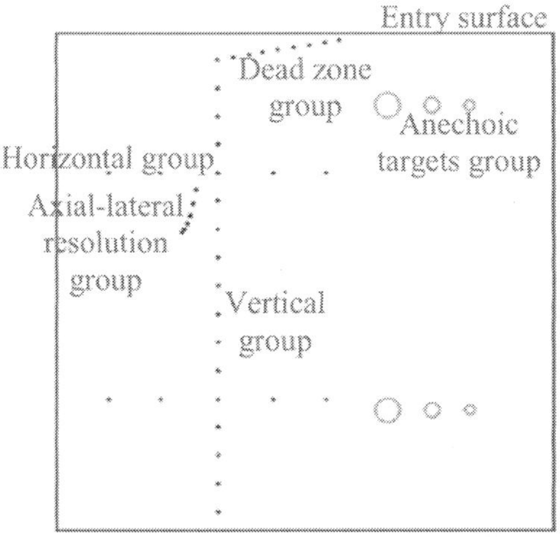

Figure 1. Ultrasound phantom (artifact).

Measurements performed during the round robin test [4, 5, 6]:

a) Measurement system accuracy. The horizontal spacing between 2 consecutive line targets from the Horizontal Group was determined (nominal distance of 20 mm). Also, a vertical distance of 170 mm (for convex transducer) from the Vertical Group was measured.

b) Dead zone. The closest line target from the Dead Zone Group to the phantom entry surface was identified and its depth was measured. Closest line target available to the entry surface of the phantom is 2 mm, nominal. During the test, no supplementary accessory was used to perform measurements (e.g. delay line bag or couplers).

c) Resolution. Focusing on the Axial-Lateral Resolution Group, system capability to resolve between 2 contiguous targets in the axial axis was determined. The group of targets is form by 6 line targets with interval spacing of (5, 4, 3, 2 and 1) mm, nominal.

d) Depth of penetration. Maximum distance from the phantom entry surface to the farthest line target of the Vertical Group was measured. The depth of the farthest line target available in phantom is 170 mm, nominal.

e) Anechoic target detection. The diameter of three anechoic targets was determined. Nominal diameters of (10, 8 and 4) mm. Measurements were performed using linear transducers.

Every participant performed the same set of measurements referred above, however due to limitations of the ultrasound machines some measurements were not able.

3. Results and discussion

Measurements from 66 medical institutions (including both public and private health centers) are presented in here as preliminary results, corresponding to 82 individual ultrasound machines.

3.1. Measurement system accuracy

Measurement deviation was determined by:

$$E_U = L_U - L_{ref} \tag{1}$$

where

E_U = Difference between the length measured by every individual ultrasound machine from the medical entity and the reference value, mm.

L_U = Length measured with the ultrasound machine, mm.

L_{ref} = Reference value taken as the average value from all measurements, mm.

Normalized error was calculated as [7],

$$e_U = \frac{|L_U - L_{ref}|}{\sqrt{U_{L_U}^2 + U_{L_{ref}}^2}} \tag{2}$$

where

e_U = Normalized error. If >1, inconsistent results.

U_{L_U} = Expanded uncertainty of the results obtained by the ultrasound machine from the medical entity participant, mm.

$U_{L_{ref}}$ = Expanded uncertainty of the reference value, mm.

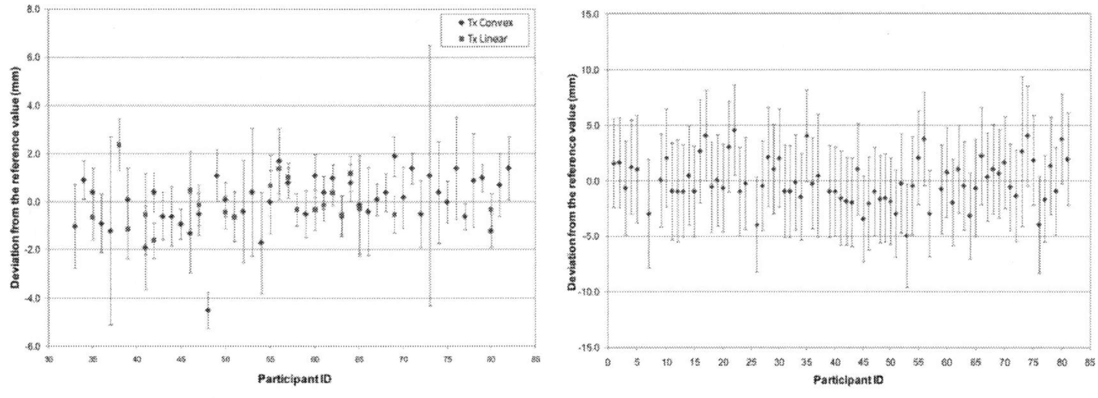

(a) Horizontal Axis *(b) Vertical Axis*

Figure 2. Measurement system accuracy on (a) horizontal axis, and (b) vertical axis.

Results for *Horizontal* and *Vertical Accuracy* are shown in figure 2. Horizontal measurements were performed with linear and convex transducers. Vertical measurements include only convex transducers. Most of the participants have a deviation value within ±2 mm (±10 %) for the horizontal axis and ±5 mm (±3 %) for the vertical axis, respectively. Regarding horizontal accuracy 51 of 68 participants obtained consistent values (normalized error ≤ 1), while for vertical accuracy 74 of 76 participants obtained consistent values.

3.2. Dead zone

Most of the participants with linear transducers reached the closest line target available to the phantom's entry surface, 2 mm. For convex transducers, widespread dispersion from 1 mm to 7 mm was obtained. This is shown in figure 3. As expected, in general, Dead Zone for linear transducers is smaller than for convex transducers. Some ultrasound machines had multi-frequency capabilities which allowed them to improve their measurements on targets corresponding to the dead zone.

3.3. Resolution

For convex transducers, 63.3 % of the individual ultrasound machines were able to resolve all the line targets of the Resolution Group, including the pair of targets separated 1 mm. The rest of the participants were able to resolve the line targets separated 2 mm.

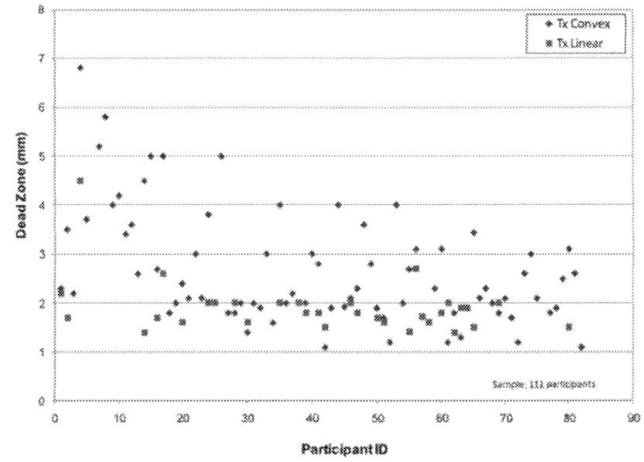

Figure 3. Dispersion of data for dead zone.

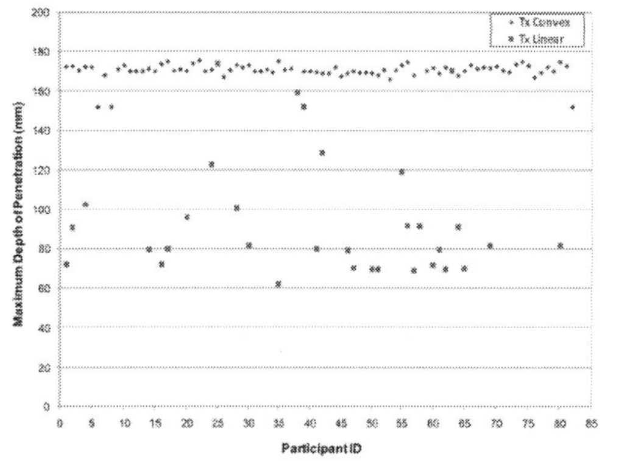

Figure 4. Dispersion of data for depth of penetration.

Journal of Physics: Conference Series **279** (2011) 012030 doi:10.1088/1742-6596/279/1/012030

For linear transducers, 79.3 % of the medical entities were able to resolve up to 1 mm. Ultrasound machine capabilities for imaging and configuration had a considerable influence in the results. As expected, linear transducers showed much better resolution capabilities than convex transducers.

3.4. Depth of penetration

Most of the participants with convex transducers identified a line target at the maximum depth available in the phantom, 170 mm. For linear transducers, a widespread dispersion is shown from approximately 60 mm up to 170 mm, some of them only reached 80 mm due to maximum available range in the ultrasound machine (scale locked). On the other hand, two linear transducers reached the maximum available depth due to relatively low frequency operation (5 MHz). As expected, convex transducers reached a larger depth of penetration than linear transducers. See figure 4.

3.5. Anechoic target detection

In the case of anechoic targets measurements with a linear transducer, most of the participants are within data dispersion values of ±0.5 mm (±12 %) for the 4 mm diameter target, ±1 mm (±12 %) for the 8 mm diameter target and ±1.2 mm (±12 %) for the 10 mm diameter target. See figure 5.

As items that may require further discussion and analysis, it is noted that measurement accuracy and equipment diagnostics capabilities may need to take into account the effect of factors, such as:

a) Equipment capabilities
- Image quality
- Spatial resolution
- Beam divergence

b) Medical specialist expertise
- Transducer alignment
- Equipment operation
- Measurement technique

c) Environmental conditions
- Electrical voltage
- Temperature
- Humidity

d) Phantom
- Geometry
- Sound speed
- Targets alignment

e) Protocol and logistic
- Limited time to perform measurements

Above factors may have led to apparent equipment limitations and/or measurement inaccuracies. A line target should produce a snap point image; however an unfocussed beam produces a distorted elliptical form (see figure 6). On the other hand, misalignment of the transducer and measurement path (using the caliper) can give an overestimated length (see figure 7).

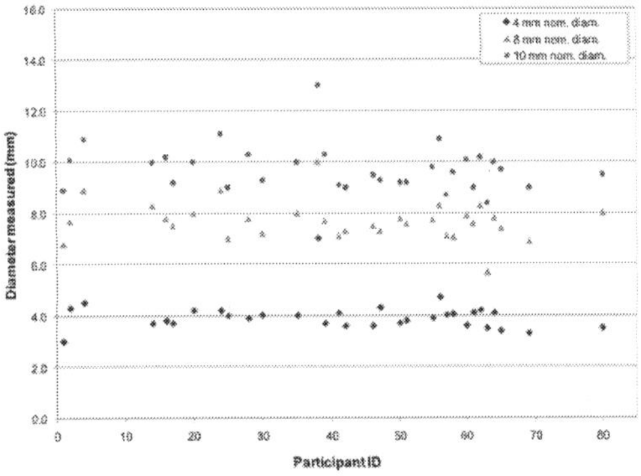

Figure 5. Dispersion of data for anechoic target size.

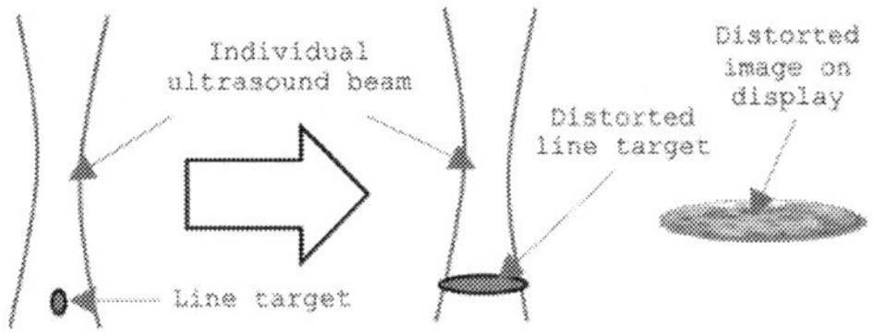

Figure 6. Capabilities and configuration may affect the image quality.

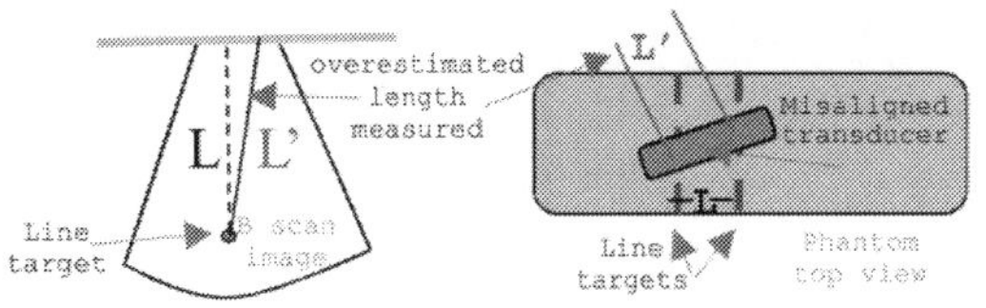

Figure 7. Misalignment on measurements may affect accuracy.

4. Conclusions

Results from this round robin test underline several aspects that need to be taken into account by a medical specialist performing ultrasound examinations or scannings, in order to improve the quality of the diagnostic. Namely, measurement capabilities may degrade as the equipment gets old, and metrological assurance is not the same as corrective maintenance. That is, ultrasound metrology and basic metrological practices may be overlooked. Measurements performed gave an objective feedback to medical specialists regarding dead zone, depth of penetration, resolution and anechoic targets detection of their ultrasound examination systems and their measurement accuracy. It is claimed that the use of an ultrasound phantom is an important tool to verify and monitor the correct operation of ultrasound machines; besides being a useful tool for training purposes.

As part of the future work, it will be convenient to find out if ultrasound machines are indeed periodically calibrated in other countries and how such calibrations are related to ultrasound measurement systems kept by the corresponding NMI. There seems to be a broken traceability chain that may need some attention. Among the remaining questions: 1) Which countries do widely have or enforce ultrasound medical diagnostics with traceability to SI units? and 2) Are there any benefits for a patient to receive ultrasound medical services with traceability to national standards?

5. References

[1] INEGI- 2003 *Instituto Nacional de Estadística y Geografía*, Mexico.
[2] INEGI- 2005 *Instituto Nacional de Estadística y Geografía*, Mexico.
[3] ATS Laboratories Phantom model 535. Operation manual.
[4] IEC 61390-96 Ultrasonics- Real time pulse-echo systems- Test procedures to determine performance specifications.
[5] AIUM Standard methods for measuring performance of pulse-echo ultrasound imaging equipment (July 13, 1990).
[6] Goodsitt, Carson, Witt, Hykes and Kofler. Real time B-mode ultrasound quality control test procedures. Report of AAPM Ultrasound task group No.1. Medical Physics, Vol. 25 No. 8, August 1998 (1385 to 1406).
[7] ISO/IEC Guide 43-1 (1997) Proficiency testing by interlaboratory comparisons-Part 1: Development and operation of proficiency testing schemes.

Acknowledgments

Authors thank the kind support of the *Consejo Nacional de Ciencia y Tecnología* (CONACYT) and the government of the State of Queretaro to perform this study (project FOMIX QRO-2008-C03-107938). Participation of all medical specialists and health institutions is also acknowledged.

The Ultrasound effects on non tumoral cell line at 1 MHz *therapeutic frequency*.

L Di Giambattista[1,2,6], P Grimaldi[1], I Udroiu[3], D Pozzi[4], G Cinque[5], MD Frogley[5], AM Cassarà[1] , A Bedini[3] , C Giliberti[3] , R Palomba[3] , S Buogo[6], A Giansanti[1] and A Congiu Castellano[1,6]

[1] Physics Department, Sapienza, University of Rome (IT).
[2] CISB, Sapienza, University of Rome (IT).
[3] DIPIA, ISPESL, via Urbana 167, Rome (IT).
[4] Experimental Medicine and Pathology Department, Sapienza, University of Rome (IT).
[5] Diamond Light Source Ltd, Didcot, Oxfordshire (UK).
[6] CNR-Institute of Acoustics O.M. Corbino, Rome (IT).

l.digiambattista@caspur.it

Abstract. The aim of this research is to investigate some bioeffects due to *Therapeutic Ultrasound* (1 MHz and $50<I_{PA}<60$ W/cm^2) which could allow to enhance drugs or genes delivery in non tumoral cells [1]. Ultrasound (US) has been demonstrated to alter the cell membrane permeability due to a biophysical mechanism, Sonoporation, and exploited as a promising non-invasive gene transfer method. We have used the NIH-3T3 cell line as a model system and exposed it to US medical equipment for 15, 30, 45, 60 minutes at distances of 10 and 15 cm from the source transducer, corresponding to the far field region where $z > z_{last} = \frac{a^2}{4\lambda} = 4.0 \pm 0.4$ cm. We have worked with the maximum power in pulsed system with 75% duty cycle. Characterization of the unfocused, planar and with a circular geometry 1 MHz source transducer, was performed and the acoustics pressure was measured by a calibrated 0.5 mm needle hydrophone; moreover, the pressure field generated by the source transducer was simulated. The US effects on cells were assessed by Fourier transform infrared (FTIR) Imaging with focal plane array (FPA) detector. By the IR analysis, the US exposure on non tumoral cells has induced a change of the intensity for CH_2 asymmetric stretching (2924 cm^{-1}) band in the lipid region (3000-2800 cm^{-1}) that it could detect an energy-dependent process. It has already shown that cells invest energy to catalyze lipid movement in order to maintain a specific transmembrane phospholipid distribution. Although asymmetry is the rule for control cells, the loss of asymmetry could be associated with the permeability change of plasma membrane inducing temporary pores.

1. Introduction

In the last years, the Ultrasound has been employed in gene delivery for the advantages over other systems as virus or nonvirus-mediated systems; the advantages are a lower invasive-method, a higher delivery efficiency, more applications and minimal cell death. Ultrasound systems are available for both *in vitro* and *in vivo* therapies [1]. The biological (*thermal, cavitation and microstreaming*) effects of ultrasound can result from physical and biological factors, such as frequency, intensity, time

exposure, duty cycle, temporal and spatial structure of sound field, the physiological state and the size-volume of a sonicated sample, and external conditions like temperature, pressure, microstreaming, shear stress . Such a great number of variables complicates the analysis of the phenomena [2]. We have introduced the FTIR microspectroscopy based on the FPA detector to study the bioeffects of ultrasound on a cellular system in the range of Therapeutic Ultrasound (1 MHz in pulsed system with 75% of duty cycle).
According to a previous test [3], we have characterized the source transducer and we have established the conditions of experimental set up, such as the size of plexiglass tank or the distances from sample to US source. Finally, we have simulated the distribution of pressure with this experimental set up through multimodules software.

2. Materials and Method

2.1. Cell culture and sample preparation
In this research, we have used a fibroblast cell line (NIH-3T3); the NIH-3T3 were grown with a solution of Dulbecco's Modified Eagle's minimum essential Medium (DMEM) without calcium with 10% fetal bovine serum (FBS), 1% penicillin-streptomycin and 1% L-glutamine at 37° C in humidified atmosphere containing 95% air and 5% CO_2. Before the US exposure, the NIH-3T3 were cultured as monolayer in 35-mm petri dish on CaF_2 windows, pre-treated with polylysine. The viability of both the untreated (control) and the US treated (sonicated) cells has been determined by Trypan blue (TB) exclusion test (over 90% of viability). After the treatment with ultrasound, the control and sonicated cells were fixed in paraformaldehyde (2% for 15 min), washed in Dulbecco's Phosphate Buffer Saline (PBS) and in distilled water to remove PBS residues, and dried in a desiccator.

2.2. Experimental set up
As shown in figure 1, we have built a plexiglass tank filled with partially degassed water and the source transducer was placed on the bottom.

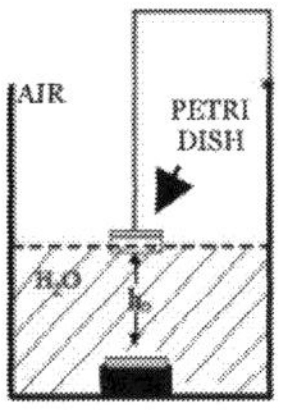

Figure 1. The general scheme of experimental setup.

A vertical directivity test of source transducer was realized and the maximum signal by transducer has shown a shift of θ=0.8° (figure 2).

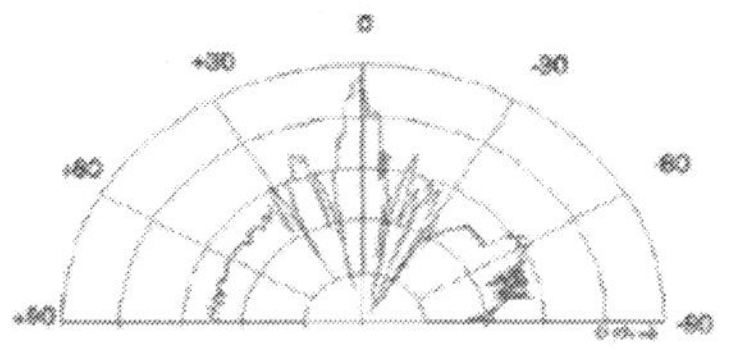

Figure 2. Polar diagram (log scale) for unfocused, planar and circular source transducer.

The excitation pressure waveform was measured at different distances from trasmitting transducer by 0.5 mm needle hydrophone (Precision Acoustics LTD, Interchangeable Probe) with a sensitivity of -272.7dB re 1V/μPa at frequency of 1 MHz (table1).

Table 1. The values of voltage and pressure at different distances.

Distance (cm) ± 0.05	$Voltage_{rms}$ (mV)	Pressure (MPa)
0.8	53.4 ± 2.8	1.62 ± 0.08
3.0	54.3 ± 2.7	1.70 ± 0.08
10.0	33.7 ± 1.7	1.03 ± 0.05
15.0	49.3 ± 2.5	1.53 ± 0.08
20.0	44.0 ± 2.2	1.34 ± 0.07

The US medical equipment has been set to 1 MHz with 100% of maximum power in pulsed system with 75% duty cycle. According to the preliminary analysis, the distances (h_D) between the cellular sample and the transducer were chosen at 10 and 15 cm (figure 2) and the US exposed times were fixed at 15, 30, 45, 60 minutes. At both distances, we have established tha values of pulse average intensity (I_{PA}): I_{PA}= (52 ± 6) Wcm^{-2} at 10 cm and I_{PA}= (58 ± 6) Wcm^{-2} at 15 cm. The temperature inside the petri dish and water of tank was monitored with a thermocouple; a maximum increase of 2°C or less was revelead during US exposed time.

2.3. FTIR microspectroscopy and data pre-processing for image analysis

An FTIR microscope (Hyperion 3000) coupled to an Vertex 80v FTIR spectrometer (both from Bruker Optics, Germany) was used for IR measurments; the Hyperion 3000 microscope was equipped with a computer-controlled x,y stage and was sealed using a specially box purged in 99.8% pure nitrogen. This microscope was equipped with a mercury cadmium telluride (MCT) based focal plane array detector (FPA) of 64x64 pixels where each pixel corresponded to an area of 40x40µm, so that the visible field corresponded to an area 170x170μm^2 with 15x magnification objective. Each pixel corresponds to a single acquired FT-IR spectrum. The FTIR images were acquired in continous with a spectral resolution of 8 cm^{-1} and by co-adding 512 scans of absorbance spectra were recorded in the range from 3000 to 900 cm^{-1} (lipid, proteins and nucleic acids regions). Every cellular sample was measured on CaF_2 window in Trasmission mode and a background spectrum/ image of CaF_2 slide was recorded before each cellular sample measurement in order to account for variations in water vapour and CO_2 levels. Three images were acquired on the CaF_2 slide, resulting 30 images with 4096 spectra for the whole experiment. The resulting FT-IR images were pre-processed by different sofwares (OPUS 6.5, OriginPro 8.0, Mathematica 7.0); to remove poor quality spectra, the data sets were subjected to a quality test such as Signal-to-Noise ratio (S/N) [3]. In the (S/N) test, the 4096 spectra were evaluated using a criterion of 100 as a threshold of the ratio in accordance with the FTIR spectroscopy; the signal with the maximum absorbance was evaluated in the frequency region of the the Amide I band (1595-1800 cm^{-1}) and the noise was calculated as the standard deviation in the spectral range 1800-1900 cm^{-1}. The spectra with a S/N ratio smaller than 100 were labeled as bad spectra and the corresponding pixels, bad pixels. The number of bad spectra inside a map is denoted with N_{bs}; the maps having more than 1000 bad pixels have been discarded. After this pre-processing analysis, the spectral data were processed by a baseline correction with a rubberband method using 64 baseline points; then, through the vector normalization, we have normalized the whole spectral region (3000-900 cm^{-1}) at CH_2 asymmetric stretching (2924 cm^{-1}) band for the analysis of the lipid region. The chemical maps were created from these spectra by plotting different spectral parameters, such as intensity peak or intensities peaks ratio, as a function of x-y pixel position.

2.4. Simulation details

In this study, 2-D simplified model has been built using multimodules software. Finite element analysis has been used to model the ultrasonic waves propagation in the plexiglass tank at distances of 10 and 15 cm from US source. The boundary condition was sound hard boundary (wall) for the whole tank; at interface between the water and air, the boudary condition was set to continuity. The mesh size has been adapted to every acoustics subdomain in order to resolve the wavelenght. The maximum dimension of the mesh has been set on a value corresponding to 2th of the acoustic wavelenght of water (IEC 61102). A frequency-dominion analysis was selected for solving the problem and a linear system solver adopted for high frequency was selected.

3. Results and discussion

3.1. Spectra pre-processing

By the results with S/N quality test, we have established the number of bad spectra in each data set and have evaluated their weight in the 4096 spectra. In figure 3, we have reported as an example, the dispersion of (S/N RMS) values around average value vs. pixels # (4096 spectra) for the control cells and the sonicated cells at 15 cm for 45 minutes (SON_{45_15}) that has shown a larger number of bad spectra (data no shown for all samples).

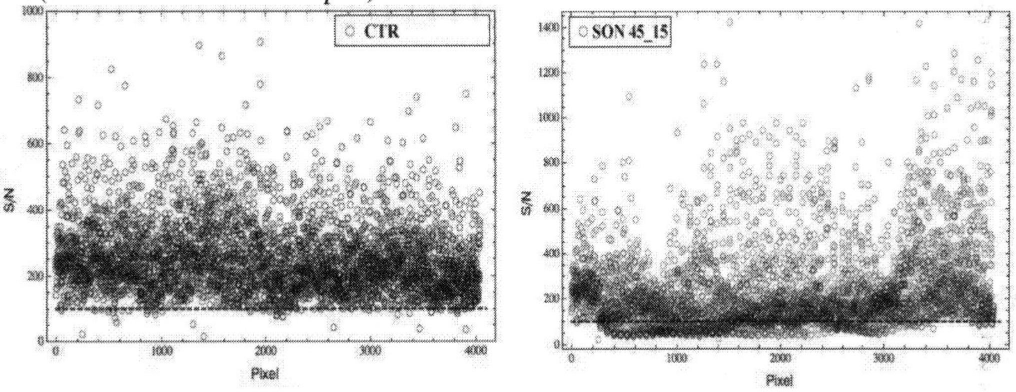

Figure 3. On the left, RMS of S/N ratio vs. pixels for control cells with 4051 good spectra and 44 bad spectra; on the right, RMS of S/N ratio vs. pixels for sonicated cells at 15 cm for 45 minutes with 3344 good spectra and 752 bad spectra.

We have not discarded the FT-IR maps with $N_{bs}<1000$ to monitor the presence of major number of empty spaces due to the US that perturbed the spatial cellular distribution on the slide forming the islands of cellular monolayers. We have compared the number of bad spectra vs. time of FTIR maps between the control and sonicated cells at both distances from the US source; in figure 4, we have reported the results only for the distance of 15 cm.

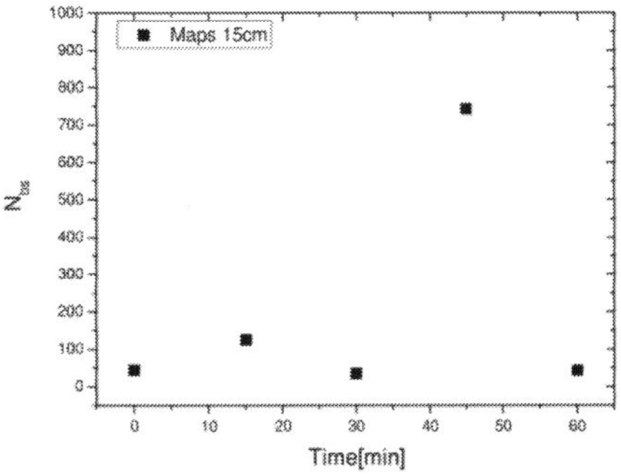

Figure 4. The number of bad spectra (N_{bs}) vs. time within the maps reported at the 15 cm from the US source.

3.2. The simulation of pressure field

We have presented a simplified 2-D geometry to study the acoustics wave propagation inside the plexiglass tank (LxH=20x25cm). The results have been validated by comparison with experimental data (figure 5).

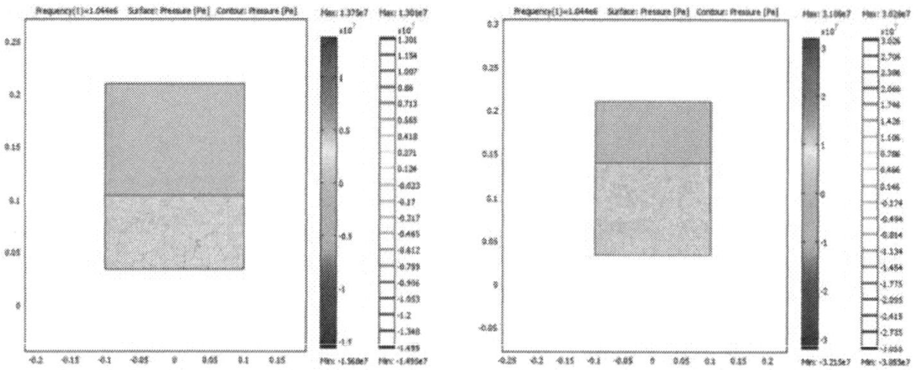

Figure 5. The distribution of pressure at distances of 10 (left) and 15 (right) cm from US source.

3.3 FTIR spectral imaging

FTIR spectral imaging enables the determination of the distribution of several molecules of interest; we have studied the chemical map due to the FTIR- imaging through the analysis of L_{2I} that indicates the intensity of the CH_2 asymmetric stretching (2924 cm^{-1}) band. The intensities of IR peaks provide quantitative analysis about sample contents, depending on the nature of molecular structure, their bonds, and their environment. As reported in figure 6, we have compared the values of this parameter vs. time obtained from the chemical map at both distances.

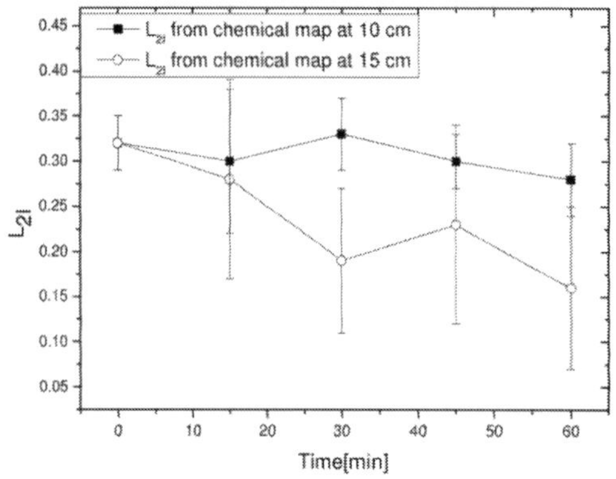

Figure 6. L_{2I} parameter vs. time at 10 cm (black point) and 15 cm (white point) due to chemical map is reported, showing differences between the values of chemical map at two distances. The error bars correspond to ± 5 % of the measured value.

From spectral data, the changes of L_{2I} parameter obtained from the map were correlated with the applied ultrasonic (US) energy (the ultrasonic energy per unit of surface). In figure 7, this correlation has a R-value of about 94% with a probability of about 99% (*Fisher test*) at the distance of 15 cm. At

10 cm from the source, the degree of correlation decreases to a R-value of about 75% with a probability of about 98% (*Fisher test*) (data not shown).

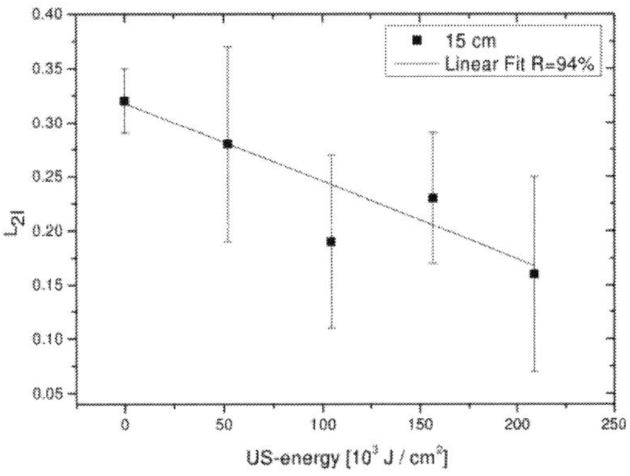

Figure 7. Correlation between the L_{2I} intensity and the US energy (the ultrasonic energy per unit surface) with a R-value of about 0,94 and a probability of about 99% (Fisher test). The error bars correspond to ± 5 % of the measured value.

4. Conclusion

In this paper, we have analyzed the effects due to US medical equipment at 1 MHz frequency in pulsed system on non tumoral cell line. The unfocused, planar and with a circular geometry source transducer was characterized through a vertical directivity test where the maximum signal has shown a shift of of θ=0.8°. At the chosen distances of 10 and 15 cm, we have established the values of pulse average intensity (I_{PA}): I_{PA}= (52 ± 6) Wcm^{-2} at 10 cm and I_{PA}= (58 ± 6) Wcm^{-2} at 15 cm. The experimental setup did not cause a thermal effect on cells; the temperature was registrated inside the petri dish and the water by thermocouple and it was increased of 2°C or less during US exposed time. The simulation of acoustics wave propagation has been validated by comparison with experimental data. The results of FTIR microspectroscopy have shown that the US experimental conditions have not induced a DNA mutations [4] whereas in the lipid region (3000-2800 cm^{-1}) there was an effect on L_{2I} parameter vs. time that could be useful in the study of the ultrasound-mediated gene transfection *in vitro*. The change of L_{2I} parameter has proved as membrane phospholipid asymmetry was perturbed; this movement of transbilayer lipid distributions can be associated to the formation of temporary pores on plasma membrane, ruled out the formation of permanent pores that can induced the cell death.

References

[1] Newman CM, Lawrie A, Brisken AF, Cumberland DC 2001 *Echocardiog.* **18** 339-347
[2] Lewin PA, Lypacewicz G, Bautista R, Devaraju V 2000 *Ultrasonics* **38** 135-139
[3] Conti L, Grimaldi P, Udroiu I, Bedini A, Giliberti C, Palomba R, Congiu Castellano A 2010 *Vibrat. Spectrosc.* **52** 79-84
[4] Di Giambattista L, Grimaldi P, Udroiu I, Pozzi D, Cinque G, Frogley MD, Giansanti A, Congiu Castellano A 2009 *Biophys. and Bioengin. Letters*, **2**, 2.

Acknowledgments
The authors are grateful Dr. M. Lattanzi for his assistance with Mathematica software and Dr. S. Belardinelli for his assistance in the laboratory experiments.

AUTHOR INDEX

Institute of Physics
Dirac House, Temple Back
Bristol BS1 6BE UK

CURRAN ASSOCIATES INC
proceedings
.com